T0291135

Evaluation of Intelligent Road Transport Systems, 2nd Edition

Other related titles:

You may also like

- PBTR007 | Meng Lu | Evaluation of Intelligent Road Transport Systems | 978-1-78561-172-8
- PBTR025 | Meng Lu | Cooperative Intelligent Transport Systems | 978-1-83953-012-8
- | PBTR001 | Michele Fiorini and Jia-Chin Lin | Clean Mobility and Intelligent Transport Systems | 978-1-84919-895-0

We also publish a wide range of books on the following topics:
Computing and Networks
Control, Robotics and Sensors
Electrical Regulations
Electromagnetics and Radar
Energy Engineering
Healthcare Technologies
History and Management of Technology
IET Codes and Guidance
Materials, Circuits and Devices
Model Forms
Nanomaterials and Nanotechnologies
Optics, Photonics and Lasers
Production, Design and Manufacturing
Security
Telecommunications
Transportation

All books are available in print via https://shop.theiet.org or as eBooks via our Digital Library https://digital-library.theiet.org.

TRANSPORTATION SERIES 046

Evaluation of Intelligent Road Transport Systems, 2nd Edition

Methods and results

Edited by
Meng Lu

The Institution of Engineering and Technology

About the IET

This book is published by the Institution of Engineering and Technology (The IET).

We inspire, inform and influence the global engineering community to engineer a better world. As a diverse home across engineering and technology, we share knowledge that helps make better sense of the world, to accelerate innovation and solve the global challenges that matter.

The IET is a not-for-profit organisation. The surplus we make from our books is used to support activities and products for the engineering community and promote the positive role of science, engineering and technology in the world. This includes education resources and outreach, scholarships and awards, events and courses, publications, professional development and mentoring, and advocacy to governments.

To discover more about the IET please visit https://www.theiet.org/

About IET books

The IET publishes books across many engineering and technology disciplines. Our authors and editors offer fresh perspectives from universities and industry. Within our subject areas, we have several book series steered by editorial boards made up of leading subject experts.

We peer review each book at the proposal stage to ensure the quality and relevance of our publications.

Get involved

If you are interested in becoming an author, editor, series advisor, or peer reviewer please visit https://www.theiet.org/publishing/publishing-with-iet-books/ or contact author_support@theiet.org.

Discovering our electronic content

All of our books are available online via the IET's Digital Library. Our Digital Library is the home of technical documents, eBooks, conference publications, real-life case studies and journal articles. To find out more, please visit https://digital-library.theiet.org.

In collaboration with the United Nations and the International Publishers Association, the IET is a Signatory member of the SDG Publishers Compact. The Compact aims to accelerate progress to achieve the Sustainable Development Goals (SDGs) by 2030. Signatories aspire to develop sustainable practices and act as champions of the SDGs during the Decade of Action (2020–30), publishing books and journals that will help inform, develop, and inspire action in that direction.

In line with our sustainable goals, our UK printing partner has FSC accreditation, which is reducing our environmental impact to the planet. We use a print-on-demand model to further reduce our carbon footprint.

British Library Cataloguing in Publication Data

A catalogue record for this product is available from the British Library

ISBN 978-1-83724-083-8 (hardback)
ISBN 978-1-83724-084-5 (PDF)

Typeset in India by MPS Limited
Printed in the UK by CPI Group (UK) Ltd, Eastbourne

Cover image: MR.Cole_Photographer via Getty Images

Contents

5   Assessment method for prioritising transport measures and infrastructure development **87**
Henk Taale and Jan Kiel

Foreword

The first edition of this book answered the pressing need for a reference source so that practitioners in national or local government, commercial organisations, research centres could quickly find the state of the art for a range of ITS activities worldwide. However the field of Intelligent Transport Systems (ITS) is a very dynamic one. Many of the earlier techniques have been improved and some new ones have moved from the laboratory to real-world service so the time was ripe for an update.

Decades ago, 'Intelligent Transport Systems' – or ITS – technologies included roadside sensors, signs that dynamically changed messaging, and coordinated traffic signals, mainly focused on improving safety and efficiency of vehicles. Now ITS technologies have expanded to include a much wider spectrum of innovation, including: automation; connectivity; digital infrastructure; sustainable technologies; mobility on demand technologies; and emerging transportation technologies like eVTOLs, delivery drones, sidewalk robots and Artificial Intelligence (AI).

Rapidly evolving innovations are being developed and deployed that promise to entirely reshape the way our transport networks operate, facilitating vast improvements to safety and overall mobility. The promise of these innovations is apparent, but the deployment and application of these technologies is not without challenges. Sustainable business models and funding remain the exception, not the rule. While technology disrupts the transport ecosystem with new innovations and ideas, government, communities, private companies, and researchers must race to catch up and harness these advances for the public good. ITS America and its global partners at ERTICO and ITS Asia Pacific have been at the centre of these innovations, advocating for transformational technology to save lives, promote sustainability and resiliency, improve access to multimodal mobility and improve the movement of goods.

I am are pleased that this book will showcase the outcomes and tangible benefits of deployments to help our ITS community scale these innovations in the years to come.

Laura Chace
President & CEO, ITS America
March 2024

It is good to see up-to-date information on ITS being developed on a global scale. The term ITS – Intelligent Transport Systems– was first used about 30 years ago. At that time, the focus was on how to utilise roads safely and efficiently. Today ITS is being discussed from the very different perspective of how it can contribute to solving social issues ranging across transport and mobility. In the twenty-first century ITS are solutions that contribute to safe and secure mobility, enjoyment, freedom of movement, carbon neutrality, disaster resilience, and learning how to improve local transport to support local economies. There are a variety of themes, and I am pleased to see that this book looks at many of those mentioned above and will be a guide for the future of ITS.

Ace Yamamoto
Secretary-General, ITS Asia Pacific
March 2024

While ITS was traditionally viewed as a catalyst for smarter, efficient and safer transport and mobility of both people and goods, we've also recognised other enablers for our sector. This includes the role of ITS in promoting sustainability, resilience and data sharing among various stakeholders and their products, ranging from vehicles, shuttles, two-wheelers, bikes, scooters, trains, ships, drones or planes, all harmonised with both physical and digital infrastructure providers. In essence we're paving the way for seamless, connected, and automated multi-modality, aligning with the UN Sustainable Development Goals. ITS are based on a wide variety of information about transport, travellers, and vehicles.

ITS technologies serve as the foundation and operating system for multimodal transport and mobility ITS have the ability to connect the dots and various elements of a large mobility ecosystem in a systemic manner. Although ITS are now a relatively mature discipline, we do not always have a robust understanding of their costs and benefits compared to the longer-established processes for building new, or upgrading, physical infrastructure. But we do know that carefully designed and implemented schemes can deliver a 30% reduction in the number of fatalities and seriously injured persons across Europe, 15% reduction of road traffic related congestion, 20% improvements in energy-efficiency, 50% increase in availability of real time traffic and travel information.

It is a pleasure to welcome this comprehensive book on the evaluation of Intelligent Road Transport Systems. ITS use electronics, information and data, combined with communications technologies, to deliver transport improvements instead of extending physical infrastructure, thereby saving money (typically 50%) and reducing environmental impact. Dr. Lu has done another excellent job and the updated book complements the work ERTICO – ITS Europe is doing to provide decision makers and related ITS stakeholders with access to reliable, understandable and coherent information on the outcomes (benefits and impacts) of existing and ongoing ITS deployment, and generally supporting them in developing fact-based policy objectives and strategies.

Albert Einstein is famous for a number of reasons but his achievements in mathematical physics are equalled by his practical daily wisdom as shown by the following two quotes:

- "True genius is knowing where to find the answer"
- "Never memorise anything you can look up"

The second edition of "Evaluation of Intelligent Road Transport Systems: Methods and Results" would be welcomed by Einstein against both of his criteria. We can recommend it unreservedly.

Joost Vantomme
CEO, ERTICO – ITS Europe

Professor Eric Sampson
Newcastle University

March 2024

Preface

The development of Intelligent Transport Systems (ITS) began in the late 1980s, driven directly by the automotive industry, and strongly supported by authorities around the world. The technical backbone of ITS is the automotive industry, with the original objectives being to substantially improve road safety, driver comfort, traffic efficiency and energy/fuel efficiency. Since then, ITS Associations have been gradually launched, such as ITS Europe, named ERTICO (European Road Transport Telematics Implementation Co-ordination Organization), ITS Japan and ITS America, followed by many other national and regional associations. The first ITS World Congress was held in 1994 in Paris, France. The congress has since been organised annually, on a rotating basis, in the three regions: Europe/Mid-East, Asia/Pacific and North America. It is the largest technical and commercial event in the domain of ITS. The Institute of Electrical and Electronics Engineers (IEEE) hosted its first IEEE Intelligent Vehicles Symposium in 1989 in Tsukuba, Japan, and the first IEEE Intelligent Transportation Systems Conference in 1997 in Boston, USA.

From a technical perspective, autonomous systems and cooperative systems can be distinguished in the domain of ITS. The term ITS covers a range of products, solutions and services based on ICT (Information & Communication Technology). Various in-vehicles systems called ADAS (Advanced Driver Assistance Systems) have been developed for decades. By far the most successful (and probably the most useful) ADAS application is the navigation system. First commercially introduced in the early and mid-1990s, its development and deployment had a boost since May 2000 when a more accurate GPS signal became available (discontinuation of Selective Availability). In the years that followed, market penetration grew rapidly through availability of both built-in systems in (initially high-end) new cars and after-market systems. Since around 2010, its adoption has continued to increase as mobile phone navigation apps entered the scene. Currently, the navigation system is widely applied as standard equipment in vehicles around the world. In addition to the development of in-vehicle systems by the private sector, the public sector has made substantial investments in intelligent infrastructure for transport network management, road pricing and provision of information to vehicles and other road users. Over the past two decades, there have been substantial developments in infrastructure-supported ITS, called C-ITS (Cooperative Intelligent Transport Systems) or Connected Vehicles, for the applications using V2X (Vehicle to Everything) communication. The rapid growth of the (C-)ITS industry was made possible by the rapid development of various technologies, such as control system engineering, absolute and relative positioning (using GPS and digital maps), communications, sensing and information processing, supported by political interest and the perception of its substantial potential impact (social, economic and environmental), and achieved to a substantial extent through intensive collaboration between government, industry and academia in many national and international projects and initiatives.

In 2016, while serving as Chair of IBEC, the International Forum for ITS Impacts Evaluation, I initiated the first edition of this book. At that stage, researchers were developing new approaches for ITS evaluation in an attempt to overcome the shortcomings of the methods generally in use; industry partners were interested in improved result-oriented technical assessment; and decision makers would like to gain better insight in investment costs and benefits. The first edition provided an overview of Intelligent Road Transport Systems, and of evaluation methods used for ITS, and presented selected evaluation results of the development and deployment of ITS.

After going through considerable technical development and reaching an advanced level of maturity, with huge investment from both the public and the private sectors, ITS has moved into the deployment phase, and has demonstrated substantial (and sustainable) economic, environmental and societal impacts. Over the past decade, enormous R&D efforts have been made in the field of high-level automated driving (see, e.g. Lu, M. (Ed.) (2019). *Cooperative Intelligent Transport Systems: Towards High-Level Automated Driving.* IET, London. DOI: 10.1049/PBTR025E). As we move towards more sophisticated and advanced C-ITS applications, the role of infrastructure is becoming increasingly important. (C-)ITS includes a broad spectrum of applications for sustainable, secure and resilient transport (all modes) to shape the future of mobility. In parallel with simultaneous other international (C-)ITS activities, the European Commission (EC), following up on the C-ITS Platform (2014-2017), established the Cooperative, Connected and Automated Mobility (CCAM) Platform in 2019 to carry out pre-deployment activities with significant public resources. Moreover, from a systems perspective, ITS evolved from a technology specific to road transport, to a more generally applicable technology for all modes of transport, especially in terms of transshipments between different surface transport modes (road, rail, air and waterborne, including inland waterways), logistics hubs (airports, dry or inland ports, railway stations, sea ports), and cross-sector multi-modal logistics and supply chains. ITS even extends into the third dimension for Urban Air Mobility (UAM) in terms of eVTOLs (electric Vertical Take-Off and Landing) aircraft and drones. Future mobility, in the context of regional development, is not only intended to improve the quality of life but also to create thriving communities, by better taking care of and even restoring nature and the environment, maintaining strong economic growth, and increasing the potential positive social impact. The second edition of the book covers current themes in contributions from international experts, and intends to provide guidance for the future of (C-)ITS deployment.

Herewith I sincerely thank all my friends and colleagues who have contributed to this book and provided very kind support. I also thank the European Commission (especially DG MOVE, DG RTD, DG CONNECT and DG GROW), the US DoT (Department of Transportation), Standards Development Organisations (such as the IEEE Standards Association, ISO – the International Organization for Standardization, and SAE International) and the ITS associations (such as ERTICO – ITS Europe, ITS America, ITS Asia Pacific, ITS Australia, ITS Canada, ITS Japan, ITS Korea, ITS China, ITS Taiwan, ITS Singapore, ITS South Africa, ITS UK, ITS Sweden, ITS Denmark, ITS Norway, ITS Finland, ITS Network Germany, AustriaTech, TTS Italia, ITS Spain, ITS Portugal, ITS France, ITS Hellas, ITS Ireland, ITS Switzerland, ITS Romania, ITS Polska, ITS Hungary, ITS Bulgaria, ITS Czech Republic, ITS Slovenia, ITS Belgium and Connekt/ITS The Netherlands), TISA (Traveller Information Services Association) as well as all (industry, authority and academia) partners that have substantially contributed to the (C-)ITS development and deployment. Last but not least, I would like to thank my parents for their excellent support during my education and further development, and my husband Kees Wevers for introducing me to the domain of ITS, and strongly supporting my (C-)ITS activities for more than two decades.

Dr. Meng Lu
Utrecht, The Netherlands
March 2024

List of acronyms

3D	3-Dimensional
A2D2	Audi Autonomous Driving Dataset
AA	Authorisation Authority
AADT	Average Annual Daily Traffic
AASE	Absolute Average Speed Error
ABS	Anti-lock Braking System
ACC	Adaptive Cruise Control
ADAS	Advanced Driver Assistance Systems
AD	Automated Driving
ADF	Automated Driving Function
ADS	Automated Driving System
ADS-DV	ADS-Dedicated Vehicles
AEB	Automated Emergency Braking
AHP	Analytic Hierarchy Process
AI	Artificial Intelligence
ALKS	Automated Lane Keeping Systems
AMoD	Autonomous Mobility-on-Demand
AMPO	Assessment Method for Policy Options
AMS	Autonomous Mobility Service
API	Application Programming Interface
ARCADIA	Architecture Analysis & Design Integrated Approach
ARGO	Algorithms for Image Processing
ARI	Autofahrer Rundfunk Information
ARTS	Automated Road Transportation Systems
ASV	Advanced Safety Vehicle
ATIS	Advanced Traveler Information System
AUC	Area Under the Curve
AV	Automated Vehicle
AWPM	Aigaleo Western Peripheral Motorway
AWS	Amazon Web Services
AIMES	Australian Integrated Multimodal EcoSystem

BCR	Benefit–Cost Ratio
BoQ	Back of Queue
BOS	Blank Out Signs
BSI	British Standards Institution
CA	Certificate Authority
CACC	Cooperative Adaptive Cruise Control
CAM	Cooperative Awareness Message
CAN	Controller Area Network
CASE	Connected, Autonomous, Shared, and Electric
CATS	Cooperative Architecture for Transportation Systems
CBA	Cost–Benefit Analysis
CBM	Context-based Matching
CCAM	Cooperative, Connected and Automated Mobility
CCTV	Closed Circuit Television
CDR	Call Detail Record
CEA	Cost-Effectiveness Analysis
CEF	Connecting Europe Facility
CEN	European Committee for Standardization
CENELEC	European Committee for Electro technical Standardization
CI	Confidence Interval
C-ITS	Cooperative Intelligent Transport Systems
C-ITS-F	Central ITS Facility
CI/CD	Continuous Integration and Continuous Deployment
CNN	Convolutional Neural Network
COMPANION	COoperative dynamic forMation of Platoons
CoMPAV	Common Mobility Platform for Automated Vehicles
CO/NO/VIS	Visibility Sensors in Tunnels
CPM	Collective Perception Message
CPU	Central Processing Unit
CORS	Continuously Operating Reference Stations
CRG-UN	C-Roads Germany Urban Nodes
CRNN	Convolutional Recurrent Neural Network
DARPA	Defense Advanced Research Projects Agency
DDPG	Deep Deterministic Policy Gradient
DDT	Dynamic Driving Task
DENM	Decentralised Environmental Notification Message
DfT	Department for Transport

DITRDCA	Department of Infrastructure, Transport, Regional Development, Communications and Arts
DL	Deep Learning
DoF	Degree of Freedom
DOVA	Distributed ODD (Operational Design Domain) Attribute Value Awareness
DOT	Department of Transportation
DSM	Design Structure Matrix
DSRC	Dedicated Short-Range Communications
EA	Enrolment Authority
EC	European Commission
ECU	Electronic Control Unit
EEBL	Emergency Electronic Brake Light
EIP	European ITS Platform
ELECTRE	ÉLimination Et Choix Traduisant la RÉalité
ENSEMBLE	ENabling SafE Multi-Brand pLatooning for Europe
EO	Egnatia Odos
ERTICO	European Road Transport Telematics Implementation Co-ordination Organisation
ESP	Electronic Stability Program
ESS	Emergency Stop Signal
ESSM	Elefsina-Stavros-Spata Airport Motorway
ESVN	Emergency Service Vehicle Notification
ETC	Electronic Toll Collect
ETL	Extract Transform Load
ETSI	European Telecommunications Standards Institute
EU EIP	European ITS Platform
euroFOT	European Field Operational Test
EVA	Emergency Vehicle Approaching
EVFT	Embedded VMS "Free Text"
eVTOL	electric Vertical Take-Off and Landing
FBX	Filmbox
FCN	Fully Convolutional Network
FCD	Floating Car Data
FIFO	first-in-first-out
FOT	Field Operational Test
FPGA	Field Programmable Gate Array
FU	Fallback User
GA	Genetic Algorithm

GAM	Goal Achievements Matrix
GAN	Generative Adversarial Networks
GDPR	General Data Protection Regulation
Gen AI	Generative AI
GHG	greenhouse gas
GLOSA	Green Light Optimal Speed Advisory
GNSS	Global Navigation Satellite System
GPS	Global Positioning System
GPT	Generative Pre-trained Transformers
GPU	Graphics Processing Unit
H3D	Honda 3-Dimensional Dataset
HD	High-Definition
HGV	Heavy Goods Vehicle
HIC	High Income Country
HLN	Hazardous Locations Notification
HMI	Human Machine Interface
HSM	Hardware Security Module
I2V	Infrastructure-to-Vehicle
ICP	Iterative Closest Point
ICT	Information and Communication Technologies
ICVP	Ipswich Connected Vehicle Pilot
IEC	International Electrotechnical Commission
IEEE	Institute of Electrical and Electronics Engineers
IEEE SA	IEEE Standards Association
IMU	Inertial Measurement Unit
INS	Inertial Navigation System
ITS	Intelligent Transport Systems
IoT	Internet of Things
IoU	Intersection over Union
IPU	Image Processing Unit
IRL	Inverse Reinforcement Learning
ISAD	Scheme for Automated Driving
ISO	International Organization for Standardization
IT	Information Technology
ITS	Intelligent Transport Systems
IVIM	In-Vehicle Information Message
IVS	In-Vehicle Signage
IVSL	In-Vehicle Speed Limit

IWPM	Imittos Western Peripheral Motorway
KAIST	Korea Advanced Institute of Science and Technology
KITTI	Karlsruhe Institute of Technology and Toyota Technological Institute
KPI	Key Performance Indicator
LCS	Lane Control Signs
LCRW	Longitudinal Collision Risk Warning
LDA	Linear Discriminant Analysis
LeGO-LOAM	Lightweight and Ground-Optimized LiDAR Odometry and Mapping
LiDAR	Light Detection and Ranging
LIME	Local Interpretable Model-agnostic Explanations
LKA	Lane Keeping Assistance
LMIC	Low- and Middle-Income Country
LOS	Level of Service
LUMS	Lane Use Management System
LWIR	Long Wave Infrared
Lx	SAE Automation Level x (0-5)
MAC	Media Access Control
MAPE	Mean Absolute Percentage Error
MAPEM	MAP Extended Message
MBSE	Model-based Systems Engineering
MCA	Multi-Criteria Analysis
METI	Ministry of Economy, Trade and Industry
ML	Machine learning
MLIT	Ministry of Land, Infrastructure, Transport and Tourism
MNO	Mobile Network Operator
MoD	Mobility-on-Demand
MPC	Model Predictive Control
MQTT	Message Queuing Telemetry Transport
MRC	Minimal Risk Condition
MRF	Markov Random Field
MRS	Monitoring and Reporting System
MSDV	Minimum Safe Distance Violation
NAP	National Access Point
NDS	Naturalistic Driving Studies
N-FOT	Naturalistic Field Operational Test
NHTSA	National Highway Traffic Safety Administration
NLP	Natural Language Processing

NMPC	Nonlinear Model Predictive Control
NPA	National Police Agency
NPU	Neural Processing Unit
NTRIP	Networked Transport of RTCM via Internet Protocol
MVDR	Mobile Digital Video Recorder
OBU	On-Board Unit
ODD	Operational Design Domain
OHVD	Over-Height Vehicle Detector
ONCE	ONe million sCEnes
OSM	OpenStreetMap
PATH	Partners for Advanced Transit and Highways
PBS	Planning Balance Sheet
PCA	Principal Component Analysis
PER	Packet Error Rate
PI	Proportional Integral
PnP	Perspective-n-Point
POC	Proof of Concept
PPP	Public Private Participation
PROMETHEE	Preference Ranking Organisation METHod for Enrichment Evaluations
PROMETHEUS	PROgraMme for a European Traffic of Highest Efficiency and Unprecedented Safety
PRT	Personal Rapid Transport
PVD	Probe Vehicle Data
R&D	Research and Development
RDS	Radio Data System
RHW	Road Hazard Warning
RICP	Robust Iterative Closest Point
R-ITS-S	Roadside ITS Station
RNM	Road Network Model
RNN	Recurrent Neural Network
ROC	Receiver Operating Characteristic
ROI	Return On Investment
RSP	Ride Sourcing Provider
RSS	Responsibility-Sensitive Safety
RSU	Road Side Unit
RTK	Real-Time Kinemetic
RTTI	Real-Time Traffic Information
RWIS	Road Weather Information Systems

RWW	Roadworks Warning
RWW-LC	Roadworks Warning – Lane Closure
SAC-IA	Sample Consensus Initial Alignment
SAE	Society of Automotive Engineers
SAP	Single Access Point
SARTE	Safe Road Trains for the Environment
SAW	Simple Additive Weighting
SCATS	Sydney Coordinated Adaptive Traffic System
SCMS	Security Credential Management System
SDO	Standards Development Organization
SEB	Speed Error Bias
SHAP	SHapley Additive exPlanations
SIP	Strategic Innovation Promotion Program
SLAM	Simultaneous Localization and Mapping
SPADE	Spatial Planning and Development Evaluation
SPaTEM	Signal Phase and Timing Extended Message
SR	Smart Routing
SRN	Strategic Road Network
SRTI	Safety-Related Traffic Information
SSD	Single Shot Multibox Detector
SSV	Stopped/Slow Vehicle
SSVS	Super Smart Vehicle Systems
SUMO	Simulation of Urban Mobility
SV	Stationary Vehicle
SVW	Slow or Stationary Vehicle Warning
SWD	Shockwave Damping
TAA	Tram Awareness Alert
TCS	Traffic Control Server
TEN-T	Trans-European Transport Network
TJW	Traffic Jam (Ahead) Warning
TLM	Trust List Manager
TM	Traffic Management
TMC	Traffic Message Chanel
TMR	Queensland Government Department of Transport and Main Roads
TOPSIS	Technique for Order Preference by Similarity to Ideal Solutions
TPW	Tram Passenger Warning
TraCI	Traffic Control Interface

TSP	Traffic Signal Priority
TTA	Time-to-Action
TWVR	Turning Warning Vulnerable Road-user
UAM	Urban Air Mobility
UN	United Nations
UTMS	Universal Traffic Management System
V2I	Vehicle-to-Infrastructure
V2N	Vehicle-to-Network
V2V	Vehicle-to-Vehicle
V2X	Vehicle-to-Everything
VDS	Variable Direction Sign
VICS	Vehicle Information and Communication System
VMS	Variable Message Signs
VPN	Virtual Private Network
VRU	Vulnerable Road User
VSL	Values of Statistical Life
VTP	Variable Text Panel
VTTI	Virginia Tech Transportation Institute
WCW	Weather Conditions Warning
WG	Working Group
xFCD	extended Floating Car Data
YOLO	You Only Look Once

Short biographies of the editor and authors

Dr. **Meng Lu** – Principal, Aeolix ITS; Member, Board of Governors, IEEE Standards Association; VP Standards Activities & Standards Committee Chair, IEEE Intelligent Transportation Systems Society; Steering Committee Member, IEEE Future Networks Technical Community. Previously, Strategic Innovation Manager, Peek Traffic (NL); Program Manager International, Dutch Institute of Advanced Logistics (NL); and Visiting Professor, National Laboratory for Automotive Safety and Energy, Tsinghua University (CN). Since 2002 active involvement in R&D and innovation. Since 2016 also contribution to standardization activities, e.g. ISO/TC 204 – Intelligent Transport Systems (also Head of Delegation (NL) Jan 2021–Apr 2023), CEN/TC 278, and IEEE. PhD at LTH, Lund University, Sweden; Master's title and degree of Engineering in The Netherlands and China.

Prof. **Johann Andersen** is Industry Associate Professor in Intelligent Transportation Systems at Stellenbosch University, South Africa. He teaches ITS principles in the graduate civil engineering programmes and guides research activities in ITS. In his capacity as CEO of Techso, a specialist consultant company, he has extensive experience in ITS planning, design and implementation in the application areas of Freeway management systems as well as advanced public transportation systems. Johann heads up the SSML, using his experience in the research and implementation of ITS applications, to lead the SSML towards becoming a renowned Knowledge Centre for development of innovative and cost-effective solutions in ITS, not only in South Africa but also developing countries.

Nicholas Brook is the Program Director of the C-ITS National Harmonisation iMOVE project (1-083) for the Department of Transport and Planning, Australia. His is a Chartered Civil Engineer with a diverse working background including road construction, traffic management, design, contract management, traffic modelling, Managed Motorways, signal optimisations, network operations planning, strategic leadership, and emerging road safety technologies. Since 2016, Nick has been working on deploying Cooperative Intelligent Transport Systems (C-ITS) with both Queensland and Victorian road authorities in collaboration with other Governments, academia, and industry. He has a passion for maximising social outcomes for road users through use of engineering and technology for improved safety, efficiency, and sustainability.

Dr. **Megan Bruwer** is a transportation engineer and Senior Lecturer at Stellenbosch University, South Africa. She joined the Civil Engineering Department in 2015, and lectures traffic engineering to both postgraduate and undergraduate students. Megan also coordinates the research activities of the Stellenbosch Smart Mobility Laboratory, a multidisciplinary unit for Intelligent Transport Systems research. Prior to joining Stellenbosch University, Megan worked as a transportation engineer in urban and rural South Africa. Her research interests are focussed on improvements to the transport field in low- and middle-income countries, including sustainable transport and land-use development, and the application of new data sources to improve transport planning and traffic management. She completed a PhD in 2023, which focused on the correct application of Floating Car Data in the sub-Saharan African context. Megan has published several journal articles and international conference proceedings.

Luke Capelli is currently a C-ITS specialist working in Toyota/Lexus Australia's Cooperative and Automated Vehicles team. With a background in transport engineering, he specialises in Cooperative Intelligent Transport Systems (C-ITS) and automated driving, spatial systems and ITS infrastructure, with a passion for future mobility and safe, accessible transport for all.

Prof. **Oliver Carsten** is Professor of Transport Safety at the Institute for Transport Studies, University of Leeds, UK. His major research focus is on driver interaction and safety with driver assistance and automation systems. He led the UK national project on Intelligent Speed Assistance and has provided advice on safety policy to the UK Department for Transport and to the European Commission. He is heavily involved in international regulatory activities on vehicle automation on the road user side as a member of the Informal Group of Experts on Automated Driving (IGEAD) under UNECE WP.1 and on the vehicle side as a member of a series of informal groups under UNECE WP.29. He is editor-in-chief of the academic journal *Cognition, Technology and Work*.

Prof. **Richard Cuerden**, Director, TRL Academy – Transport Research Laboratory. Richard is a member of TRL's Executive Team and responsible for the company's Technical Strategy and the associated forward-looking investment and thought leadership activities. He ensures the technical quality of science and engineering outputs, supports the academic development of staff and manages engagement with stakeholders on programmes of collaborative Innovation, Research and Development. Richard has a strong track record of using real-world evidence to identify safety and environmental priorities and developing transport solutions, supporting policies and regulations. He led the technical work that in 2024 transformed EU Vehicle Safety Type Approval, the UK's HGV Platooning Trials and co-created London's Smart Mobility Living Lab.

Daniel Cullern (BA Hons, MA) has been a transport consultant since 2005 and has worked in the field of European ITS initiatives on behalf of National Highways since

2008. In this capacity, he was responsible for the coordination of the multi-Member State Arc Atlantique ITS deployment corridor from its initiation through to its completion (2012–2021). In parallel, he has also worked extensively on the European ITS Platform (EU EIP), most notably as coordinator of the EU EIP Evaluation Activity, which developed a consensus-built and commonly adopted harmonised approach to evaluating the benefits of ITS services on the European road network.

Kia Eisinga is a data scientist at TomTom, based in Amsterdam, The Netherlands. She obtained a Master's degree in Data Science from Tilburg University, The Netherlands (2016). During her master's degree, she worked on runway allocation prediction for flights at Amsterdam Airport Schiphol. At TomTom, her focus is on the interpretation of Floating Car Data (FCD), particularly for travel time estimations and various other traffic use cases.

Dr. **Catherine M. Elias** received her MSc degree (2018) in Mechatronics Engineering in the field of cooperative control of multi-robot systems, German University in Cairo (GUC), Egypt. She received her PhD degree (2022) in the design of V2X-based cooperative architectures for transportation systems agents. Catherine is currently holding a lecturer position in the Computer Science Engineering Department, Faculty of Media Engineering and Technology, GUC, Egypt, and Coordinator of the Multi-Robot Systems (MRS) Research Group. She is a member of the Board of Governors (2023–2025) of the IEEE Intelligent Transportation Systems Society.

Dr. **Stig E.R. Franzén** is Adjunct Professor Emeritus, formerly at Design & Human Factors, Chalmers University of Technology, Sweden. With a multi-disciplinary background in engineering and behavioural science, his interest in the interaction between humans and technology has had a particular focus on the human in complex socio-technical systems. He was one of the authors of the book *Driving Future Vehicles*, published in 1993, and bringing together work on driver behaviour, traffic safety and human-machine interface design. He has been involved in several national and international R&D projects on drivers, vehicles, advanced driving support systems as well as information support systems for information officers and traffic management. Examples of projects include FESTA, TeleFOT, Bus System of the Future and AEOLIX. He was a key actor in the establishment of the Swedish national ITS research school as well as in Chalmers Research Area of Advance Transport.

Dr. **Bolin Gao** received the BS and MS degrees in Vehicle Engineering from Jilin University, Changchun, China, in 2007 and 2009, respectively, and the PhD degree in Vehicle Engineering from Tongji University, Shanghai, China, in 2013. He is currently Associate Research Professor with the School of Vehicle and Mobility, at Tsinghua University, Beijing, China. His research interests include the theoretical research and engineering application of the dynamic design and control of intelligent and connected vehicles, especially collaborative perception and tracking methods in cloud control systems, intelligent predictive cruise control systems on commercial trucks with cloud control mode, and the test and evaluation of intelligent vehicle driving system.

Dr. **Jiaxin Gao** received the BS and PhD degrees from the University of Science and Technology Beijing in 2017 and 2023, respectively. He is currently working as an assistant researcher at the School of Vehicle and Mobility, Tsinghua University. His current research interests include vehicle dynamics and control, reinforcement learning, approximate dynamic programming and the alternating direction method of multipliers.

Divya Garikapati is a Senior IEEE Member and a Systems and Safety Engineering Expert. Her current work focuses on Systems and Safety research and development for Level 2, 3 and 4 automated vehicles research and development. She has over 11 years of experience in the automotive industry. She currently serves as the standards committee member within the IEEE Intelligent Transportation Systems Society and a peer reviewer for IEEE ITS conferences. She actively participates in several industry standards discussions within IEEE and SAE, and also the working group Chair for the IEEE Vehicular Technology Society. Currently she serves as the SAE Ground Vehicle AI Committee Member. She worked as an Associate Editor for IEEE ICHMS, IEEE IV, IEEE ACC and other conferences. She received her Master's degree in Electrical Engineering from the University of Michigan, Ann Arbor, MI, USA.

Orestis Giamarelos holds a Diploma in Civil Engineering from the National Technical University of Athens (NTUA), Greece and a MSc in Transportation Engineering and Management from the University of Twente, The Netherlands. Since 2018, he has been working as a research engineer at the Federal Highway Research Institute (BASt) in Germany. His research activities include the fields of traffic management, Intelligent Transport Systems and freight transport. He is engaged in European projects pertaining to ITS, most notably the European ITS Platform (EU EIP) and C-Roads. Within EU EIP, he led the Expert Groups Traffic Management and Freight & Logistics of the Monitoring and Dissemination activity, while he contributed his expertise to the evaluation activities of both EU EIP and C-Roads.

Prof. **Susan Grant-Muller** is Chair in Technologies and Informatics at the Institute for Transport Studies and Co-Director of the pan-Faculty Leeds Institute for Data Analytics (LIDA), University of Leeds, UK. Her research is at the multidisciplinary interface between digital technologies, data analytics and sustainable transport futures. She provides leadership to the "Digital Futures" research theme at the Institute for Transport Studies, building on rapid and recent developments in pervasive technology. Her specific interests are in the role of data at scale in sustainable transport paradigms, the evaluation of new technology-based transport schemes, transport-health interactions, the resilience of ICT enhanced transport, and influencing technologies to reduce the carbon, health, energy and other burdens of the transport sector.

Dr. **Frances Hodgson**, provides leadership in the area of Transport Planning as the Director of Student Education for the Institute for Transport Studies, University of Leeds, UK. She has over 25 years of experience in transport research. Her work encompasses the area of behaviour, planning and policy focusing on the evaluation and design of schemes for sustainable travel. Most recently, her research has been on the effectiveness and social impact of the use of social innovation or new sources of data in transport. She has worked in the UK, Europe, Asia and Africa and published in international journals.

Dr. **Javier Ibanez-Guzman** received the MSEE degree from the University of Pennsylvania, Philadelphia, PA, USA, as a Fulbright fellow, and the PhD degree from the University of Reading, UK on a SERC fellowship. He was a Visiting Scholar at the University of California, Berkeley, CA, USA. He was a Senior Scientist with SimTech, A-Star Research Institute, Singapore, where he spear-headed work on autonomous ground vehicles for defense applications. He is currently a Corporate Expert of autonomous systems with Renault SAS and the Co-Director of the SIVALab Common Laboratory between the CNRS, UTC Compiegne, and Renault, working on intelligent vehicle technologies. He is also an Expert to the EU and Eureka Research Program. He is a Chartered Engineer (C. Eng.) and a fellow of the Institute of Engineering Technology, UK. Currently, he is a Senior Editor and an Associated Editor of related IEEE Transactions and country representative to ISO groups associated to automated vehicles and AI.

Dr. **Satu Innamaa** works as Principal Scientist at VTT Technical Research Centre of Finland Ltd., Finland. She has over 25 years of experience in research on transport and mobility, connected and automated driving, impact and quality assessment, and user needs. She has wide experience in field operational tests and other evaluations. Her work has also had a strong focus on evaluation methodologies. Currently, she is leading methodology work packages and tasks in the EU-funded projects Hi-Drive and FAME, related to automated driving. She is Certified Senior Project Manager, IPMA Level B.

Dr. **I.C. MariAnne Karlsson** holds a professorship (chair) in Human-Technology Systems at Design & Human Factors, Chalmers University of Technology, Sweden. Her research aims to develop knowledge of the multidimensional relationship between people and technical artefacts. Central themes include methods for eliciting user requirements for new technical products and systems, prerequisites for individuals' adoption of new technology (ICT in particular) and its effects on everyday life, and the development of new technical solutions to support safe and sustainable behaviour. A more recent research interest has been drivers' understanding of automated vehicles. She has been involved in several national and international projects, for example, FESTA, TeleFOT, Bus System of the Future, and MeBeSafe. She has been actively involved in the development of Chalmers Research Area of Advance Transport as well as in SAFER Vehicle and Traffic Safety Centre at Chalmers.

Jan Kiel is Senior Advisor at Panteia, The Netherlands. He is specialised in projects concerning both passenger and freight transport for a variety of national and international clients. With a background that spans 35 years, he assisted diverse organisations in different types of projects such as transport modelling, assessments, evaluations, procurement, training, capacity building and stakeholder engagement. Jan's work is based on data and modelling, not only conventional transport models but also including the development of instruments such as the integration of Multi-Criteria Analysis and Cost-Benefit Analysis for AMPO and SPADE. Jan is member of the GTiT committee of the European Transport Conference, as well as co-organiser of the Dutch Platos colloquium. Jan graduated in Economic Geography (University of Groningen, The Netherlands), with minors on Civil Engineering (Delft University of Technology, The Netherlands) and Transport Economics (University of Groningen, The Netherlands).

Areti Kotsi has been working since July 2017, as a Research Associate at the Hellenic Institute of Transport (HIT), Centre for Research and Technology Hellas (CERTH), Greece. She is a graduate of the School of Civil Engineering of the National Technical University of Athens (NTUA) and in 2016 she acquired a MSc in Design, Organization and Management of Transportation Systems from the Aristotle University of Thessaloniki (AUTH), Greece. She is a PhD candidate in the School of Civil Engineering AUTH. Her main research activity focuses on ITS and C-ITS. Since joining HIT, she has been involved in Research & Innovation projects related to (C-)ITS. She is the (co-)author of scientific papers and has participated in many international congresses.

Prof. **Risto Kulmala** works as Principal Advisor on ITS at Traficon Ltd., Finland, with past positions at Finnish Transport Agency (Finland), VTT (Finland) and University of Lund (Sweden). He has been a coordinator of several major national and international R&D, innovation and deployment programmes and projects, with more than 300 publications. He has been a member or chair of various international ITS bodies, as well as scientific and technical committees. His areas of expertise include connected and automated driving, physical and digital infrastructures for CCAM, ODDs for highly automated driving, ITS evaluation, deployment road maps, road safety, statistical modelling and field studies of road user behaviour.

Prof. **Shengbo Eben Li** received his M.S. and Ph.D. degrees from Tsinghua University in 2006 and 2009. Before joining Tsinghua University, he has worked at Stanford University, University of Michigan, and UC Berkeley. His active research interests include intelligent vehicles and driver assistance, deep reinforcement learning, optimal control and estimation, etc. He is the author of over 190 peer-reviewed journal/conference papers, and co-inventor of over 40 patents. Dr. Li has received over 20 prestigious awards, including Youth Sci. & Tech Award of Ministry of Education (annually 10 receivers in China), Natural Science Award of Chinese Association of Automation (First level), National Award for Progress in Sci & Tech of China, and best (student) paper awards of IET ITS, IEEE ITS, IEEE ICUS, CVCI, etc. He also serves as Board of Governor of IEEE ITS Society, Senior AE of IEEE OJ ITS, and AEs of IEEE ITSM, IEEE TITS, IEEE TIV, IEEE TNNLS, etc.

Ada Lin has 18-year career in the automotive industry, which is distinguished by her commitment to crafting safer, more accessible and sustainable mobility solutions. Since earning her Master of Engineering from the University of Melbourne (Australia), she has led diverse teams across the globe. Her areas of expertise include hybrid vehicles, connected automated vehicles, sensors and safety assist technologies. In recent years, Ada has spearheaded several award-winning Cooperative Intelligent Transport Systems (C-ITS) and Automated Vehicle (AV) programs. Ada is also a passionate advocate for road safety, dedicating her efforts to the development of real-world solutions. Her work strives to ensure safer roads and vehicles, aiming to secure the safe return of individuals to their homes.

Chris Lodge is Consultancy Manager of Digital and Data at TRL, UK. He is an experienced team leader and consultant with a strong commercial skill set and research background. He is the Consultancy Manager for TRL's Digital and Data teams responsible for over 40 staff working in Technology, Automation, Data & Evaluation and Behaviour. Chris is focused on leading high performing teams and projects in the Connected and Autonomous Vehicle space and beyond. The core of his professional experience is in CAV trials and evaluation, traffic management and modelling, and pedestrian safety. This breadth of experience allows Chris to understand the needs of clients on a technical and commercial level.

Dr. Evangelos Mitsakis is Researcher Director of the Hellenic Institute of Transport (HIT), Centre for Research and Technology Hellas (CERTH), Greece. He is Head of Laboratory "B3 – Infrastructure and traffic management in land transport", as well as Deputy Head of "Department B – Infrastructure, Networks, Mobility and Logistics". He holds a Diploma in Civil Engineering and a PhD in Transport Engineering. He participates in several national and international R&D projects, and he is the author of several scientific publications in peer-reviewed journals and conferences. He is a member of committees and expert groups related to topics of Intelligent Transport Systems abroad (e.g. ERTICO, ECTRI, TRB, IEEE and others) and in Greece, including the expert committees for the National ITS Strategy and National ITS Architecture.

Prof. Elsayed I. Morgan received his PhD degree in Mechatronics Engineering in the field of Mechanics in 1972. His main study field is Robotics, Mechanics and Vibrations.

Dr. **Fjollë Novakazi** is a postdoctoral researcher at the Machine Perception and Interaction Lab at Örebro University, Sweden. Her research concerns the interplay between people and technology. Specifically, she focuses on the cognitive aspects of people's interaction with automated systems and robotics, and how human perception influences these complex interactions. The main application areas are transportation and industrial manufacturing plants. She approaches her studies through empirical mixed-methods research, employing surveys, in-depth interviews, naturalistic studies and observations.

Dr. **Arunkumar Ramaswamy** received his PhD in Computer Science and Robotics from ENSTA ParisTech, France, in 2017. He has served as an architect and systems engineer for multiple automated driving-based mobility projects at Renault's research department. Additionally, he was the product owner for the development of the automated driving software stack. His research interests span AI, mobile robots, robot architectures, computer vision, human-robot interaction and systems engineering. Arunkumar obtained dual Master's degrees in Robotics from Ecole Centrale de Nantes in France (2012) and Warsaw University of Technology in Poland (2011).

Jan Schappacher studied civil engineering with a focus on transportation at the Karlsruhe – University of Applied Sciences (Germany) and the Ryerson University in Toronto (Canada) and obtained the Master of Engineering degree from both universities. From 2018 onwards, he has been working as a Research Associate at the Federal Highway Research Institute (BASt). In his field of expertise on urban traffic management, his responsibilities include the technical supervision of research projects as well as committee work in the German Road and Transportation Research Association (FGSV e.V.) and the Open Traffic Systems City Association (OCA e.V.). Furthermore, he is an expert for Cooperative Intelligent Transport Systems (C-ITS) and is actively involved in the European C-Roads Platform.

Matthias Seidl is Principal Consultant in the area of Vehicle Regulation at TRL. He joined the Safety Group in 2012, after graduating in Mechanical Engineering/Automotive Safety from the Dresden University of Technology. He has 12 years' experience working on and leading projects in the areas of vehicle safety, type-approval regulation and technology road-mapping. Matthias represents TRL regularly at regulatory and stakeholder meetings and has presented as technical expert on behalf of the European Commission to working groups at European and at international levels (UNECE WG29). Matthias has authored scientific papers and presented his research at conferences.

Dr. **Omar M. Shehata** received his first MSc degree in Mechatronics Engineering in the field of multi-cooperative systems, at Ain Shams University (ASU), Cairo, Egypt, in 2014. His second MSc degree was completed in the field of Micro-Robotics used in medical applications, German University in Cairo (GUC), Egypt, in 2015. His PhD degree was finished in the field of Intelligent Transportation Systems from ASU in 2018. Dr. Shehata is currently an Associate Professor at the Mechatronics Engineering Department, GUC, Egypt.

Sneha Sudhir Shetiya is a Senior IEEE Member and a Staff SW Engineer at Torc Robotics Inc.. She received her Master's degree in Electrical Engineering with a major in Computer Vision and Signal Processing from North Carolina State University, USA in 2021. Her work involves middleware topics for embedded development of automated driving stack, automotive diagnostics, systems engineering and functional safety. She has two patents filed for the SOTIF standard. She is an active volunteer with IEEE Region 4, part of the committee for senior member evaluation at IEEE for 2024 and has been a proctor for IEEExtreme 24 hour coding competition. Sneha is a pioneer of Women in Technology and leads several mentorship programs.

Chaitanya Shinde is AV Systems and Safety Expert with a demonstrated history of 8+ years of working in the automotive and robotics industry. Also, he is an active WG member of the SAE, IEEE and ISO Standards Committee for Autonomous Driving and AI Safety. He serves on an Advisory Board for the SAE Young Professional Committee and is an acting industry advisor on CMU's Racing team. He serves as a safety architecture expert for The Autonomous, an international AV organization. Chaitanya has a Master's degree from Purdue and he is skilled in robotics, AI, and autonomous systems design, systems architecture, safety architecture, algorithms design & development and new product development and is interested in solving challenging real-life safety critical problems in the automated driving and mobile robotics world.

Zhiying Song received the BEng degree in Vehicle Engineering from the School of Automotive Studies, Tongji University, Shanghai, China, in 2021. He is currently working toward a Master's degree in Vehicle Engineering at Tsinghua University, Beijing, China. He is a member of the Tsinghua Intelligent Vehicle Design and Safety (IVDAS) Research Institute. His research interests include the robustness and safety of cooperative perception for automated driving vehicles.

Prof. **Christoph Stiller** received the Diploma degree in Electrical Engineering and the PhD degree from the RWTH Aachen University, Aachen, Germany, in 1988 and 1994, respectively. He held a Postdoctoral position with INRS, Montreal, QC, Canada. In 1995, he joined the Research of Robert Bosch GmbH in Germany. Since 2001, he has been a full Professor with the Karlsruhe Institute of Technology, Karlsruhe, Germany. He has served the IEEE Intelligent Transportation Systems Society in numerous positions, including as an Editor-in-Chief for the IEEE ITS Magazine, and the Vice-President and the Society President in 2012 and 2013, respectively.

Dr. **Henk Taale** is a senior consultant employed by Rijkswaterstaat, The Netherlands. He has over 30 years of experience in the fields of traffic management, traffic models and evaluation. Part of his job is the development and dissemination of knowledge and therefore he has been a member of several conference committees. Currently, he is the chair of the "Intelligent Mobility – Management and Operations" committee of the European Transport Conference He is also a part-time researcher at the Delft University of Technology to supervise MSc and PhD research and to continue his own research on traffic management. Henk graduated in Applied Mathematics and did his PhD on the subject of anticipatory control of road networks, dealing with the interaction between traffic management and road users.

Dr. **Fuxi Wen** (Senior Member, IEEE) received a PhD degree in Electrical and Electronic Engineering from Nanyang Technological University, Singapore, in 2013. He held research positions in Singapore and UK, from 2013 to 2017. He was a Researcher and a Marie Sklodowska-Curie Fellow with the Department of Electrical Engineering, Chalmers University of Technology, Gothenburg, Sweden, from 2017 to 2020. In 2020, he joined the School of Vehicle and Mobility, at Tsinghua University, Beijing, China, as an Associate Professor (research track). His research interests include vehicular communications and V2X-enabled cooperative driving techniques.

Tenghui Xie received the BEng degree in Intelligent Manufacturing Engineering from the School of Mechanical Engineering, Tongji University, Shanghai, China, in 2023. He is currently working toward the Master's degree in Vehicle Engineering at Tsinghua University, Beijing, China. He is a member of the Tsinghua Intelligent Vehicle Design and Safety (IVDAS) Research Institute. His research interests include automated driving and cooperative perception.

Prof. Dr.-Ing. **Adrian Zlocki** studied mechanical engineering at the RWTH Aachen University, Germany. From 2004 to 2012, he was employed at the Institute for Automotive Engineering, RWTH Aachen University (ika) as scientific assistant. Since 2010, he is leading the Automated Driving department. Currently, he is employed by fka GmbH. He is working in the fields of ADAS, active safety systems and automated driving.

Part I

Introduction

Chapter 1

History and deployment of (Cooperative) Intelligent Transport Systems

Adrian Zlocki[1] and Meng Lu[2]

1.1 Introduction

Increasing comfort and safety have been development goals of transport since its early beginning. Efficiency and traffic flow are also in focus as climate change and environmental and ecological impacts shape the further development of modern societies. In the last century, the value of road-based transport gained increasing importance within these modern societies. Individual mobility in particular and therefore the automobile, which offers everyone such an individual transport option, represents this development.

In order to increase safety, efficiency and comfort, different systems and technologies have been developed (e.g. automatic starter, chassis systems and restrain systems, start/stop functions). Various ideas and concepts for modern transport already emerged in the first part of the twentieth century, but at that time they could not be fully implemented and deployed due to technical limitations. A good example in the automotive sector is the idea of vehicle automation and automated driving. For instance, visions of automated driving and futuristic vehicle concepts were already displayed at the 1939 New York World's Fair Futurama [1].

The rapid development of electronics, micro-electronics and computer technology in the 1980s made measures available to provide such a comfortable and safe method of transport. These were first implemented at a research level by means of demonstrators and proofs of concept, followed by a switch to market introduction and deployment. Vehicle technology, intelligence for individual mobility and infrastructure-based systems for mass transport (including individual mobility) have since experienced significant achievements such as the introduction of ABS (anti-lock braking system), ESP (electronic stability program), routing technology and driverless public transport systems.

Technologies and measures to enhance comport, safety, efficiency and effectiveness of public transport and individual mobility are referred to as ITS

[1]fka GmbH, Steinbachstr. 7, Germany
[2]Aeolix ITS, Utrecht, The Netherlands

(intelligent transport systems). ITS describes a broad range of diverse technologies in the area of transport. These technologies include domains such as sensors and perception, communications systems, information and data processing and vehicle control. Applications of these technologies range from vehicle-manufacturer-dependent functions up to large-scale traffic management networks.

ITS applications are classified into the main areas of vehicle-based (on-board) and infrastructure-based (off-board) systems. Examples for vehicle-based ITS are advanced driver assistance systems (ADAS), active safety systems and vehicle automation, and include technologies for perception, positioning and (vehicle-to-vehicle) communication. Examples of infrastructure-based ITS are tolling, traffic monitoring, traffic control (vehicle-to-infrastructure), communication and all backend systems.

The term ITS is fairly generic and used in context of different disciplines such as traffic engineering, automotive engineering and communication engineering. A clear definition of ITS is not available. The following definitions are provided by different stakeholders.

EU Directive 2010/40/EU (7 July 2010): 'ITS are advanced applications which without embodying intelligence as such aim to provide innovative services relating to different modes of transport and traffic management and enable various users to be better informed and make safer, more coordinated and "smarter" use of transport networks' [2].

European Telecommunications Standards Institute (ETSI): 'ITS include telematics and all types of communications in vehicles, between vehicles (e.g. car-to-car), and between vehicles and fixed locations (e.g. car-to-infrastructure). However, ITS are not restricted to Road Transport – they also include the use of Information and Communication Technologies (ICT) for rail, water and air transport, including navigation systems' [3].

Federal Highway Administration: 'ITS improves transportation safety and mobility and enhances American productivity through the integration of advanced communications technologies into the transportation infrastructure and in vehicles. Intelligent transportation systems (ITS) encompass a broad range of wireless and wire line communications-based information and electronics technologies' [4].

ITS Japan: 'ITS offers a fundamental solution to various issues concerning transportation, which include traffic accidents, congestion and environmental pollution. ITS deals with these issues through the most advanced communications and control technologies. ITS receive and transmit information on humans, roads and automobiles' [5].

In general, it can be concluded that ITS applications aim to improve road-based traffic and transport by means of smart communications and automation technologies.

1.2   Classification of ITS

Different classification schemes can be applied to structure the wide variety of ITS applications available today. A methodical approach for the classification of ITS is a formal description of these systems. In the following, different classification schemes are provided.

Classification of ITS with regard to system architecture

This classification scheme clusters infrastructure based and vehicle based. The main distinguishing feature for this type of classification is on the architecture-dependent collection of ITS relevant information. Examples for infrastructure-based ITS range from traffic management systems, cloud-based systems such as tolling to even infrastructure-guided vehicles. Infrastructure-based automated vehicles are self-driving vehicles operated in a dedicated infrastructure such as dedicated lanes or closed areas (e.g. airport taxi, underground rail transport). These vehicles do not need a driver or operator. Vehicle-based ITS applications are defined according to different automation levels [6] or different types of active safety systems and rely mainly on on-board sensor information. A link between data collected from on-board and off-board sensors is the basis for combined systems. Next to the automation levels also the degree of connectivity is a scale for classification of ITS applications. These range from autonomous systems, which are not connected to the environment, up to highly connected systems with high data transfer rates.

Classification of ITS with regard to the driving tasks

ITS applications can also be classified using a three-level model for the driving task. This model distinguishes three operating levels: navigation, guidance and stabilisation [7]. At the navigation level, a route is determined inside an existing road network, based on origin and destination, and other selected route parameters. At the guidance level, the driver adjusts all relevant control parameters to the selected route in the network and the surrounding traffic. The stabilisation level is characterised by setting all necessary control parameters and changes thereof within the chosen driving strategy e.g. steering torque, accelerator pedal position or brake force.

Classification of ITS with regard to the type of support

The level of support provided by ITS can range from information to warning, and even extend to active intervention. An informational ITS application provides support by supplying exactly the information that is required in the current situation, e.g. a navigation system, which informs about a necessary motorway change at the next intersection. A warning system provides support by giving an acoustical, optical and/or haptic warning in a critical situation. This can be a critical state of the vehicle (e.g. low tyre pressure), of the driver (e.g. drowsiness) or of the overall situation (e.g. a close distance to another vehicle). An intervening system takes over a part of the driving task and thus supports or relieves the driver, e.g. ACC (adaptive cruise control) or AEB (automated emergency braking).

Classification of ITS according to the level of safety improvement
This classification, with a main focus on traffic safety, results from consideration of the (level of) safety improvement provided by ITS applications. The classification [8] distinguishes the following levels: safe traffic, risk avoidance (minimising risks like small time gaps between vehicles, high decelerations), collision prevention (e.g. emergency braking), protection of passengers (e.g. active and reversible restraint systems) and rescue management (e.g. eCall – electronic call).

ITS in passenger vehicles or lorries enables new functions to increase vehicle safety and vehicle automation. C-ITS (Cooperative Intelligent Transportation Systems) focuses on the connectivity of each single traffic participant in an overall connected network, facilitating the exchange of messages with the aim to increase comfort, safety and traffic and environmental efficiency.

1.3   History of ITS

Ideas for modern ITS systems have emerged since the invention of the automobile as given in [1]. A first mechanical implementation of an ITS comfort system installed in a passenger vehicle was the mechanical cruise control in the Chrysler Imperial as the so-called 'auto pilot' in 1958 [9]. This mechanical system was able to keep a constant set speed and therefore could take over the longitudinal vehicle control on a motorway. This system was designed by means of mechanical components. In the 1980s, these could be replaced by a computer controlled unit due to the developments in the field of electronics and information technology [10].

With regard to communication technologies and infrastructure, the German Autofahrer-Rundfunk-Information system (ARI, motorist broadcasting information system) was one of the first systems in Europe, active from 1974 to 2008 [11]. The system was developed by the German company 'Blaupunkt' in collaboration with national radio broadcasters. ARI added a special modulated sound ('Hinz-Triller') to radio-based traffic messages, which could be detected by special equipped receivers in order to determine the availability of a radio station sending out traffic messages, a geographic area to determine the relevance of the radio station and finally the traffic message itself, which could be replayed to inform the driver.

Since the introduction of these first systems, information and communication technologies have been applied in many fields of road transport. A systematic research of such technologies was started in several research programs with national or international funding. The goal of these programs was joint development of systems and standards, independent of one single company. Such research programs were implemented in different countries such as PATH (Partners for Advanced Transit and Highways) and IVHS (Intelligent Vehicle Highway Systems) in the United States, ARTS (Automated Road Transportation Systems) in Japan and PROMETHEUS (PROgraMme for a European Traffic of Highest Efficiency and Unprecedented Safety) in Europe.

In 1986, the California PATH program was established. The program was sponsored by the California Department of Transportation. The aim of PATH was to develop long-term strategies in order to cope with the immense traffic in California. The University of California at Berkeley and other university partners from California investigated solutions for public and private transport involving ITS [12]. Between the 7 and 10 August 1997, a vehicle platoon consisting of eight Buick LeSabre vehicles was demonstrated as one of the outputs of PATH. The distance between each longitudinally and laterally controlled vehicle was 6.5 m at a velocity of up to 96 km/h. The distance accuracy was measured in the range of 10 cm at constant driving and 20 cm at acceleration and deceleration manoeuvres [13]. At national level, the IVHS program was authorised by the Intermodal Transport Surface Transportation Efficiency Act of 1991. Within this IVHS program, five broad, interrelated areas were addressed: advanced traffic management systems, advanced traveller information systems, advanced vehicle control systems, commercial vehicle, operations and advanced public transportation systems [14]. The program initiated several IVHS operational tests in different parts of the United States.

In Japan, funded research in the field of ITS was started in the 1980s by different ministries and agencies. Vehicle-oriented projects, such as SSVS (Super Smart Vehicle Systems) and Energy ITS, were sponsored by METI (Ministry of Economy, Trade and Industry). ASV (Advanced Safety Vehicle) was sponsored by the MLIT (Ministry of Land, Infrastructure, Transport and Tourism) in addition to infrastructure-oriented projects like VICS (Vehicle Information and Communication System), ETC (Electronic Toll Collect) and Smartway. The infrastructure-oriented project UTMS (Universal Traffic Management System) was sponsored by the NPA (National Police Agency) [15]. In 2008, the Energy ITS project was aiming at energy efficiency and CO_2 emission reduction by means of truck platooning [15]. At the ITS World Congress 2013, a platoon of three automated trucks driving at 80 km/h with the gap of 15 m was demonstrated as one of the results of the project. In 2014, the Japanese government started the Cross-Ministerial Strategic Innovation Promotion Program (SIP), which refers to, inter alia, the next generation of ITS infrastructure [16]. A main project in this research program is the SIP-adus project (Automated Driving for Universal Service), which aims to deploy automated driving in Japan. Within the SIP-adus project, research is conducted in the field of automated driving, on topics such as dynamic map, connected vehicle, security, impact assessment, human factors and next-generation transport [17].

In Europe, following preliminary research activities of the European automobile industry on automated driving, the PROMETHEUS project, initiated in 1985, and lasting 1986–1994, was a research initiative of Eureka, an intergovernmental organisation for pan-European research and development funding and coordination. It was the first Pan-European cooperation of 14 European vehicle manufacturers and over 100 suppliers. The initiative was a direct reaction to the joint research activities in the United States and Japan [18]. The aim of PROMETHEUS was the improvement of traffic safety and traffic management on the three levels of the driving task: navigation, guidance and stabilisation. The PROMETHEUS project consisted of seven sub-projects. Within the sub-project PRO-CAR [19] first advanced driver assistance functions and necessary ICT were developed and demonstrated, including new sensor concepts, algorithms for image processing and control functions [19]. Automated driving functions were implemented for the very first time by the project. Today, many of the elaborated ideas and demonstrations are well established and available in vehicles on the market. Prominent examples are integrated navigation support, Adaptive Cruise Control (ACC) and Lane Keeping Assistance (LKA).

The positive results of these first research actives and the resulting market introductions led to continuous research programs and the founding of member organisations active in the field of ITS such as ERTICO (European Road Transport Telematics Implementation Coordination Organisation) – ITS Europe, founded in 1991, and similar ITS Associations in Asia Pacific, North America and Europe, as well as the Intelligent Transportation Systems Society (ITSS), launched in 1999 within IEEE (the Institute of Electrical and Electronics Engineers). Since the 1980s, European framework programmes, up to the current programme Horizon Europe (running until 2027) have funded a wide range of research projects and deployment measures in the field of ITS.

1.4   Infrastructure-based ITS

Infrastructure-based ITS applications use off-board sensors and communication technology. Due to the wide spread availability of state-of-the-art communication technology, the deployment of these systems is increasing.

1.4.1   *Emergency vehicle notification systems, eCall*

After a long struggle, eCall was finally adopted through the European directive 2007/46/EG EC, which required it to be operational from 2018 [20]. This ITS application can determine the exact location in case of an accident and initiate an automated emergency call to the nearest emergency call centre independent of the location in the EU. Due to the ambitions to collect and communicate data, including personal information such as location, and the continuous technical advances in communication technology, the introduction of a unified eCall system and its design was discussed and investigated for many years. The eCall system shall be free of charge and use the international emergency number 112. To prevent this data from being available to third parties, all location-related data may not be recorded and stored. Meanwhile, vehicle manufacturer-dependent systems are being introduced as services, providing additional comfort for intelligent routing and navigation to points of interest.

1.4.2   Automatic road law enforcement

Since 1970s, vehicle speed has been measured by law enforcement using ITS technology like laser or radar. These speed controls are infrastructure-based at selected locations or mobile by means of so-called laser guns. The collected data are transferred to a central server in order to issue a ticket to the driver, if the traffic rules were violated. In addition to the observance of speed limits, automatic road law enforcement is also used for the detection of red light runners at crossings.

1.4.3   Tolling

Tolls for using certain roads or waterways exist long before the invention of the automobile. A first system for electronic tolling was proposed in 1959. By means of Dedicated Short-Range Communication (DSRC), data are transmitted from a transponder in the vehicle to the infrastructure. A computer calculates the toll for the vehicle according to the driven distance and the time of day. Electronic toll collection systems rely on four components: automated vehicle identification (mainly by radio-frequency identification via transmitter), automated vehicle classification (classification of vehicle), transaction processing and violation enforcement (e.g. police patrols, physical barriers and automatic number plate recognition) [21].

1.4.4   Variable speed limits

Standard speed limits signs do not take the given conditions into account. Variable speed limit signs adapt depending on the environmental situation, traffic flow and time of day. Infrastructure-based ITS applications in the form of sensors (e.g. cameras, induction loops and laser) measure traffic flow, weather condition and determine events such as heavy congestion. Data are collected in traffic management centres, which transmit relevant information to the driver by means of spoken radio broadcast messages, the Traffic Message Channel of the Radio Data System (RDS-TMC) and variable traffic signs.

1.4.5   Automated driving supported by infrastructure

Vehicles using infrastructure-based automated driving provide new mobility concepts. Especially public transportation systems such as trains, buses and Personal Rapid Transit (PRT) are operated on different automation levels even up to completely driverless. Automated trains and transportation systems in dedicated areas (e.g. railways, special driving lanes, closed environments) have been introduced since many years. PRT comprises a new modes of automated transportation. This concerns fully automated driverless vehicles, which navigate automatically along a network of dedicated guide ways providing on-demand service. An example of such a system is the Ultra at Heathrow Airport, which uses four-seater vehicles the size of a small car that run on four wheels with rubber tires [22]. Another PRT system in operation is the CyberCab in Masdar, Abu Dhabi [23]. In the recent years, fleets of automated vehicles are operated in different cities around the globe. Most prominent is the city of San Francisco in the United States, in which different companies operate their so-called 'Level 4' vehicles (partly driverless). In order to provide a safe and reliable service, all vehicles are connected via digital infra-structure to a control centre. In these control centres, human operators are able to interact with the passengers and control the vehicle remotely in case of any failure, emergency or unexpected event.

Digital infrastructure for automated driving is becoming more important. Various test fields and even entire cities are investing in vehicle automation through connectivity. For automated shuttles, such digital infrastructure with an infrastructure-based control centre can be mandatory (see the German legislation on Level 4 vehicle operation [24]). In 2019, an infrastructure categorisation auto-mated driving was introduced in the EU-funded research project INFRAMIX [25]. This ISAD scheme (Infrastructure Support Levels for Automated Driving) distin-guishes five different levels of infrastructure support for road automation, as listed in the table below.

A: **Cooperative driving:** Infrastructure is capable of perceiving vehicle trajectories and guide single AVs (or AV groups)
B: **Cooperative perception:** Infrastructure is capable of perceiving microscopic traffic situations
C: **Dynamic digital information:** All static and dynamic information can be provided to the AVs in digital form
D: **Static digital information/map support:** Digital map data (including static road signs) complemented by physical reference points
E: **Conventional infrastructure/no AV support**

1.5   Vehicle-based ITS

Since the introduction of research and deployment programs for ITS, a large number of projects have been carried out. The technical development, made pos-sible by the availability of high resolution perception systems (see Figure 1.1), computational power and wireless communication, is driven by two different approaches: automation and connectivity.

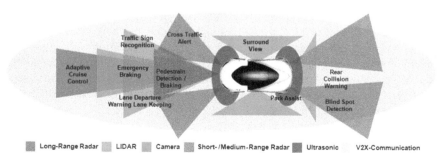

Figure 1.1   Overview on available perception systems and resulting ADAS

With regard to the application of ITS technologies in vehicles, fundamental results have been achieved in research projects over the past 25 years. In Europe, the first test drive was undertaken in the VAMP activity at the end of the PROMETHEUS project. In 1995, a journey of approximately 1,600 km took place from Munich (Germany) to Odense (Denmark), with most of the driving performed by the VaMoRs (Versuchsfahrzeug für autonome Mobilität und Rechnersehen) system [26]. In Italy, the ARGO (Algorithms for Image Processing) project demonstrated vehicle automation during a journey of approximately 2,000 km in the MilleMiglia tour in 1998 [27].

A new mobility concept in terms of 'Cybercars' was created, which uses the advantages of automobiles combined with ITS technology. The idea of Cybercars was developed in the European-funded research project Cybermove [28]. Cybercars are slow-moving automated driving vehicles on existing road infra-structure. The first systems were put in operation in the Netherlands at the end of 1997 and have been running successfully 24 hours a day. Several other systems have been implemented in European research projects on this topic: the Cybermove project, 2004, the Cybercars project, 2006 and the CityMobil projects 1 and 2 (cities demonstrating automated road passenger transport), 2011 and 2016.

In the US demonstration NavLab 5, a journey of approximately 4,587 km from Pittsburgh (United States) to San Diego (United States) was conducted, also already in 1995 [27]. The foundation for much of today's self-driving vehicle activities of various companies and institutions was laid by the DARPA (Defense Advanced Research Projects Agency) Grand Challenges. The DARPA Grand Challenge 1 was conducted in 2004, followed by the DARPA Grand Challenge 2 in 2005. Two years later in 2007, the DARPA Urban Challenge took place in an urban environment. In all three challenges, robot vehicles had to complete a test track (in the desert and in an urban city scenario) without any human interaction or control [29].

Shortly after the DARPA Grand Challenges, Google announced its self-driving car program. Since then, various vehicle manufacturers and an automotive supplier have demonstrated milestones in vehicle automation such as the VW Golf 53+1, the BMW Track Trainer, the Daimler Berta Benz drive, Audi's Jack and Delphi's Coast to Coast drive [30]. Nowadays, experience and data are collected in large-scale field tests around the world and vehicle automation is part of every development roadmap in the transportation sector.

The first systems that came onto the market and were sold to customers focused on supporting of the driver through comfort, information and warning functions. Active safety systems, capable of taking over either the longitudinal or the lateral vehicle control, use sophisticated sensor systems and fusion of different sensor principles. The systems are classified on guidance level (according to [7]) into three categories: A – informing and warning systems; B – vehicle automation; and C – temporary intervention in accident prone situations [31]. Category B is divided into different levels of automation that a vehicle-based ITS application can provide. These levels of automation are defined by SAE (Society of Automotive Engineers) [6]. In the following, the different automation levels according to the SAE definition are described.

1.5.1   Automation Level 1 – driver assistance

At Level 1, ADAS applications execute driving mode-specific driving tasks by either steering or acceleration/deceleration using information about the driving environment. The human driver is in the driving loop and monitors the driving environment. Such systems have been introduced in the market since the mid-1990s. Figure 1.2 shows the history of the market introduction of these systems.

1.5.2   Automation Level 2 – partial automation

Specific for Level 2 is the execution by one or more driver assistance systems of both steering and acceleration/deceleration using information about the driving environment. The driver remains still in the driving loop and needs to monitor the environment. He is supported in lateral and longitudinal vehicle control. One of the first systems on the market is, for example, the traffic jam assist, which can be activated in a specific condition (traffic jam) and supports the driver in the dynamic driving task. A concept known from railways is to combine single vehicles into so-called platoons. Research activities like PATH [13], KONVOI (Entwicklung und Untersuchung des Einsatzes von elektronisch gekoppelten Lkw-Konvois) [32], Energy ITS Japan [15], SARTRE (Safe Road Trains for the Environment) [33], COMPANION (Cooperative dynamic formation of Platoons safe and energy-optimised goods transportation) [34] or ENSEMBLE (ENabling SafE Multi-Brand pLatooning for Europe) [35] develop and demonstrate these platoons. Within the SARTRE project, a platoon of five vehicles (two lorries and three passenger vehicles) was built up and demonstrated on public road. Currently, the logistics of how platoons are formed during a trip is under research [34].

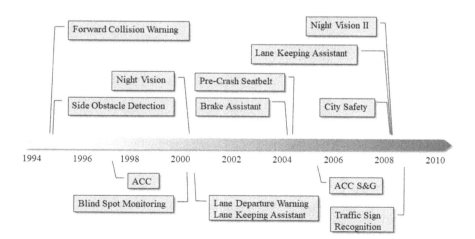

Figure 1.2   History of market introduction of Level 1 driver assistance systems focusing on guidance and navigation level up to 2010

1.5.3   Automation Level 3 – conditional automation

Starting at automation Level 3, the system monitors the driving environment. The system performs the dynamic driving task in every aspect. If necessary, the human driver will respond appropriately to a request to intervene. Such systems were first introduced in 2021 in Germany and 2023 in California by Daimler as a traffic jam pilot for specific traffic jam conditions. The driver can concentrate on secondary tasks in addition to the driving task without having to monitor the traffic in front of the vehicle.

1.5.4   Automation Levels 4 – high automation, and Level 5 – full automation

At Level 4, all aspects of the dynamic driving task are performed by the automation in a limited ODD (operational design domain). Even if a human driver does not respond adequately to a request to intervene, the system needs to be able to solve the situation. The driver is not to be foreseen as a fallback for the performance of the driving task. In Europe, automated shuttles are the focus of Level 4 automation. At Level 5, the full-time performance of all aspects of the dynamic driving task under all road and environmental conditions (unlimited ODD) is taken over by the automated system. The driver does not need to be available. This automation level is not reached by any known public research activity.

1.5.5   Teleoperation

Wireless communication improves vehicle automation by providing additional data from the outside, e.g. from road side units (RSUs) or other vehicles, to the automation function. Full connectivity of the vehicle also makes remote operation (teleoperation) possible. Teleoperative driving can also be divided into different levels [36]: (1) *remote driving*: the system is completely controlled remotely; (2) *remote assistance*: the system receives event-driven remote assistance from the operator, while still being responsible for the driving task; (3) *remote monitoring*: the system is monitored remotely with very limited intervention possibilities.

1.6   Deployment of ITS

Large-scale deployment with high penetration rates is desirable in order to maximise the impact of ITS. Especially in recent years, implementation and deployment has been the focus of funding activities and AD-companies due to the high level of maturity of ITS technology. Cross-border and cross-company standards are a prerequisite for a successful deployment. Standards pave the way for market introduction.

1.6.1   Standardisation

Standardisation of new technologies in the field of transport allow safe operation of systems and exchange of information, regardless of design and manufacturer. Different institutions are involved in the standardisation of ITS.

CEN (European Committee for Standardization) was established in 1974. Its Technical Committee TC 278 – Road Transport and Traffic Telematics, now bearing the name Intelligent transport systems, was established in 1992 and consists in 2024 of eight active working groups. ITS-relevant standards in Europe are managed in CEN/TC278 [37]. ETSI (European Telecommunication Standards Institute) was created on 29 March 1988, and has developed a number of standards related to ITS [38]. The ISO (International Organization for Standardization) Technical Committee TC204 – Intelligent transport systems has been developing standards on the overall system and infrastructure aspects of ITS. ISO/TC 204 oversees the coordination of the ISO program in this field, which includes setting the schedule for standards development and taking into account the work of existing international standardisation bodies. Work in ISO/TC 204 includes the standardisation of information, communication and control systems in the field of urban and rural surface transportation, traveller information, traffic management, public transport, commercial transport and emergency and commercial services [39]. Standardisation work in the IEEE Standards Association (SA) on communication and ITS, SAE on vehicle automation, CENELEC (European Committee for Electrotechnical Standardization), IEC (International Electrotechnical Commission) and other national institutions as well as standardisation from robotics, industrial automation and communication are relevant in the field of ITS. In addition to these different standardisation organisations, the United Nations Economic Commission for Europe (UNECE) paved a way to vehicle automation. This resulted in UN regulations such as Automated Lane Keeping Systems (ALKS), UN Regulation No. 157 [40]. ALKS defines the control of the lateral and longitudinal movement of the vehicle for extended periods of time without further command from the driver. In a first step, the maximum operational speed was limited to 60 km/h, but an extension to 130 km/h was proposed by the Working Party on Automated and Connected Vehicles (GRVA).

1.6.2   Testing and demonstration

Following the design and development of ITS, the final steps before market introduction are testing and development. In addition to track tests and field tests, so-called field operational tests (FOTs) constitute an important approach for large-scale test on public roads. These FOTs, which analyse the developed and available technologies in the target environment, are conducted at manufacturers and/or in collaborative dedicated projects. ITS technologies have been investigated in FOTs since many years. Examples of important FOT testing activities that have taken place in the last decades are the 100-car study performed by VTTI (Virginia Tech Transportation Institute) in the United States [41], and European projects such as euroFOT (the first large-scale European Field Operational Test) [42] and Drive C2X (comprehensive assessment of cooperative systems through Field Operational Tests) [43], amongst others. In Europe, the L3Pilot project demonstrated Level 3 or Level 4 systems in several European countries [44]. In addition, system manufacturers can obtain tests permits to conduct evaluations and collect market readiness sign-off data. Especially tests of vehicle automation Levels 3 and 4 have been the focus of recent activities. The importance of intelligent infrastructure led to the introduction of several physical test fields. These test fields range from closed proving grounds, via single infrastructure measures like intersections, and test highway sections, up to whole test cities. In addition to the physical infrastructure, test fields consist of digital infrastructure, which is capable of providing information to an intelligent vehicle. Several states and cities in different countries issue driverless permits and allow driverless testing.

1.6.3   Market introduction

A successful market introduction is the last step of deployment. In order to max-imise the impact of a technology, a high penetration rate is desirable. For market introduction, different approaches need to be distinguished:

1. A technology is introduced for new infrastructure or vehicles as an option. In this case, it depends on the customer whether this option is chosen. It can be expected that the customer will go for the option in case a benefit (monetary, comfort, safety) from the system is expected. Typically to this approach is the slow increase of the penetration rate, since only new infrastructure or vehicles are equipped. This implies that existing infrastructure or vehicles are not involved and need to be replaced over time.
2. A technology is introduced for new infrastructure or vehicles as an option, and a retrofitting of existing infrastructure or vehicles is possible. As with the first approach, it is up to the customer whether a technology is successful or not. In contrast to the previous approach, at least theoretically, higher penetration rates can be reached faster.
3. A technology is introduced for new infrastructure or vehicles as standard equipment. This could happen, for instance, due to a change in a law that requests a certain technology. In this case, the customer cannot choose and the penetration rate entirely depends on the market introduction of the new infra-structure or vehicles. Due to the enforced introduction of the technology, it can be hard to charge the customer extra.

Figure 1.3 provides an overview on the penetration of different vehicle-based ITS applications in Germany. The penetration rates of the systems were derived from 5,006 interviews and vehicle inspections of ordinary drivers. A further increase of the penetration rate is expected if such systems will become required in the coming years by independent vehicle assessment programs, such as Euro NCAP, and also European legislation. This will foster rapid introduction by all relevant vehicle manufacturers.

First Level 3 vehicle automation systems are also already being introduced to customers of the automotive industry. In addition to these advances, some technology companies are considering a fleet-based introduction of Level 4 vehicles. The first driverless test fleets are being operated in selected cities. These Level 4 vehicles are true ITS-based applications, as they combine automation and connectivity with a control centre. The operation results in experience gained, identification of issues and safety cases, and both positive and negative media coverage.

System Name	Penetration rate of fitted vehicles [%]	System Name	Penetration rate of fitted vehicles [%]
Navigation and driver information		**Lane Keeping and Lane Change Assist**	
Navigation Device	85	Lane Change Warning	13
Speed Warning	22	Blind Sport Detection	20
Attention Assist	23	Lane Departure Warning	18
Traffic Sign Detection	20	Lane Change Assist	3
Vehicle Dynamics and Distance Control		**Parking system**	
Brake Assist	86	Park Distance Control	60
ESC	90		
Forward Collision Warning	18	Rear View Camera	25
Collision Warning	25		
Secondary Collision Mitigation	17		
AEB up 30 km/h	22	Park Assist	14
AEB above 30 km/h	12		
Cruise Control		**Others**	
Cruise Control	59	eCall	23
Speed Limiter	36		
ACC (Adaptive Cruise Control)	17	Emergency Assist	2
Traffic Jam Assist	5		

Figure 1.3   Overview of the penetration rate of different vehicle-based ITS applications in Germany in 2021 on basis of [45]

1.7   Outlook

Today, ITS technologies have already a long history and are being introduced into the transportation sector as a wide range of different systems. A disruptive impact of these technologies is forecasted with the introduction of automated driving technologies for higher automation levels. Currently, a phased introduction versus a direct introduction of dedicated automated vehicles is considered by several companies from different industry branches.

Roadmaps and ITS action plans indicated a further deployment of ITS technology in each upcoming new transport generation, regardless of vehicle type and manufacturer. Electronics and software will play the most important role. A shift from hardware to software development is one of the essential transformations that can be observed in current research and development. In addition to the software-defined vehicle, the role of large data sets offers new technological approaches.

1.8   Outline of the book chapters

The book is composed of four Parts: I. Introduction; II. Needs and Methods; III. Intelligent Transport Systems Evaluation Results; IV. Discussion and Conclusions. In this second edition, the chapters are either entirely new or substantially updated chapters from the first edition. Table 1.1 presents an outline with brief descriptions of each chapter.

Table 1.1   Outline of the book chapters

Part	Chapter	Description
I	1. History and Deployment of (Cooperative) Intelligent Transport Systems	Up-to-date technical review of (C-)ITS in terms of relevant research programmes, challenges, market introduction and recent developments of automated driving. Infrastructure-based and vehicle-based key technologies are comprehensively reviewed. Standardisation and testing of (C-)ITS are summarised.
II	2. The Evolution towards Automated Driving – Classification of Impacts, Review of Assessments of Automated Driving Functions, Challenges for Evaluation	Update of an evolutionary approach towards higher level automated driving. Introduction of automation levels and use cases. Review of social, environmental and economic impact evaluation frameworks and practices. Discussion of infrastructure impacts. And analysis of challenges for the evaluation of automated driving.
	3. Field Operational Tests (FOTs) – (Still) the Ultimate Answer to Impact Assessment?	Update of the evaluation of Field Operational Tests (FOTs) by analysing whether a FOT is (still) the best approach for impact assessment, for example for studying drivers, driving behaviour, use of different functions and the impact of these functions; and for assessing the impact of more advanced automated systems.
	4. Driving the Future: The Role of Artificial Intelligence in Road Vehicles	Introduction of the role of and use cases for Artificial Intelligence (AI) in Automated Vehicle (AV) applications. Extensive discussion of data processing techniques, AV datasets, software package size, processing power and AI challenges in automotive. In addition, a review of relevant standardisation and regulatory activities is provided.
	5. Assessment Method for Prioritising Transport Measures and Infrastructure Development	Introduction and discussion of the Assessment Method for Policy Options (AMPO) – a framework addressing decision-making processes by integrating cost–benefit analysis and multi-criteria analysis while promoting stakeholder engagement. Discussion of a use case of transport policy and infrastructure planning by using AMPO.
	6. Evaluation of ITS: Opportunities and Challenges in the Era of New Pervasive Technology	Discussion of the potential for new data forms to contribute to the evaluation of a variety of established ITS infrastructure schemes, of approaches for evaluation of the impact of ITS schemes based on

(Continues)

Table 1.1   (Continued)

Part	Chapter	Description
		pervasive connected technologies, and whether the needed approach differs from the approach used to evaluate established ITS infrastructure schemes.
III	7. The Potential Benefits of Heavy Goods Vehicle (HGV) Platooning	This chapter targets cooperative road freight transport, especially regarding the benefits of decarbonisation, safety and efficiency. It brings together recent research results from real world platooning, and discusses how once an impact assessment is completed, such ITS technologies can be effectively scaled to realise commercial success and societal benefits.
	8. C-ITS Deployment in Australia – Achievements & Key Learnings	Comprehensive outline of C-ITS infrastructure and services. Review of the application of various C-ITS services in Australia. Technical assessment thereof, and in addition evaluation of impacts, especially on user experience and road safety, using a solid evaluation approach.
	9. C-ITS Evaluation on C-ROADS – Findings from C-ROADS Germany	Introduction of the C-Roads Platform, and presentation of the method and the processes of the impact assessment of the C-ITS deployment in Europe. Summary of the results of the C-ITS services deployed at German C-Roads Pilots, especially regarding Green Light Optimal Speed Advisory (GLOSA) and RoadWorks Warning (RWW).
	10. Impact Assessment of Large-Scale C-ITS Services in Greece	Impact assessment and evaluation of C-ITS services implemented in Greece, under C-Roads Greece. Overview of the implementation of the C-ITS services, and of the specific use cases. Presentation of subjective and objective evaluation methods, and of evaluation results, with detailed analysis.
	11. System Architecture for the Deployment of Autonomous Mobility On-Demand Vehicles	Introduction of a system architecture for on-demand mobility services based on the automated vehicles (such as shuttles and robotaxis), developed using a systems engineering approach. Focus on ensuring flexibility and scalability of the Autonomous Mobility on-Demand (AMoD) system.

(Continues)

Table 1.1   (Continued)

Part	Chapter	Description
	12. Robust Cooperative Perception for Intelligent Transport Systems	Discussion of the challenges of cooperative perception, especially regarding obtaining precise positioning information with a low-cost and robust approach. Context-based matching is proposed as such, and details are provided. Account of the extensive experiments that were conducted using this method on both simulated and real-world datasets, and the excellent results that were obtained.
	13. Cooperative Architecture for Transportation Systems (CATS): Assessment of Safety and Mobility in Vehicular Convoys	Convoy in the context of ITS refers to multiple cooperative vehicles spreading over multiple lanes, traversing with a constant speed while maintaining a pre-designed formation. The chapter studies the convoy system and its potential positive impact on safety, mobility and the environment, and explores some important behaviours of the vehicular convoy.
	14. Traffic Management using Floating Car Data in Low- and Middle-Income Countries	Exploration of FCD applications in traffic planning and management, ensuring maximum benefit from this data source, while maintaining the integrity of traditional traffic engineering methods, in the context of low/middle-income countries. With a case study in South Africa and a proposal for the deployment of xFCD.
	15. European ITS Platform: Evaluating the Benefits and Impacts of ITS Corridors	Introduction of the European ITS Platform (EU EIP), the EU EIP Evaluation Group and the related evaluation approach. Overview of the evaluation results of the five ITS Road Corridor projects, and consolidation of a harmonised and substantiated impact evaluation of the socio-economic benefits of ITS services.
IV	16. Key Findings and the Future of Intelligent Transport Systems	Overview of the development and deployment of (C-)ITS and services, critical review of the evaluation methods, and presentation of the evaluation results of (C-)ITS applications. Some further thoughts (regarding the evaluation, research, ITS deployment and decision making challenges) and recommendations.

References

[1]   Herman, A. (2012). *Freedom's Forge: How American Business Produced Victory in World War II*, pp. 58–65, Random House, New York, NY, 2012.

[2]   Buzek, J. and Chastel, O. (2010). Directive 2010/40/EU of the European Parliament and of the Council. 26 August 2024. https://eur-lex.europa.eu/eli/dir/2010/40/oj

[3]   N.N. (2016). ETSI Automotive Intelligent Transport Systems (ITS). 26 August 2024. https://highways.dot.gov/safety/other/intelligent-transportation-systems-safety

[4]   N.N. (2016). Intelligent Transportation Systems Joint Program Office. www.its.dot.gov/about/its_jpo.htm

[5]   N.N. (2016). ITS Japan. 26 August 2024. www.its-jp.org/english/about_e/

[6]   N.N. (2021). Taxonomy and Definitions for Terms Related to Driving Automation Systems for On-Road Motor Vehicles, SAE Standard J3016, USA, 2021.

[7]   Donges, E. (1982). Aspekte der aktiven Sicherheit bei der Führung von Personen-kraftwagen, Automobil-Industrie, Heft 2.

[8]   Eckstein, L. (2014). Automotive Engineering – Active Vehicle Safety and Driver Assistance Systems, Lecture Notes, Institute für Kraftfahrzeuge (ika), RWTH Aachen University, Aachen, Germany.

[9]   Rowsome, F. (1958). What It's Like to Drive an Auto-Pilot Car, Popular Science Monthly, USA, April 1958.

[10]   Shaout, A., and Jarrah, M. A. (1997). Cruise Control Technology Review, *Computers Electric Engineers* 23(4), 259–271.

[11]   Weiskopf, B. (2001). ARI-Technik. UKW/TV-Arbeitskreis der AGDX e.V., Mannheim, Februar 2001.

[12]   Shladover, S. (1992). The California PATH Program of IVHS Research and Its Approach to Vehicle-Highway Automation, IEEE Intelligent Vehicle '92 Symposium, Detroit, USA.

[13]   N.N. (1997). *PATH Program – Vehicle Platooning and Automated Highways, PATH Fact Sheets*, University of California, Berkeley, CA.

[14]   Mammano, F. J., and Bishop, J. R. (1992). Status of IVHS Technical Developments in the United States, *IEEE Vehicular Technology Conference*, Denver, USA.

[15]   Tsugawa, S. (2011). The Current Trends and Issues on ITS in Japan: Safety, Energy and Environment, *IEEE MTT-S International Microwave Workshop Series*, Daejeon, South Korea.

[16]   Harayama, Y. (2014). The National Program for Innovation, Cross-Ministerial Strategic Innovation Promotion Program (SIP), *1st SIP-adus Workshop on Connected and Automated Driving Systems*, Tokyo, Japan.

[17]   Kuzumaki, S. (2015). Innovation of Automated Driving for Universal Service (SIP-adus) – Mobility Bringing Everyone a Smile, *The 2nd SIP-adus Workshop on Connected and Automated Driving Systems*, Tokyo, Japan.

[18] Nagel, H. H. (2008). *EUREKA-Projekt PROMETHEUS und PRO-ART (1986–1994), Informatikforschung in Deutschland*, Springer, Berlin, pp. 151–202.

[19] Hofflinger, B., Conte, G., Esteve, D., and Weisglas, P. (1990). Integrated Electronics for Automotive Applications in the EUREKA Program PRO-METHEUS, *ESSCIRC '90. Sixteenth European Solid-State Circuits Conference*, Vol. 2, pp. 13–17.

[20] N.N. (2015). Regulation (EU) 2015/758 of the European Parliament and of the Council (29 April 2015) Type-approval Requirements for the Deployment of the eCall In-vehicle System Based on the 112 Service and Amending, Directive 2007/46/EC.

[21] Kelly, F. (2006). Road Pricing: Addressing Congestion, Pollution and the Financing of Britain's Road. *Ingenia (The Royal Academy of Engineering)* 39, 36–42.

[22] Bly, P., and Lowson, M. (2009). Outline Description of the Heathrow Pilot PRT Scheme, Deliverable D1.2.2.2, CityMobil project.

[23] N.N. (2011). 2getthere. 26 August 2024. https://www.2getthere.eu/

[24] N.N. (2022). Verordnung zur Genehmigung und zum Betrieb von Kraftfahrzeugen mit autonomer Fahrfunktion in festgelegten Betriebsbereichen (Autonome-Fahrzeuge-Genehmigungs-und-Betriebs-Verordnung – AFGBV), Autonome-Fahrzeuge-Genehmigungs-und-Betriebs-Verordnung vom 24. Juni 2022 (BGBl. I S. 986), die durch Artikel 10 der Verordnung vom 20. Juli 2023 (BGBl. 2023 I Nr. 199) geändert worden ist 24.06.2022.

[25] Lytrivis, P., Manganiaris, M., Reckenzaun, J. *et al.* (2019). Infrastructure Classification Scheme. INFRAMIX Deliverable D5.4 (Grant Agreement No: 723016). 26 August 2024. www.inframix.eu

[26] Maurer, M., Behringer, R., Fürst, S., Thomanek, F., and Dickmanns, E. D. (1996). A Compact Vision System for Road Vehicle Guidance, *Proceedings of IEEE ICRP 1996*, IEEE, Piscataway, NJ.

[27] Bertozze, M., Broggi, A., and Fascioli, A. (2000). Vision-Based Intelligent Vehicles: State of the Art and Perspectives, *Robotics and Autonomous Systems,* 32(1), 1–16.

[28] Parent, M., and Gallais, G. (2002). Intelligent Transportation in Cities with CTS, *Proceedings of IEEE 5th International Conference on Intelligent Transportation Systems*.

[29] Urmson, C., Anhalt, J., Bagnell, D. *et al.* (2008). Autonomous Driving in Urban Environments: Boss and the Urban Challenge, *Journal of Field Robotics* 25(8), 425–466 (2008), DOI: 10.1002/rob.20255, Wiley Periodicals, Inc.

[30] Zlocki, A. (2014). *Automated Driving, Encyclopedia of Automotive Engineering*, John Wiley & Sons, West Sussex, PO19 8SQ, United Kingdom, DOI: 10.1002/9781118354179.auto023.

[31] Gasser, T. M., Seeck, A., and Smith, B. W. (2015). Rahmenbedingungen für die Fahrerassistenzentwicklung, *Handbuch Fahrerassistenzsysteme*, Springer, Berlin.

[32]   Deutschle, S. (2008). Das KONVOI Projekt – Entwicklung und Untersuchung des Einsatzes von Lkw-Konvois, Aachener Kolloquium Fahrzeug und Motorentechnik 2008, Aachen, Germany.

[33]   Chan, P. Gilhead, P. Jelinek, P. Krejci, and Robinson T. (2012). Cooperative Control of SARTRE Automated Platoon Vehicles, *Proceedings of the 19th ITS World Congress*, Vienna, Austria.

[34]   Farokhi, F., and Johansson, K. H. (2015). A Study of Truck Platooning Incentives Using a Congestion Game, *IEEE Transactions on Intelligent Transportation Systems*, 16(2), 581–595.

[35]   Pont, J. (2021). Public Demonstration. ENSEMBLE Deliverable D5.6 (Grant Agreement No: 769115). 26 August 2024. https://platooningensemble.eu

[36]   Majstorovic, D., Hoffmann, S., Pfab, F. *et al.* (2022). Survey on Teleoperation Concepts for Automated Vehicles. *2022 IEEE International Conference on Systems, Man, and Cybernetics (SMC)*, IEEE, Piscataway, NJ.

[37]   CEN/TC278 (2016). 26 August 2024. www.itsstandards.eu/index.php

[38]   ETSI TC 278 (2024). 26 August 2024. www.etsi.org/index.php/technologies-clusters/technologies/intelligent-transport

[39]   ISO/TC 204 – Intelligent Transport Systems (2016). 26 August 2024. https://www.iso.org/committee/54706.html

[40]   N.N. (2021). UN Regulation No. 157, Uniform Provisions Concerning the Approval of Vehicles with regard to Automated Lane Keeping Systems. E/ECE/TRANS/505/Rev.3/Add.156

[41]   Dingus, T. A., Klauer, S. G., Neale, V. L. *et al.* (2006). The 100-Car Naturalistic Driving Study Phase II – Results of the 100-Car Field Experiment National Highway Traffic Safety Administration, Report Nr. DOT HS 810 593, Washington, DC, USA.

[42]   Benmimoun, M., Zlocki, A., Kessler, C., *et al.* (2011). Execution of a Field Operational Test within the euroFOT Project at the German1 Test Site. *The 18th World Congress on Intelligent Transport Systems*, Orlando, USA.

[43]   Brakemeier, A. (2012). Co-operative ITS – Standardization, Improving Road Safety and Traffic Efficiency. *The 18th European Wireless Conference*, Poznan, Poland.

[44]   Stoyanov, H., Metzner, S., and Stehlik, P. (2021). Showcases. L3Pilot Deliverable D6.6 (Grant Agreement No 723051). 26 August 2024. https://l3pilot.eu

[45]   Gruschwitz, D., Hölscher, J., van Nek, L. *et al.* (2023). Marktdurchdringung von Fahrzeugsicherheitssystemen 2021, Report of Bundesanstalt für Straßenwesen (BASt), Mensch und Sicherheit, Heft M 339, Bergisch Gladbach, Germany.

Part II

Needs and methods

Chapter 2

Challenges in the evaluation of automated driving

Risto Kulmala[1], Oliver Carsten[2] and Satu Innamaa[3]

2.1 Introduction

In recent years, the concept of automated driving has received substantial interest from various sectors, including the automotive industry and its suppliers, the IT and telecom industries, public authorities and road operators. The interest has resulted in various efforts (example references are provided but do not aim for completeness):

- to define research needs and agendas for automated driving [1],
- to define cross-sector roadmaps for closing the research gaps [2],
- to develop industry strategies and roadmaps towards automated driving [3,4],
- to develop national and European strategies towards automated driving [5–9] and
- to set up adequate infrastructure for testing and deployment of automated driving functions [10–13].

Most of the roadmaps and strategic plans for automated driving stress the urgent need to have reliable evidence of the impacts, benefits and costs of the investments to be made by the different stakeholders.

2.2 Levels of automated driving

There are different levels of automated driving. The most frequently used levels are those specified in the SAE classification, described in the SAE J3016 Information Report 'Taxonomy and Definitions for Terms Related to On-Road Motor Vehicle Automated Driving Systems' [14].

[1]Traficon Ltd, Espoo, Finland
[2]Institute for Transport Studies, University of Leeds, UK
[3]VTT Technical Research Centre of Finland Ltd, Espoo, Finland

The levels of driving automation describe the extent of the dynamic driving task (DDT*) being performed by the human driver and the extent being performed by the driving automation system. The DDT represents the operational and tactical aspects of driving, but not the strategic tasks such as planning routes or choosing destinations. The DDT tasks include basic steering and speed control plus identifying and tracking hazards in the driving environment, manoeuvring around obstacles and hazards and planning and selecting local paths. The levels of driving automation are as follows: [14,15]

Level 0 – No driving automation: The human driver performs the complete DDT but may be assisted by collision warning systems, collision avoidance and collision mitigation systems, or systems that act intermittently in response to specific road or hazard conditions, without changing the driver's fundamental driving tasks.

Level 1 – Driver assistance: The system performs either lateral (steering) or longitudinal (acceleration and braking) control on a sustained basis, while the human driver performs all other DDTs and therefore is required to remain fully engaged in driving.

Level 2 – Partial driving automation: The system simultaneously performs lateral and longitudinal control on a sustained basis, while the human driver continues to perform object and event detection, recognition and response tasks. Therefore, although the driver's hands and feet may be off the steering wheel and pedals, he or she still needs to remain fully engaged in the driving task and needs to continuously monitor the performance of the system to be prepared to intervene when necessary.

Level 3 – Conditional driving automation: The system is capable of performing the complete DDT under certain limited conditions (within its ODD), but it depends on a human 'fallback user' (FU) in the driver's seat to intervene when it requests help to contend with situations that it cannot handle by itself. The human FU can shift attention to other activities while the Level 3 ADS is performing the DDT but needs to be alert enough to respond promptly to any requests to intervene. So, he or she could be conducting work or leisure activities online but should not go to sleep while the system is driving.

Level 4 – High driving automation: The system is capable of performing the complete DDT under certain limited conditions (within its ODD), and it does not need an attentive driver or FU to ensure safety. It must be capable of bringing the vehicle to a stable, stopped condition (a 'minimal risk condition' or MRC) as

---

*The DDT or dynamic driving task contains all of the real-time operational and tactical functions required to operate a vehicle in on-road traffic [14].

necessary to respond to internal failures or external hazards in the driving environment. Level 4 automation may be applied on ADS-dedicated vehicles (ADS-DV) that are only intended to be driven by the ADS and therefore do not need conventional human driver control interfaces (steering wheel and pedals) or it may be applied on vehicles that are also intended to be driven by human drivers and therefore have conventional driver control interfaces.

Level 5 – Full driving automation: The system is capable of performing the complete DDT under all conditions in which humans are capable of driving, so it has no ODD constraints. Similar to Level 4 automation, it must be capable of bringing the vehicle to an MRC as necessary to ensure safety, and it may be applied on ADS-DVs or on vehicles that can also be driven by human drivers using conventional control interfaces. Because of the technical challenges, this is unlikely to become reality until many decades in the future, so it is only a long-term dream rather than reality.

Level 1 and 2 systems are considered driving assistance or driver support systems because the human driver is still in charge and needs to make all safety-critical decisions. The higher-level automation systems above L2 are considered automated driving systems (ADS) because they are capable of operating without continuous human supervision under at least some conditions.

The Level 4 and 5 systems are the ones that could have revolutionary impacts on travel behaviour and urban form by eliminating the disutility of travel time, decoupling parking locations from travellers' origins and destinations, providing individual mobility to all users including those without a driving license, facilitating vehicle sharing as well as ride sharing, and breaking down the boundaries between public and private transportation. At Level 4, these impacts are restricted to specific roads and conditions that allow the use of the automated mode.

Automated driving systems can only perform the DDT in their operational design domains (ODD). The ODD is determined by the set of conditions in which each driving automation system is capable of performing the DDT. This could mean, e.g. motorway links or specific routes in cities or specific areas in transport terminals in good weather conditions.

One important principle underlying all the technical discussions is that the descriptions of systems for automating road transport are focused on specific driving automation features rather than vehicles, because an individual vehicle may be equipped with multiple automation features that are capable of different kinds of automated operations under different conditions. For example, a passenger vehicle may be equipped with a SAE Level 3 Motorway Chauffeur System as well as a SAE Level 4 Automated Valet Parking System. Therefore, a vehicle cannot generally be defined in terms of a unique automation level, cooperation class, or ODD, because its driving automation features could differ from each other in each of these dimensions [13].

2.3   Case of Level 1 and 2 driver support systems

As defined by SAE [14], also Level 1 and 2 driver support systems operate within their ODD, e.g. on highways with well distinguishable lane markings.

It is increasingly common for new road vehicles to provide drivers with driver support systems that can support the driver in the control and manoeuvring aspects of the driving task. These systems can analyse the vehicle environment, inform or warn the driver and increase driving comfort by actively stabilising or manoeuvring the vehicle. Some of these systems can be enabled in appropriate circumstances in order to substantially reduce or even supplant the driver's input in both longitudinal and lateral vehicle control, relieving the driver of interaction with the pedals and assisting, to various degrees, in steering control. Such functionalities, however, still leave the driver responsible for the safety of the DDT and therefore conform to SAE Level 2 automation [16].

In the case of Level 2 driving, the driver is expected to immediately resume their input into the driving controls upon realisation that the driving assistance system is not capable of handling a situation, or more generally, whenever necessary to maintain their safety and that of other road users. Such technology therefore constitutes 'driver support', in contrast with 'automated driving' at Levels 3 and above, where the responsibility for the safety of the DDT is assumed by the automated driving system. With L2 driver support, continued driver attention to the roadway and traffic scene is required [16].

Some Level 2 automated driving systems have been named in a way that the vehicle owners may be led to believe that their vehicle is capable of Level 3 or higher automation. Examples of such product names are Tesla's Autopilot and Full Self Driving. Even without misleading names, the use of Level 2 systems increases driver inattention to the DDT as shown in [16]. Recent road crashes involving Tesla vehicles suggest that the misleading names for Level 2 automated driving systems promote inattention or the use of the systems outside their nominal ODD leading to increased crash risks [17].

2.4   Use cases of higher-level automated driving systems

This book chapter focuses on the evaluation of higher-level automated driving systems, i.e. SAE Level 3 or 4 only. As SAE Level 5 will likely not be realistic within decades, it will not be addressed by us. Currently about 20 use cases have been identified [11,18]. Some examples are listed below:

- highway and urban chauffeurs (L3)
- highway autopilot and convoy (L4)
- highly automated freight vehicles (L4) in confined areas / dedicated roads / open roads
- delivery robots (L4)
- automated public transport vehicles (L4) in confined areas / dedicated roads / open roads
- Commercial driverless vehicles (L4) as taxi services / robotaxi (L4)

- Driverless maintenance and road works vehicles (L4)
- Valet parking (L4)

The use cases can be classified in various ways. Some are for freight vehicles, public transport vehicles, vehicles operated as fleets and some for personal passenger vehicles. Some are designed to work in confined areas (such as terminals, parking establishments, industrial, agricultural fields), on dedicated roads (hub to hub, bus routes) and some on open roads (urban street, highways, rural roads). Some provide automated driving systems which have been evolved from driver support systems by the vehicle manufacturers, whereas some can be classified as driverless vehicles, developed sometimes by IT companies. Each use case will have different impacts on mobility, road network efficiency, road safety, environment, the transport system and society depending on their use and operating domain.

2.5   Impact evaluation of higher-level automated driving

2.5.1   *Impact assessment frameworks for automated driving*

Several impact assessment frameworks exist to be used for automated driving. Which one is the most suitable framework depends on the project type, technology under evaluation and resources available for evaluation.

FESTA [19] is a methodology developed specifically for field operational tests (FOT), but it is applicable for also other types of field tests. It was published originally already in 2008, with the latest update in 2021. An FOT is defined as: 'A study undertaken to evaluate a function, or functions, under normal operating conditions in road traffic environments typically encountered by the participants using study design so as to identify real-world effects and benefits'. Thus, FESTA is most applicable for driving automation with high technology readiness level when it can be tested with real users in real traffic without additional safety protocols such as supervision by a safety driver. In 2021, the ARCADE project published Micro-FESTA [20], which is a condensed version of FESTA to support small pilot projects of connected, cooperative and automated mobility (CCAM).

The Trilateral Impact Assessment Sub-Group for Automation in Road Transportation made with their framework the first effort to harmonise impact assessment studies globally at a high level [21]. Their framework includes, for example, definitions for direct and indirect impact areas and how they link, and adaptation of nine impact mechanisms to all impact areas, specifically for systematic assessment of the different impacts of automated driving covering intended and unintended, direct and indirect, short-term and long-term impacts of both users and non-users of driving automation.

The L3Pilot and Hi-Drive projects under the EU's Horizon 2020 Framework Programme made extensive efforts to adapt the FESTA methodology for large-scale automated driving pilots. Their full impact assessment methodologies [22,23] are publicly available as examples of how certain specific aspects of these pilots can be addressed in evaluation. Note that these methodologies apply to projects with multiple test sites and significant resources for evaluation. The scope of these innovation actions was to pave way for higher-level driving automation of passenger cars.

At the time of publication of this book, the FAME project under the EU's Horizon Europe Framework Programme was developing a Common Evaluation Methodology for CCAM (EU-CEM Handbook) [24], which will provide guidance on how to set up and carry out an evaluation or assessment of direct and indirect (wider socio-economic) impacts of different CCAM solutions, to overcome gaps identified in FESTA and to share best-practices. This approach was needed due to the complexity and the novelty of CCAM systems and services, and the wide-ranging societal impacts of CCAM. The EU-CEM Handbook will be part of European framework for testing (of CCAM) on public roads.

2.5.2   *Examples of evaluations carried out*

2.5.2.1   L3Pilot

L3Pilot was a flagship project for automated driving in EU's Horizon 2020 research and innovation programme from 2017 to 2021. The overall motivation in L3Pilot was to test and study in large scale the viability of automated driving, focusing on SAE level 3 and passenger cars, as a safe and efficient means of transportation. The project also aimed to develop a knowledge base for exploring and promoting new service concepts to provide inclusive mobility to all user groups. L3Pilot also included an extensive evaluation activity, with impact assessment which focused on mobility, safety, efficiency, environment and socio-economics. The L3Pilot methodology was built by applying the FESTA methodology in adapted form for a large-scale pilot study of automated driving [25].

In L3Pilot, personal mobility behaviour was determined by individual travellers' choices, which are constrained by their available time budget and means of travel and influenced by personal and cultural attitudes. Consequently, the assessment was focused on understanding people's attitudes and beliefs, and how those can evolve into future behaviour. The methods included user questionnaires, global surveys and focus groups. Examples of the main results include that automated driving functions (ADF)[†] are likely to increase the quality of travel, but the experience of travel quality depends on the individual traveller. The main results showed that, on average, increased travel quality may decrease the perceived costs of travelling by car. In addition, drivers may switch to routes within the operational design domain of the automated driving system or travel mode [26].

The safety impact assessment focused on estimation of effects on the number and the likelihood of road accidents and on their severity, with breakdowns by driving and traffic scenarios. The evaluation approach was to cluster driving situations into typical driving scenarios, to simulate automated and baseline driving for them with tools developed specifically for this purpose, to compare accident outcomes and to scale-up results at the European level using accident statistics. The

---

[†]Automated driving function (ADF) is '*a common feature addressed by a group of ADSs*'.

main results indicated that automated driving would be capable of reducing the number of injury accidents for all severities and that this reduction is expected for most scenarios. For example, overall, on the motorways, the number of fatal accidents on motorways was estimated to be reduced by 2.0% (at a 5% penetration level) to 13.1% (at a 30% penetration level). In relation to the target accidents of the motorway ODD defined for the mature ADF addressed in the impact assessment of L3Pilot, this would be equivalent to 3.8% to 25.1%. In the urban environment, the corresponding estimates were a 2.0% to 12.2% reduction overall, and a 4.3% to 25.9% reduction within the urban ODD [26].

The efficiency and environmental impact assessment assessed the potential impacts in terms of changes in travel time, delay, average speed, energy demand, fuel consumption and CO_2 emissions. Impacts were assessed on the basis of traffic simulations. Motorway simulations covered a large proportion of different motorway layouts in Europe with a range of traffic volumes. The urban ADF was studied at a more general level using two examples of urban road networks without scaling-up. Results of simulations and scaling up showed that benefits are possible in situations with high traffic volumes, where ADFs have potential to improve efficiency and reduce emissions, especially on motorways. When scaled up to the European level, the expected impacts were overall positive but small (less than 1% decrease of CO_2 emissions and less than 3% decrease of energy demand, for all penetration rates) due to most vehicle kilometres being travelled in low traffic volumes, where no large impacts were predicted. At a local level, larger impacts could be observed, for example on regularly congested urban motorways [26,27].

The assessment of the socio-economic impacts addressed the net effect of ADFs on social welfare: Do the benefits outweigh their costs? A snapshot approach was chosen for this with a time perspective narrowed down to one year, with today's traffic situation as the baseline. In the assessment, the estimated impacts from the mobility, safety, efficiency and environmental assessments were evaluated quantitatively by applying standard unit costs to the scaled-up impacts. The costs of the technology were estimated by independent experts from the car manufacturing industry. The results showed that for the motorway function, the benefit–cost ratio for the quantified impacts is less than 1 for all penetration rates studied (5%, 10% and 30%) but in an urban environment, the expected net social benefits from accident prevention clearly exceed the social costs of this function for all penetration rates. The additional impacts regarding safety impacts of inbuilt sensors on accidents occurring outside the ODD defined for the mature ADFs, and the impacts on the cost of travel time, indicate even higher monetary benefits. The main conclusion was that it is likely that a package of ADFs for motorways, urban environment and parking together, would generate social benefits, which exceed the social costs [26].

2.5.2.2 Waymo

Waymo has carried out a retrospective analysis of the crash involvement rate of their L4 ride-hailing fleet of robotaxis operating without either a safety driver or a remote human driver in Phoenix, San Francisco and Los Angeles with a cumulative mileage of 7.14 million miles (11.49 million km) [28]. The crash involvements of the Waymo vehicles, as reported in compliance with the incident reporting requirements of NHTSA for ADS driven vehicles, were compared with crash involvements of light-duty passenger vehicles driven on surface streets in the same geographic areas as the Waymo service. The involvement rates of the background population were adjusted for underreporting using correction rates calculated from the SHRP-2 naturalistic driving study [29]. Comparisons were made both for property-damage-only involvements and for involvements in injury crashes. To calculate involvement rates per million miles travelled, the observed mileage of the Waymo vehicles was used alongside data on vehicle miles travelled on surface streets by all light-duty passenger vehicles in the Waymo areas of operation [28].

The calculations for property-damage only crashes indicated that the human-driven vehicles operating in the same areas as the Waymo vehicles had a crash involvement rate that was 2.3 times as high as that for the Waymo vehicles (ADS to human rate ratio 0.43, 95th percentile CI 0.22 to 0.76). The comparable calculation for involvement in injury crashes indicated a rate that was 6.8 times higher for human-driven vehicles (ADS to human rate ratio 0.15, 95th percentile CI 0.02 to 0.49). Results for both differences were significant at the $p \leq 0.05$ level [28].

These findings indicate a very substantial improvement in the safety of driving for the Waymo vehicles as compared to vehicles driven by the general driving population, with particularly impressive results for involvement in injury crashes. It cannot be concluded that all the improved performance is attributable to the ADS — the general vehicle population will include older vehicles that are less well maintained and have fewer advanced safety features and in all probability more occupants who are unbelted (Waymo passengers are reminded to buckle up) — but nevertheless one can conclude that the results are impressive.

2.5.3 Current knowledge of the impacts of the changes in user and driving behaviour

2.5.3.1 Review of the impacts on vehicle kilometres travelled, modal split, traffic flow, environment and traffic safety

VTT conducted an extensive review of scientific journal articles which address the potential impacts of automation in road transportation by use case, and how these results apply to Nordic conditions [30]. Specifically, the study focused on the impacts on vehicle kilometres travelled, modal split, traffic flow, environment, traffic safety and interaction between road users, addressing privately owned automated passenger cars, public transport, robotaxis and logistics solutions. The focus was on driving automation of SAE levels 3–5 but addressed for some questions also the lower levels.

Overall, there has been lot of research on the potential impacts of automation. However, all the impacts cannot be assumed to be realisable in real traffic. This is due to the availability of only a limited selection of possible research methods. To date the number of highly automated vehicles in traffic has been small and their operational domain has been limited; consequently in most studies, the impacts of driving automation have not been measured but assessed in virtual environments and via surveys using a variety of simplifications and assumptions [30].

The literature indicated that automation of road transportation may affect **vehicle kilometres travelled and modal split** in a number of different ways. These impacts can be assessed via the effects of automation on the perceived travel cost of different modes, and on the other hand, via the mobility solutions enabled by driving automation [30].

In an automated passenger car, the driver may be freed from the driving task and be allowed to use travel time for other purposes. Consequently, the perceived time cost is reduced which makes car use more attractive. Models show how this leads to more and longer car trips. Such effects are likely to lead to an increase in total kilometres travelled by car and a shift from public transport to automated passenger cars. Automation of passenger cars does not likely change the fact that an increase in vehicle kilometres travelled may increase traffic congestion, and the worse congestion, the longer the travel time. Even though automation of passenger cars may lead to somewhat more efficient traffic flow, the benefits of better efficiency may be cancelled by the increased vehicle kilometres travelled by car [30].

The impacts of robotaxis will depend on many factors such as their fleet size and whether the rides are shared with other passengers or not. The larger the fleet, the better is their service offering and, likely, the more common is their use. Shared rides mean fewer vehicle kilometres travelled than in private rides, but from a single customer perspective the ride is not as short as possible which reduces the attractiveness of the service. Robotaxis will also drive empty. Consequently, this service is likely to increase the total vehicle kilometres travelled. Naturally, the pricing of robotaxi rides also affects their use. This service may compete mostly with public transport and traditional ride services, but may also make it easier to give up car ownership [30].

If an automated fleet makes the operation of public transport and logistics services cheaper, this may make it easier to maintain a level of service which provides a worthy alternative for the less space-efficient travel modes (passenger car, robotaxi). It may enable public transport service provision to new areas. In goods logistics, lower operation costs may increase the attractiveness of road transport over rail or ships [30].

The literature shows how automated driving can improve **traffic flow**. However, the impacts e.g. on travel time have been studied mostly for certain specific scenarios which may not be representative for Europe or other regions. These studies have mostly focused on motorways with a high traffic volume even if those represent only a small fraction of European motorway traffic. Thus, the large-scale impact cannot be assessed based on such studies [30].

Impacts of automated passenger cars have been studied mostly with traffic simulation. The studies have used very different assumptions of the capabilities of the automated fleet, and not all studies report their assumptions which makes it hard to apply their results. In addition, current driver and car following models available in microsimulation tools may not work well for modelling the differences between a human driver and an automated vehicle. Conditions in simulations are also free of disturbances such as poor weather conditions and longitudinal road gradients, and are limited in representing different preferences and abilities of drivers [30].

The results in the literature indicate that it is possible that automation of passenger cars leads to longer travel time in mixed traffic. The results on throughput depend directly on the assumption used regarding the desired headway in car following. Research shows that benefits for traffic flow can be achieved only with high penetration of connected automated vehicles, with driving behaviour that does not compromise string stability and with shared rides so that the total vehicle kilometres travelled do not increase [30].

Very little is known of the impact on traffic flow of other use cases than the automated passenger car. It is likely that these impacts are mostly consequence of the changes they lead to in travel behaviour [30].

The results of the **environmental** impacts of automated passenger cars are in line with the traffic flow impacts above. These impacts are usually consequence of changes in traffic flow dynamics and total vehicle kilometres travelled, and therefore the results depend on assumptions made on these. In addition, the assumptions on fleet composition and engine type affect results [30].

Field experiment trials on commercial ACCs indicate that in steady-state flow, the fuel consumption and CO_2 emissions of automated vehicles can be lower than of human drivers, assuming that automated cars will obey the speed limit. However, in the presence of even small deviations in speed, the result may also be the opposite. The results are less clear in conditions and environments common in real traffic such as the presence of intersections, congestion and slopes. The impact of the energy needed for sensors and algorithms is usually excluded from the studies and related estimates vary greatly [30].

The impact of robotaxis on emissions depends greatly on the scale of deployment and modal shift that they cause. Truck platooning may reduce the emissions of heavy-duty vehicles, but more research would be needed to study the impacts of the platoons on traffic flow in total. Delivery robots may reduce emissions of delivery services for the last kilometres at least in densely populated areas if they are part of effective transport chains [30].

Most research on the **traffic safety** impact of automation focuses on the accident risk of the automated vehicle or its user. In addition, most research is based on simulations, and the simulation tool employed, assumptions used and parametrisation of driving behaviour affect the outcome. Many simulation tools have been developed for research on traffic flow, and their applicability directly for research on safety critical situations is questionable. They also neglect the impact of e.g. bad weather conditions [30].

Research results indicate that the impact of automation depends on the use case and studied environment. The literature shows rather large effects on the number of conflicts for motorways but there is more variability in the magnitude of impact for single intersections or for an entire city. Results indicate that truck platooning may reduce the number of rear-end collisions but reduce the safety of lane changes or make entering the motorway more difficult [30].

According to studies using real-word data from California, robotaxis have a higher accident risk than human drivers [31,32]. However, these results may be consequent of the differences in reporting of incidents of automated vehicles and reporting of accidents of manually driven cars. Only recently have robotaxis been used in real-world operation without a safety driver. The strongest research evidence is on the safety benefits of driver assistance systems. Yet, their performance in more challenging environmental and weather conditions should also be further investigated to determine the safety impact [30].

The implications of the changes in travel behaviour on traffic safety depend on modal shifts and changes in routing, i.e. whether people shift to more or less safe modes and routes, and whether their exposure to traffic, in terms of kilometres or time, increases or decreases. If the deployment of automation leads to new and longer journeys and a shift from public transport to less safe travel modes, transport safety benefits may be lost. The ODD of ADS also affects the impact potential. If the most dangerous conditions are outside ODD, the safety improvement potential of automation is limited [30].

The penetration rate of automation affects the outcome. The results on ADAS indicate that the relative benefit of automation in conflicts is largest in low penetration rates. This is likely because in some conflicts, it is enough that one of the vehicles involved has a relevant support system. For the conflicts with vulnerable road users, the effect of increased penetration is more linear [30].

The VTT study [30] concluded that automation in road transportation can contribute to many of the objectives set for the transport system, specifically for efficiency, environment, accessibility, safety and health. However, progress towards these objectives cannot be taken for granted. Research suggests that the promotion of public transport and active travel modes, in particular shared rides in automated vehicles, is important. Yet, more research is needed on all of these impacts.

2.5.3.2   Review of the safety implications of higher levels of automation

Tafidis *et al.* (2021) carried out a scoping review of the studies on the safety impacts of vehicles with higher levels of automation [33]. They defined such higher levels as L4 and L5 in the SAE classification. They adopted the approach of a scoping review, as opposed to a systematic review, on the grounds that the topic was an emergent one with limited evidence. The exclusion of L3 vehicles was justified on the grounds that this technology might never reach deployment because of the unclear boundary between system and human responsibilities. Their methodology was built upon the PRISMA Extension for Scoping Reviews (PRISMAScR) which provides a 22-item checklist for systematic reviews [34]. The initial search of bibliographic databases revealed 4185 studies, which after screening were reduced to 24; the excluded studies typically focused on levels of automation lower than L4 or on connectivity [33].

The included studies applied a variety of methods to derive their conclusions: 15 used traffic simulation, 8 applied accident analysis and 1 used accident prediction modelling. The traffic simulation studies were applied on limited road networks: five on rural roads, four on urban networks and four on suburban inter-sections. The reviewers determined that only a minority of studies (11 out of 23 with no information from one study) included pedestrians and cyclists in their considerations; even for the urban investigations, only two-thirds addressed pedestrians and cyclists. The simulation studies tended to estimate impacts at 100% penetration of automated vehicles (AVs) into the vehicle fleet, and therefore did not address the safety impacts of mixing with human-driven vehicles [33].

The vast majority of the studies reviewed (20) reported positive safety benefits from AV deployment, with eight studies concluding negative impacts and six neutral implications. All the studies that reported negative findings applied microsimulation, with several studies finding that intersections and roundabouts were especially problematic, while others found negative effects at low levels of penetration. However, only one such study found solely negative impacts on safety. All the studies that applied analysis of existing crash types came to positive con-clusions, typically by assuming that AVs would perform faultlessly when addres-sing current safety problems. They ignored the potential of AVs to introduce new types of error into the traffic system [33].

The microsimulation studies used surrogate safety measures – traffic conflicts, time headway, speed, etc. – as indicators of changes in risk. The authors of the review note that these results are highly dependent on the behaviour of the AVs assumed in the models. They also points out that the modelling omits consideration of performance in poor weather conditions or on steep gradients. Also omitted is the behavioural adaptation of other road users when interacting with AVs [33].

The review concludes that both the accident studies and the microsimulation approach have considerable limitations. Challenges to the safe operation of AVs from system faults or errors are not addressed and interactions with pedestrians and cyclists tend not to be covered. Equally, there is little empirical evidence on the interaction of human drivers with AVs. For example, aggressive drivers might take advantage of the presumed caution of AVs. The authors make a plea for the acquisition of more real-world data on the driving performance of AVs in order to prove the reliability of AV functionality, to permit evidence-based safety assess-ment and to provide input into future microsimulation modelling [33].

2.5.4   *Current knowledge of the infrastructure impacts*

Highly automated driving can affect road infrastructure in primarily two ways. First, the operation of specific automated driving use cases might affect infra-structure in a different manner than human driven vehicles or transport are doing. Second, the need to provide driving environments that fulfil the requirements of ODDs set for ADSs may result in changes made in the infrastructure.

2.5.4.1   Physical infrastructure

The operation of automated driving use cases is expected to have impacts on the physical infrastructure. Some of the most likely short- and medium-term impacts expected are briefly described below.

Automated vehicles are likely designed to drive in a much more controlled way compared to human beings. That is why their impact on the pavement structure is also more focused and will most likely cause premature damage and other problems unless the vehicles better resemble the driving performance of manually driven vehicles in this regard [35].

In the future, lanes can be much narrower allowing the provision of increased pedestrian, cyclist or public transport space or wider shoulders or an additional lane in the road corridor. Reducing lane width will, however, lead to more canalised traffic loading patterns that will reduce the lifetime of the pavement structure. The negative impacts of automated driving to the pavement structure can be roughly classified to (a) impacts due to reduced tyre wander, and (b) impacts of platoon driving [35]. If the ADS developments do not provide sufficient 'artificial tyre wander' to mitigate the problem, a possible solution is to change road paving so that the narrow strips, where the vehicle wheels run, will be equipped with material tolerating wear better. This would increase the costs for paving and re-paving [36].

Shoulders or emergency parking bays are needed for situations when the ODD is ending, and there is a need to accommodate a 'minimal risk condition'. The situation is challenging on sections, where hard shoulder running is allowed, since the emergency lane cannot be used as an emergency parking space. When and if automated vehicles are used on high-speed roads, the road operator or traffic manager may need to be proactive in instructing vehicles on what to do in minimum risk condition mode and/or understanding what provisions may need to be designed into the road network. Future road projects will need to consider, on a case by case basis, the requirements to ensure minimal risk conditions [37].

Road operators and cities will need to begin turning their attention to kerbside management. Automated public transport vehicles, robotaxis and delivery robots need kerbside space to pick up or drop off passengers and goods. Space is also needed for parking. Demand for kerbside space will inevitably increase as the deployment and use of automated vehicles increases [37].

Stabling requirements will rise as uptake of shared services increases and vehicles become automated. As services emerge and capture passengers, the demand for stabling facilities will likely grow at a faster rate. At higher market shares of travel, stabling requirements are likely to plateau [37].

Parking space savings may become associated with shared automated vehicle operations. The expected decrease in parking demand enables a decrease in parking supply. It may even be appropriate in some cases to not allow any parking to be provided. The decrease in non-residential parking demand (and therefore locations for decreased supply) will vary by location. The largest decrease is likely to occur in high-density urban areas, with high-value land. It is likely to become increasingly feasible for parking to be relocated to the fringes of these areas – although there may be social implications of relocating parking to low-value areas [37].

Automation of light vehicle passenger transport will create a range of opportunities to increase the efficiency of the layout of parking structures. Some gains relate to the application of automated valet functionality. When passengers can be dropped at a remote pick up and drop off facility, parking layouts can be altered and a higher bay yield is possible. This can be due to reduced bay dimension, aisles and ramps becoming one-way, tandem parking arrangements, and the precise manoeuvrability assumed of automated driving systems [37].

Finally, road and winter maintenance operations can be improved to ensure that conditions are kept such that they fulfil the criteria set for ODDs of the ADSs and for instance to maintain space for minimal risk manoeuvres at the kerb or shoulder of the roads.

2.5.4.2   Digital infrastructure

The key aspects of the digital infrastructure are related to ODDs:

- connectivity
- accurate positioning
- real-time data on ODD attribute values
- rules of the road

Connectivity is essential to automated driving systems operating at high speeds due to the reasonably short range (30–300 m) of current vehicle sensors. At low speeds, the ADS can operate the vehicle and perform safe stops within the vehicle sensors' range. At high speeds, it needs data about the conditions outside its sensors' range, the so-called electronic horizon to be aware of the situation and availability of its ODD on its immediate route ahead to be able to proceed safely. The connectivity can be realised via a cellular network, satellite network or short-range connectivity, and can require additional investments when such connectivity established for other purposes is not sufficient for higher level automated driving [2,12,35].

The ADS has to be aware of its exact position on the lane and with regard to the traffic environment to be able to plan its route and proceed towards its destination using a safe trajectory. GNSS satellite positioning supported by land stations can provide the required positioning accuracy when combined with the data from vehicle's sensors with regard to the location and distance to nearby landmarks and other objects in the digital map used by the ADS. This might require additional investments e.g. in land stations in areas with GNSS problems or additional landmarks in areas with no 'natural' landmarks. The establishment and real-time maintenance of HD maps of the transport system is also quite costly [2,12,35].

The ADS can operate safely only within its ODD, and for that reason it has to be aware of the value of all of its ODD attributes. If it does not know the value of just one of its ODD attributes, it must cease operation and transfer the vehicle control to a human vehicle occupant or move to a minimal risk condition. ODD attribute lists have been provided in [12,13] and [38].

The ODD attribute information can be provided by other vehicles based on their sensor data, infrastructure operators based on their monitoring system data, or information service providers such as meteorological offices. The concept of Distributed ODD Attribute Value Awareness (DOVA) is detailed in [13].

In order to provide real-time data on ODD attribute values, such data must be available in digital form concerning the road infrastructure, traffic and any incident, events or other situations affecting the attribute values. This requires a digital twin, shadow or model of the road infrastructure, traffic and environmental conditions as well as an access point to the digital data via a data cloud [2,12].

The rules of the road determine whether and how the ADS can operate on a specific road section. The rules to be provided to the ADS contain the digital version of the regulations of the traffic law/code as well as the additional regulations imposed in the dynamic traffic and incident management plans set by traffic management centres, the police and/or the rescue authorities [13].

2.5.4.3   Operational infrastructure

The operational infrastructure includes the traffic management centres, the traffic management and information systems and services, and fleet management and operation services.

Traffic management systems will not actively manage the tactical or operational decision making of ADS, i.e. activate and de-activate automation; instead its added value to ADS and thereby road safety lies in improving the situational awareness of ADS and providing strategic guidance. Thereby, higher level automation does not necessitate major changes in the operation of traffic management centres [12].

However, the local traffic management at road works, incident and event sites needs to be harmonised in such a way that the ADS can easily detect the existence of these sites, and even could navigate in automated mode through these sites. The latter would likely require up-to-date digital maps of the sites unless the ADS can receive detailed information of safe trajectories through the sites based on the experiences of other vehicles that have already passed the sites. Naturally, even careful marking of available paths by using standard road cones would help both the ADS and human drivers [12].

The provision of C-ITS services is also part of the operational infrastructure. In addition to various types of information of hazardous locations, and signal control phases and timing, collective perception services are useful for the ADS especially at locations of restricted visibility for the vehicle sensors. In such locations, infrastructure-based sensors can be deployed for detecting approaching road users otherwise invisible to the ADS [12,13].

A necessary operational infrastructure element is fleet management and operation. This is especially so for automated public transport and robotaxi services as well as driverless freight services, where a fleet manager has the responsibility for the safe operations of a fleet of ADS-operated vehicles. In case of emergencies and ODD departures, the fleet operator can give remote assistance and support to the ADS otherwise incapable of continuing its operation. The remote operation of the ADS or the vehicle is not currently regarded as safe other than in strictly con-fined areas such as mines [13].

2.6   Challenges for the evaluation of automated driving

In the evaluation of driving assistance systems and ADS, the evaluation paradigm is to utilise different methods in the various phases of the system development life-cycle as illustrated in Figure 2.1.

There is a long tradition of evaluation of ADAS systems with the gold standard being naturalistic FOTs in which the users are provided with equipped vehicles and allowed to use those vehicles in everyday driving. This allows an experimental design in which driving performance, behaviour and safety without the relevant system can be compared with driving performance, behaviour and safety with the system active (or at least available to the user). Such evaluations have most frequently been carried out on prototype systems, e.g. the FOT of Adaptive Cruise Control conducted by the University of Michigan in 1996–1997 [40] or the various FOTs of Intelligent Speed Assistance [41–43]. They have also sometimes been carried out on production systems, which approach by definition reduces the burden on the FOT organisers of obtaining approvals to use non-homologated or pre-production systems in real-world driving. An example here is the euroFOT project which evaluated the safety and acceptance of a range of production ADAS systems [44]. In recent years as L2 assistance systems have come onto the market, there have also been naturalistic investigations of driver behaviour when using such functionality. An example here is the study of driver responses to automation-initiated disengagement when driving with the Cadillac hands-off SuperCruise system [45]. An analogous study, examining driver pro-pensity to inattention when using L2 assistance, was also carried out with a nat-uralistic methodology. A particular focus of this study was how propensity to inattention evolved over time as drivers became familiar with the assistance of the automation system [46].

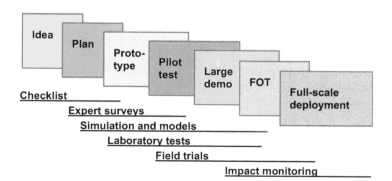

Figure 2.1   Assessment setups for different phases in system lifecycle (adapted from [39])

Naturalistic trials with vehicles that permit multiple levels of automation, i.e. which at L3 and L4 have fallback users, are a very attractive proposition. As opposed to pilots with drivers often recruited from manufacturer staff or controlled trials with a safety driver present, they would allow investigation of propensity to use the automated driving features in various naturally occurring circumstances, alongside travel behaviour impacts and usability as well as acceptance, not to mention user willingness to pay. Moreover, with sufficient trial length, learning effects could be ascertained. However, currently (at least in Europe), it is not feasible to organise such FOTs for ADSs which are not yet on widely on market. Indeed, vehicle manufacturers and public authorities have difficulty in conducting their own tests of vehicles under development on real roads in Europe. The European Commission has identified this as a significant obstacle to vehicle development by vehicle manufacturers and to real-road testing for type approval by relevant national authorities [47].

In the absence of naturalistic FOTs with vehicles equipped with L3 or L4 ADS, on-road trials have been far more limited in scope. Thus, the European L3Pilot project stated: 'The idea in Piloting was to ensure that the functionality of the systems is exposed to variable conditions'. Motorway drives in which members of the public had the role of the fallback driver were conducted with a Safety Driver present. For urban drives, all members of the public had the role of passengers [48]. The same restriction has been imposed in the successor project to L3Pilot, Hi-Drive: 'The technological readiness level of the highly automated driving systems and the safety requirements of public road tests do not yet allow for FOTs with unsupervised ordinary drivers or naturalistic driving studies' [49].

With such limited piloting of functions, evaluation has to be restricted to the technical performance of the function alongside scaling-up through simulations to examine impacts. Such scaling-up requires extensive and unverifiable assumptions about the system, usage and the quality of user interaction with the ADS, as regards for example user responsiveness in the absence of a Safety Driver to requests to intervene from an L3 ADS. The implication is that the lessons learned from those trials are somewhat limited with consequent harm to understanding of the impacts of higher levels of automation on traffic efficiency, safety and the environment, and especially on personal mobility.

Simulations have specific issues with validity. This is especially the case with safety-related simulation even for human-operated vehicles. In real life, road accidents tend to result from errors in human perception, judgement and action as well as the failed interaction between the participating humans, vehicles and the infrastructure. So far, we are far from the valid modelling of human errors and interaction failures, let alone the failures of automated driving systems.

Current driver and car following models available in microsimulation tools may not work well for modelling the differences between a human driver and an automated vehicle. The behavioural adaptation of other road users, including vulnerable road users when interacting with AVs, is also not usually addressed in any way. So far, there is little empirical evidence on the interaction of human drivers with AVs. The models so far do not seem able to deal with all frequently occurring driving conditions nor situations including for instance adverse weather conditions or steep gradients.

It is essential to have more real-world data on the driving performance of AVs in order to prove the reliability of ADS functionality, to permit evidence-based evaluations, and to provide input into future microsimulation modelling of ADS behaviour in traffic.

References

[1] CCAM Partnership (2024). The Strategic Research and Innovation Agenda (SRIA). Accessed 31 January 2024. www.ccam.eu/our-actions/sria/

[2] ERTRAC (2022). Connected, Cooperative and Automated Mobility Roadmap. Version 10. Accessed 18 January 2024. www.ertrac.org/wp-content/uploads/2022/07/ERTRAC-CCAM-Roadmap-V10.pdf.

[3] ACEA (2022). Automated Driving. Roadmap for the Deployment of Automated Driving in the European Union. Accessed 31 January 2024. www.acea.auto/publication/roadmap-for-the-deployment-of-automated-driving-in-the-european-union/

[4] 5GAA (2022). A Visionary Roadmap for Advanced Driving Use Cases, Connectivity Technologies, and Radio Spectrum Needs. 5GAA Automotive Association White Paper. Accessed 31 January 2024. https://5gaa.org/content/uploads/2023/01/5gaa-white-paper-roadmap.pdf

[5] Die Bundesregierung (2015). Strategy for Automated and Connected Driving, September 2015, Accessed 31 January 2024. www.bmvi.de

[6] Ministry of Transport and Communications (2021). Action Plan on Legislation and Key Measures of Transport Automation [Liikenteen automaation lainsäädäntö- ja avaintoimenpidesuunnitelma; Finnish with Abstract in English]. Accessed 31 January 2024. https://julkaisut.valtioneuvosto.fi/bitstream/handle/10024/163629/LVM_2021_28.pdf

[7] Centre for Connected and Autonomous Vehicles (2022). Connected and Automated Mobility 2025: Realising the Benefits of Self-driving Vehicles. Accessed 31 January 2024. www.gov.uk/government/publications/connected-and-automated-mobility-2025-realising-the-benefits-of-self-driving-vehicles

[8] DGITM (2023). Development of Automated and Connected Road Mobility. State of Play, Challenges and Actions for the National Strategy. French Ministries of Ecological Transition and Territories, Economy, Interior and Overseas Territories, Transport, Industry and Digital. Accessed 31 January 2024. www.ecologie.gouv.fr/sites/default/files/dgitm-strategie-vehicule-automatise-et-connectee-2022-2025-EN.pdf

[9] Trafikverket (2022). Roadmap – Digitalisation of the Road Transport System. Accessed 31 January 2024. www.diva-portal.org/smash/get/diva2:1651954/FULLTEXT01

[10] Austroads Research Report AP-R665A-22 (2022). Minimum Physical Infrastructure Standard for the Operation of Automated Driving Part A – Infrastructure Investment. Accessed 31 January 2024. https://austroads.com.au/publications/connected-and-automated-vehicles/ap-r665-22

[11] EU EIP (2020). Road Map and Action Plan to Facilitate Automated Driving on TEN Road Network – Version 2020. EU EIP SA4.2 Facilitating Automated Driving Deliverable Task 3. Accessed 31 January 2024. www.its-platform.eu/wp-content/uploads/ITS-Platform/AchievementsDocuments/AutomatedDriving/EU%20EIP_SA42_%20Deliverable%20_Task_3_2020_Road_Map_and_Action_Plan_v6.0.pdf

[12] FTIA (2021). Automated Driving on Motorways (AUTOMOTO). Study of Infrastructure Support and Classification for Automated Driving on Finnish Motorways. Finnish Transport Infrastructure Agency. Helsinki 2021. Publications of the Finnish Transport Infrastructure Agency 21/2021. Accessed 31 January 2024. www.doria.fi/handle/10024/182620

[13] Khastgir, S., Shladover, S., Vreeswijk, J., Kulmala, R. and Wijbenga, A. (2022). Report on ODD-ISAD Architecture and NRA Governance Structure to Ensure ODD Compatibility. TM4CAD Deliverable D2.1. March 2022. Accessed 31 January 2024. https://tm4cad.project.cedr.eu/deliverables/TM4CAD%20D2.1_submitted.pdf

[14] SAE (2021). Taxonomy and Definitions for Terms Related to Driving Automation Systems for On-Road Motor Vehicles. Standard J3016_202104. Society of Automotive Engineers. Accessed 31 January 2024. www.sae.org/standards/content/j3016_202104/

[15] SAE (2021). SAE Levels of Driving Automation[TM] Refined for Clarity and International Audience. Accessed 31 January 2024. www.sae.org/blog/sae-j3016-update

[16] Carsten, O., Perrier, M. and Jamson, S. (2023). Driver Attentiveness to the Driving Task During ADAS Use. Institute for Transport Studies, University of Leeds. Accessed 31 January 2024. https://environment.leeds.ac.uk/downloads/download/5070/adas_user_attentiveness_report

[17] NHTSA (2023). US DOT NHTSA ODI Resume on Tesla Autopilot and First Responder Scenes. National Highways Transport Safety Agency Office of Defects Investigation (ODI). Accessed 31 January 2024. https://static.nhtsa.gov/odi/inv/2022/INOA-EA22002-3184.PDF

[18] ERTRAC (2017). Automated Driving Roadmap 2017. Accessed 31 January 2024. www.ertrac.org/wp-content/uploads/2022/07/ERTRAC_Automated-Driving_Public-Consultation-April-2017.pdf

[19] FOT-Net, CARTRE & ARCADE (2021). FESTA Handbook, Version 8. Accessed 31 January 2024. www.connectedautomateddriving.eu/wp-content/uploads/2021/09/FESTA-Handbook-Version-8.pdf

[20] Barnard, Y., Koskinen, S., Wilmink, I., Sanz, L. and Mesones, A. (2021). Micro-FESTA for CCAM Pilot Projects. Accessed 31 January 2024. www.connectedautomateddriving.eu/wp-content/uploads/2021/08/Micro-FESTA-v2.pdf

[21] Innamaa, S., Smith, S., Barnard, Y., *et al.* (2018). Trilateral Impact Assessment Framework for Automation in Road Transportation: Version 2.0. Accessed 31 January 2024. www.connectedautomateddriving.eu/wp-content/uploads/2018/03/Trilateral_IA_Framework_April2018.pdf

[22] Innamaa, S., Aittoniemi, E., Bjorvatn, A., *et al.* (2020). L3Pilot Deliverable D3.4: Evaluation Plan. Volkswagen Group Research, Wolfsburg, Germany. Accessed 31 January 2024. https://l3pilot.eu/fileadmin/user_upload/Downloads/Deliverables/Update_28042021/L3Pilot-SP3-D3.4-Evaluation_plan-v1.0_for_website.pdf

[23] Vater, L., Aittoniemi, E., Bjorvatn, A., *et al.* (2023). Hi-Drive Deliverable D4.5: Effects Evaluation Methods. *Volkswagen Group Research*, Wolfsburg, Germany. Accessed 31 January 2024. www.hi-drive.eu/app/uploads/2023/09/Hi-Drive-SP4-D4.5-Effects-evaluation-methods-v1.0-.pdf

[24] FAME (2024). European Common Evaluation Methodology: EU-CEM Handbook. Public draft, version 0.6. www.connectedautomateddriving.eu/methodology/common-evaluationmethodology/, accessed on September 2, 2024.

[25] L3Pilot (2021). *L3Pilot Deliverable D1.7: Final Project Results*. Volkswagen Group Research, Wolfsburg, Germany. Accessed 31 January 2024. https://l3pilot.eu/fileadmin/user_upload/Downloads/Deliverables/Update_10082022/L3Pilot-SP1-D1.7-Final_project_results-v1.0_for_website.pdf

[26] Bjorvatn, A., Page, Y., Fahrenkrog, F., *et al.* (2021). *L3Pilot Deliverable D7.4: Impact Evaluation Results*. Volkswagen Group Research, Wolfsburg, Germany. Accessed 31 January 2024. https://l3pilot.eu/fileadmin/user_upload/Downloads/Deliverables/Update_14102021/L3Pilot-SP7-D7.4-Impact_Evaluation_Results-v1.0-for_website.pdf

[27] Aittoniemi, E., Itkonen, T., and Innamaa, S. (2023). Travel time, delay and CO_2 impacts of SAE L3 driving automation of passenger cars on the European motorway network. *European Journal of Transport and Infrastructure Research*, 23(1): 1–29. https://doi.org/10.18757/ejtir.2023.23.1.6553

[28] Kusano, K.D, Scanlon, J.M., Chen, Y.H., *et al.* (2023). Comparison of Waymo Rider-Only Crash Data to Human Benchmarks at 7.1 Million Miles. arXiv e-print archive. Accessed 31 January 2024. https://doi.org/10.48550/arXiv.2312.12675

[29] Scanlon, J.M., Kusano, K.D., Fraade-Blanar, L.A., McMurry, T.L., Chen, Y.H. and Victor, T. (2023). Benchmarks for retrospective automated driving system crash rate analysis using police-reported crash data. arXiv e-print archive. Accessed 31 January 2024. https://doi.org/10.48550/arXiv.2312.13228

[30] Aittoniemi, E., Itkonen, T., Lehtonen, E., Maasalo, I., Malin, F. and Innamaa, S. (2024). *Tieliikenteen automaation vaikutuksia*. Traficom Research Reports.

[31] Favarò, F.M., Nader, N., Eurich, S.O., Tripp, M. and Varadaraju, N. (2017). Examining accident reports involving autonomous vehicles in California. *PLoS ONE*, 12(9). Accessed 31 January 2024. https://doi.org/10.1371/journal.pone.0184952

[32]   Goodall, N.J. (2021). Comparison of automated vehicle struck-from-behind crash rates with national rates using naturalistic data. *Accident Analysis & Prevention*, 154(106056). Accessed 31 January 2024. https://doi.org/10.1016/j.aap.2021.106056

[33]   Tafidis, P., Farah, H., Brijs, T. and Pirdavani, A. (2021). Safety implications of higher levels of automated vehicles: a scoping review. *Transport Reviews*, 42(2): 245–267. https://doi.org/10.1080/01441647.2021.1971794

[34]   Tricco, A.C., Lillie, E., Zarin, W., *et al.* (2018). PRISMA extension for scoping reviews (PRISMA-ScR): checklist and explanation. *Annals of Internal Medicine*, 169(7): 467–473. www.acpjournals.org/doi/10.7326/M18-0850

[35]   Kulmala, R., Jääskeläinen, J. and Pakarinen, S. (2018). The Impact of Automated Transport on the Role, Operations and Costs of Road Operators and Authorities in Finland. EU EIP Activity 4.2 Facilitating Automated Driving. Accessed 31 January 2024. www.traficom.fi/sites/default/files/media/publication/EU_EIP_Impact_of_Automated_Transport_Finland_Traficom_6_2019.pdf

[36]   Törnqvist, J. (2018). Telephone Interview of Senior Research Scientist Jouko Törnqvist, a Specialist in Road Engineering at VTT Technical Research Centre of Finland, 16 February 2015.

[37]   Johnson, B. and Rowland, M. (2018). Automated and Zero Emission Vehicles. Transport Engineering Advice. Infrastructure Victoria & ARUP. REP/261257, Issue 1 July 2018. Accessed 31 January 2024. www.infrastructure-victoria.com.au/wp-content/uploads/2019/04/Infrastructure_Victoria_Automated_and_Zero_Emissions_Vehicles_Transport_Engineering_Advice.pdf

[38]   BSI (2020). Operational Design Domain (ODD) Taxonomy for an Automated Driving System (ADS) – Specification. The British Standards Institution, BSI PAS 1883. Accessed 31 January 2024. www.bsigroup.com/globalassets/localfiles/en-gb/cav/pas1883.pdf

[39]   Kulmala, R. (2013), Assessment Methods in ITS Lifecycle. Powerpoint Presentation. Finnish Transport Agency, 12 March 2013.

[40]   Fancher, P., Ervin, R., Sayer, J., *et al.* (1998). *Intelligent Cruise Control Field Operational Test (Final Report). Report DOT HS 808 849.* National Highway Traffic Safety Administration, U.S. Department of Transportation, Washington, DC. https://rosap.ntl.bts.gov/view/dot/3135/dot_3135_DS1.pdf?

[41]   Carsten, O., Fowkes, M., Lai, F., *et al.* (2008). Final Report. Intelligent Speed Adaptation Project. Institute for Transport Studies, University of Leeds. Accessed 31 January 2024. www.its.leeds.ac.uk/projects/isa/deliverables/Final%20Report%20080626.pdf

[42]   Biding, T. and Lind, G. (2002). Intelligent Speed Adaptation (ISA): Results of Large-scale Trials in Borlänge, Lidköping, Lund and Umeå during the period 1999-2002. Vägverket, Borlänge, Sweden. Accessed 31 January 2024. www.diva-portal.org/smash/get/diva2:1363740/FULLTEXT01.pdf

[43]   Carnet de Route du LAVIA: Limiteur s'adaptant à la vitesse autorisée. Ministère de la Transition écologique et solidaire, Paris, France. Accessed

31 January 2024. https://temis.documentation.developpement-durable.gouv.fr/docs/Temis/0073/Temis-0073937/PREDIT0072.pdf

[44]   Benmimoun, M., Pütz, A., Ljung Aust, M., *et al.* (2012). *euroFOT Deliverable D6.1: Final Evaluation Results*. Ford Research & Advanced Engineering Europe, Aachen. Accessed 31 January 2024. www.eurofot-ip.eu/en/library/deliverables/sp6_d61_final_evaluation_results.htm

[45]   Gershon, P., Mehler, B. and Reimer, B. (2023). Driver response and recovery following automation initiated disengagement in real-world hands-free driving. *Traffic Injury Prevention*, 24(4): 356–361. https://doi.org/10.1080/15389588.2023.2189990

[46]   Reagan, I.J., Teoh, E.R., Cicchino, J.B., *et al.* (2021). Disengagement from driving when using automation during a 4-week field trial. *Transportation Research Part F: Traffic Psychology and Behaviour*, 82: 400–411. https://doi.org/10.1016/j.trf.2021.09.010

[47]   European Commission (2023). Implementation of EU 2022/1426: Summary Report on Policy-related Topics. Accessed 31 January 2024. https://circabc.europa.eu/sd/a/4e1e0bc9-bcad-4bc6-a42d-4a7040c14512/PolicyTopics_summary%20report_v1.2%20rev%20EC_10-2023_CLEAN.docx

[48]   Andreone, L., Borodani, P., Pallaro, N., *et al.* (2021). *Deliverable D6.5 of L3Pilot: Pilot Reporting Outcomes*. Volkswagen Group Research, Wolfsburg, Germany. Accessed 31 January 2024. https://l3pilot.eu/fileadmin/user_upload/Downloads/Deliverables/Update_14102021/L3Pilot-SP6-D6.5-Pilot_Reporting_Outcomes-v1.0_for_website.pdf

[49]   Sintonen, H., Bolovinou, A., Guelsen, B., *et al.* (2023). *Deliverable D4.3 of Hi-Drive: Experimental Procedure*. Volkswagen Group Research, Wolfsburg, Germany. Accessed 31 January 2024. www.hi-drive.eu/app/uploads/2023/05/Hi-Drive-SP4-D4.3-Experimental-procedure-v1.1.pdf

Chapter 3

Field operational tests (FOTs) – (still) the ultimate answer to impact assessment?

I.C. MariAnne Karlsson[1], Stig E.R. Franzén[1] and Fjollë Novakazi[2]

3.1 Introduction

During the past decades, various smart functions have been introduced into vehicles – passenger cars, trucks, and buses. One group of systems encompasses, for example, blind spot detection, lane departure warnings, and adaptive cruise control. Other types of systems include navigation support and eco-driving support. The primary motives behind the implementation of these and similar systems have been to increase traffic safety, improve efficiency, enhance convenience, and promote an overall smarter utilisation of vehicles and roads.

Conclusions regarding the effectiveness of the systems have been based on an abundance of driving simulator studies, studies on test tracks, and limited trials in real traffic. However, more extensive trials are necessary to understand more in-depth the impact of different functions and systems and how the systems are used by ordinary people in real traffic and over a more extended period. Several such field studies – or field operational tests (FOTs) – have been conducted over the years in, for example, Australia, Canada, Japan, and the USA, as well as in Europe, with the large European FOTs conducted between 2012 and 2015.

In an overview of some of these European FOTs, their respective methodologies and outcomes, the authors raised the question: Are FOTs the ultimate answer to impact assessment? [1]. We believe that the question is still valid. While FOTs offer significant potential for studying the impact of different ITS solutions, conducting large-scale studies of this nature requires substantial resources and significant efforts related to project management and coordination, as well as data collection and analysis. Are FOTs and the associated experimental approach (still) the answer to studying drivers, their driving behaviour, their usage of different functions, and the impact of these functions? Is it also the answer to assessing the impact of future ITS including more advanced automated functions, automated driving and cooperative and connected systems?

[1]Design and Human Factors, IMS, Chalmers University of Technology, Sweden
[2]School of Science and Technology, Örebro University, Sweden

3.1.1   What is a field operational test?

To address the question, it is essential to define what we mean by a FOT. One definition of a FOT reads:

> a study undertaken to evaluate a function, or functions, under normal operating conditions in road traffic environments typically encountered by the participants using study design so as to identify real world effects and benefits. [2]

The definition does not specify that an FOT is necessarily large scale, but in the context of the European FOTs, the term has been used to refer to large-scale testing and thorough assessment of what should be ICT solutions, encompassing stand-alone in-car systems as well as cooperative systems [3]. It should engage a larger sample of potential users using the functions and systems over a longer period of time, thereby covering a wide range of scenarios involving different types of drivers. The participants involved should drive their own cars as they normally do and there should be minimal intervention from any outside experimental body. 'Normal operating conditions' imply that the FOT participants use the systems during their daily routines, that data logging works autonomously and that the participants do not receive special instructions about how and where to drive. However, an important feature is that it must be possible to compare the effects of a specific function, for example, when activated compared to when deactivated. To accomplish this, the research team should be able to manipulate the participants' control over or interaction with the function(s) being tested.

3.1.2   The FESTA FOT methodology

At least from a European perspective, it is not possible to discuss FOTs without discussing also the FESTA methodology. The methodology was developed within the FESTA project (Support Action Field Operational Test) conducted between 2007 and 2008, that is before the large-scale European FOTs were funded. The project resulted in extensive recommendations for developing and implementing an experimental procedure for FOTs.

The steps that typically should be carried out during an FOT were represented in a V-shaped diagram – the *FESTA V* – with a correspondence between the levels on the left-hand and right-hand sides (Figure 3.1).

The steps were grouped into three phases; (i) Preparing, (ii) Using, and (iii) Analysing. The 'Preparing' phase involves:

Identification and description of functions to be tested;

Defining use cases;

Formulation of research questions and hypotheses that can be statistically tested;

Defining qualitative and quantitative performance indicators and study design;

Deciding on measures and sensors to be used.

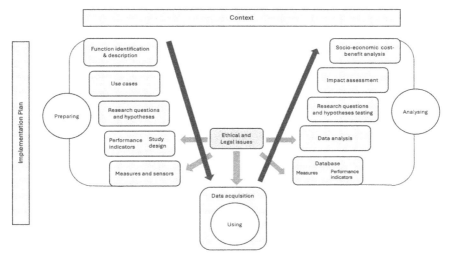

Figure 3.1   The FESTA V. Illustration adopted from the FESTA Handbook Version 5 [4] and the following versions. This differs slightly from the original description in [2].

'Using' includes data acquisition, collecting quantitative, and qualitative data from the FOT, and the 'Analysing' phase is divided into:

Database design and implementation;

Data analysis including data quality analysis, etc.;

Research questions and hypotheses testing;

Impact assessment regarding safety, efficiency and/or environment;

Scaling up of results to assess socio-economic impacts –if the system(s) was fully deployed in a large proportion of vehicles.

The methodology was documented in the first version of the *FESTA Handbook* [2], in which the steps described in Figure 3.1 were developed in detail. This includes how to set up an FOT, formulate hypotheses, choose performance indicators, experimental procedures, data acquisition guidelines, guidelines for databases and analysis tools, data analysis and modelling, impact assessment, as well as legal and ethical issues.

3.2   The European large FOTs

To address the questions, we begin by looking into some of the large-scale European FOTs, which were conducted between 2008 and 2015, and which adhered to the experimental FESTA methodology–although to different degrees.

euroFOT – European Field Operational Test on Active Safety Systems – assessed the impact of in-vehicle support (e.g., forward collision warning) [5,6].

TeleFOT – Field Operational Tests of Aftermarket and Nomadic Devices in Vehicles – investigated the implications of nomadic and aftermarket devices (e.g. speed alert) [7].

DRIVE C2X – Driving Implementation and Evaluation of C2X Communication Technology in Europe – carried out assessments of cooperative systems (e.g., road works warning), considering in-vehicle systems and roadside equipment [8,9].

FOTsis – Field Operational Test on Safe Intelligent and Sustainable Road Operation – focused on road infrastructure management systems [10].

In addition to the type of functions and systems being studied, another difference between the FOTs was that whereas euroFOT and TeleFOT involved systems already available on the market, the systems tested in the DRIVE C2X and FOTsis projects were described as 'close to market'. An overview is provided in Table 3.1.

Despite the differences, the projects had common aims: (i) assessing the impacts of the respective functions and systems on traffic safety, traffic efficiency, driver comfort and the environment, and (ii) demonstrating the overall benefits of

Table 3.1   Overview of European large FOTs

FOT	Countries where the FOTs took place	Number of vehicles/drivers	Functions tested
EuroFOT	Germany, France, Italy, Sweden	790 Cars, 80 heavy trucks	Forward collision warning, advanced cruise control, speed regulation system, blind spot information, lane departure warning, impairment warning, curve speed warning, fuel efficiency advisor
TeleFOT L-FOTs	Finland, Greece, Italy, Sweden, Spain, the UK	Private cars, approx. 2800 drivers	Navigation support, green driving support, speed alert, speed limit information, traffic information
DRIVE C2X	Finland, France, Germany, Italy, Spain, Sweden	>200 Vehicles 750 Drivers	Road works warning, traffic jam ahead warning, car breakdown warning, weather warning, approaching emergency vehicle warning
FOTsis[*]	Germany, Greece, Spain, Portugal,	Systems implemented at nine highways	Emergency management, safety incident management, intelligent congestion control, dynamic route planning, special vehicle tracking, advanced enforcement, infrastructure safety assessment

*Not commented on further.

the systems. However, from an EU Commission perspective, another important aim was (iii) contributing to the deployment of ITS technologies on the market.

3.2.1 Impact assessments

FOTs are conducted with the intention of studying impact.

In euroFOT [5,6], some of the conclusions regarding the impacts of the functions tested were, for example:

> A bundle of Advanced Cruise Control and Forward Collision Warning had a positive impact on driving safety; average time-headway increased, and the relative frequency of harsh braking manoeuvres decreased. There were also positive effects on the environment due to reduced fuel consumption.

> The Speed Limiter function led to a reduction in over-speeding and harsh braking events, while Cruise Control resulted in an increase in over-speeding but a reduction in harsh braking. No safety effects were found, but an increase in traffic efficiency was anticipated as an effect of the increase in average speed.

> Overall, the drivers had a positive attitude towards Curve Speed Warning and felt it increased safety. The same was true for drivers' perceptions of blind spot information.

In TeleFOT [7], the analyses concluded that not all tested functions affected or played the same role in different impact areas. For example:

> Access to green driving support was found to have a positive impact on the environment, resulting in significant decreases in fuel consumption and CO_2 emissions.

> Access to static and dynamic navigation support had primarily an impact on mobility, as journey distances and journey durations (for comparable journeys) were shorter.

> Access to traffic information could have a negative impact on safety as it resulted in an increase in the use of rural roads (higher risks) and a decrease in the use of city roads.

> Participants reported increases in perceptions of safety and comfort and decrease in stress.

In DRIVE C2X [8,9], the assessment found overall improvements in traffic safety with some adverse effects on traffic efficiency and positive effects on the environment. For example:

The In-Vehicle Signage (IVS) function had positive impacts on driver behaviour, especially in areas where special attention should be paid to vulnerable road users. An IVS child sign and a 'pedestrian crossing ahead' sign reduced speed in relevant areas.

The IVS on speed limit and weather warnings showed the most potential to decrease fatalities.

Functions such as the IVS on speed limit and green-light optimal speed advisory indicated significant effects on the environment and traffic efficiency.

User acceptance was high; they expressed a willingness to use functions if available in the vehicle. Specifically, journey quality was perceived to be improved in terms of decreased user uncertainty and stress and a feeling of safety and travel comfort.

3.2.2   Methodological challenges

When planning and preparing the trials, collecting data during and analysing the data from the trials, all projects faced challenges. Considerable efforts were made to adhere to the guidelines outlined in the *FESTA Handbook* and maintain a common study design approach, nevertheless various practical and logistical obstacles were encountered. These challenges required new developments and modifications to the original plans, some of which are described below.

In euroFOT, for example, the trials should have lasted 12 months, including a three-month baseline during which the function was not available and a nine-month phase when the functions were available but not all sites could follow this plan. In TeleFOT, the trials with access to the function(s) should last a minimum of six months to allow for a 'steady state' to be reached, but some trials were shorter even so.

Furthermore, in euroFOT, the recruited drivers should be between 30 and 50 years old to ensure homogeneity, but the age boundaries had to be changed considerably to reach the desired number of participants [6]. In TeleFOT, the participants should mirror the intended user (or customer) population, but – again – this was achieved to different degrees [7].

Formulating and prioritising between research questions and hypotheses were not without problems. In the TeleFOT project, a way forward was found by means of a modified and more elaborate procedure, integrating hypotheses formulated top-down, i.e. based on an underlying theoretical framework, and bottom-up, i.e. based on the function to be tested. The research questions finally addressed were then chosen based on three criteria: (i) importance in relation to the context of the impact assessment, (ii) feasibility of collecting performance indicator data, and (iii) cost of collecting data to answer research questions and test hypothesis [7].

Data collection also faced difficulties. In euroFOT the need to install data loggers required adaptations to the participants' cars and this led, in turn, to difficulties in recruiting participants [6]. Similarly, in TeleFOT, only a limited amount of logged data could be collected due to the restricted access to the participants' private vehicles. Common data acquisition equipment was not achieved across all trials [7].

The FOTs generated large amounts of objective data, and the analysis leading up to answering the research questions and addressing the hypotheses demanded considerable resources for pre-processing raw data files, checking data quality, data reduction, etc. In the DRIVE C2X project, coordination and communication problems led to non-harmonised developments of logging procedures, resulting in unconventional logging styles, which made analyses difficult and required months of additional work to sort out [9].

3.3   The future of FOTs

Planning a (large) FOT, running the trials and analysing the data, whether adhering to the FESTA methodology or not, are thus challenging. We began the chapter by asking if FOTs are, or rather, will FOTs be the ultimate answer to impact assessment in the future. The question must be divided into two, one focusing on FOTs per se: Are FOTs the ultimate answer to impact assessment in the future? and a second question focusing on the FESTA methodology: Is the FESTA methodology the ultimate answer to impact assessment in the future?

3.3.1   *Will FOTs be the ultimate answer to impact assessment?*

In addressing research questions and hypotheses related to the impacts of ITS, the type of large FOTs referred to implies, enabled by in-vehicle electronics and advanced sensor technology, the collection of large amounts of real-world data 'on real drivers on real roads...' [11] in various driving conditions, road types, and traffic situations. This, together with the experimental approach, composes the particular strengths of FOTs. At the same time, large FOTs are expensive, logistically challenging, and require significant resources for administrative, managerial, technical, procedural and legal matters [12–15]. There have been doubts voiced about how cost-effective large FOTs are, and it has been suggested that the same results could be achieved with at least a smaller number of partners [9].

Focusing on the period from 2015 and onwards, there are few larger FOTs. One example, but with only a few partners, is the Australian CAVI 500-vehicle C-ITS FOT, which was performed between 2020 and 2022, maintaining the experimental approach in terms of a larger group of participating drivers, baseline and treatment phase, and collection of qualitative and quantitative data. The 355 public participants in 1 Australian city drove their own vehicles retrofitted with connected vehicle technology (C-ITS), such as turning warnings for vulnerable road users [16,17].

It is easier to find smaller FOTs with a more limited number of participants, vehicles, and sometimes duration. One example is the vulnerable road user detection system FOT, conducted in Canada in 2018. This study involved 49 drivers and

14 vehicles, differing in size and design, and equipped with different camera/sensor systems and driver-vehicle data loggers to capture the performance of the detection systems under natural driving conditions but still over a period of one year [18]. Another example is the *Older Driver Support System FOT*, a 12-month study of an app, which examined the baseline driving behaviour of 28 older drivers in Minnesota and Kansas, USA, their driving behaviour with RoadCoach feedback, and their driving behaviour during a follow-up, no-feedback period [19]. However, even more common are field studies with a very small number of participants and vehicles, with a fairly short duration but, at the same time, often with purposes other than impact assessments.

With a focus on real-world benefits and assessments, the methodological landscape is evidently broader and also includes, for example, Naturalistic Driving Studies (NDS), such as UDRIVE (27, 28) or SHRP2 [20], and Naturalistic FOTs (N-FOTs), such as SeMiFOT [20] and SeMiFOT2. Although complementary in terms of focus and perspectives [10], there important differences [21] and with their more experimental approach, FOTs remain a fundamental component in systematic assessments of impacts of ITS.

3.3.2   *Will FESTA be the ultimate methodology to impact assessment?*

The FESTA methodology was developed to ensure a systematic and scientific approach to FOTs, but it was also developed to allow for comparisons between trials, within and between FOTs [22].

Already in 2016, when the question was first posed: 'Are FOTs the ultimate answer to impact assessment?' the authors raised the question of whether the FESTA methodology was appropriate for FOTs testing of automated vehicles or whether there is a need for modifications or a completely new methodology [1]. This has since then been a recurring theme [22–24] with the argument that impact assessments of automated systems and services require an adapted methodology.

The FESTA methodology advocates, for example, that a baseline should be established so that conclusions can be drawn as to the impact of a function when activated compared to when it is not, but such a situation may be difficult to create for automated vehicles [1,12]. In vehicles with high driving automation, the role and responsibilities of drivers differ compared to when driving with no or partial automation. Consequently, the research questions and hypotheses will also differ, as will the data necessary to address them. Furthermore, the impact analysis has primarily concerned traffic safety, efficiency, and the environment. An impact assessment of higher-level automated vehicles may involve other and longer-term societal impacts such as travel behaviour and mode choice that are related to the need for travel and the costs associated with the different mode alternatives [1,22].

The *FESTA Handbook* is a dynamic document. Since the FESTA project, the handbook has been updated several times, first by the FOT-Net network and more recently by CARTRE and ARCADE, a coordination and support action for planning, updating and dissemination of the FESTA methodology. The modifications made have been based on feedback from the FOT community, including discussions, workshops and seminars with users and experts. Topics addressed include data analysis, legal and ethical issues, impact assessment, scaling up, etc. Data sharing has been a particular theme.

The eighth version of the handbook was released in September 2021 [25]. The updated version includes sections on how FESTA can be used in projects testing and evaluating automated vehicles, but the modifications have been considered insufficient for FOTs on Cooperative and Connected Automated Mobility (CCAM) [26]. CCAM is argued to require 'innovative methods and tools' [26] and, therefore, the development of a new, common evaluation methodology to address methodological gaps covering, for example, new impact areas, new research questions, and a more agile evaluation process. The European FAME project was therefore set up to develop such a methodology, and a final version is expected mid-2025.

3.3.3 Conclusion

There is no doubt a further need for FOTs. Studies, where technical function(s) or system(s) are used by ordinary drivers (or users) under normal operating conditions in real traffic and transport environments, are important for developing more in-depth knowledge of drivers' (users') driving (using) behaviour and the impacts that different ITS solutions have. Other types of studies conducted in more controlled environments, such as driving simulator studies, or in real-world settings, such as naturalistic driving studies (NDS) and/or Naturalistic FOTs (N-FOTs), will still be important complementary tools for generating knowledge in the interaction between people and ITS solutions.

All FOTs can be expected to come with administrative, managerial, legal, technical, and/or methodological challenges. However, size is an additional complicating factor, and it is possible that future FOTs will be smaller and more focused. Furthermore, independent of FOT size, new digital tools will play an important role in facilitating the collection and analysis of data.

Nevertheless, FOTs must rely on scientific, systematic, and robust methodology. The FESTA methodology can form a fundamental framework also for future FOTs. However, although modifications have been made to the *FESTA Handbook* to adhere to the requirements posed by evaluations of automated systems, the complexity of trials with CCAM has been found to require the development of a new, common evaluation methodology for CCAM FOTs. This work is anticipated to be finished in the summer of 2025.

References

[1] Franzén, S.E.R., and Karlsson, I.C.M. 'Field Operational Tests (FOTs) – the ultimate answer to impact evaluation?' in Lu, M. (ed.). *Evaluation of Intelligent Road Transport Systems. Methods and Results.* London: IET; 2016, pp. 141–160.

[2] *FESTA Handbook Version 2. Deliverable D6.4 of Field Operational Support Action.* FESTA Consortium, 2008.

[3]   *FOT-Net. Field Operational Test. Evaluating ITS-Applications in a Real World-environment.* 2011 Available from https://erticonetwork.com/wpcontent/uploads/2011/03/www.fot-net.eu_download_Print_material_j02193_pol_fotnet_brochure_design_09_web.pdf. [Accessed 2024-03-01].

[4]   *FESTA Handbook Version 5.* Revised by FOT-Net. 2014.

[5]   Benmomoun, M., Pütz, A., Zlocki, A., and Eckstein, L. euroFOT: Field operational test and impact assessment of advanced driver assistance systems – final results. *Proceedings of the FISITA 2012 World Automotive Congress. Lecture Notes in Electrical Engineering*, vol 197. Berlin. Heidelberg: Springer; 2013, pp. 537–547.

[6]   Kessler, C., Etemad, A., Alessandretti, G., *et al. Deliverable D11.3. Final Report.* Deliverable to the euroFOT project, 2012.

[7]   Mononen, P., Franzén, S., Pagle, K., *et al. TeleFOT Deliverable D1.15. Final Report.* Deliverable to the TeleFOT project, 2013

[8]   Malone, K.M., Innamaa, S, and Hogema, J.H. Impact assessment of cooperative systems in the DRIVE C2X project. *Proceedings of the 22nd ITS World Congress.* France, Bordeaux, Oct 2015, pp. 1–13.

[9]   Schulze, M., Mäkinen, T., Kessel, T., Metzner, S., and Stoyanov, H. 2014. *DRIVE C2X. Accelerate cooperative mobility. Deliverable D11.6. Final Report.* https://www.eict.de/fileadmin/redakteure/Projekte/DriveC2X/Deliverables/DRIVE_C2X_D11_6_Final_report__full_version_.pdf. (Accessed 2024-04-08)

[10]   Alfonso, J., Sánchez, N., Menéndez, J., and Cacheiro, E. (2015). Cooperative ITS communications architecture: the FOTsis project approach and beyond. *IET Intelligent Transport Systems*, 2015; **9**(6): 591–598.

[11]   Flament, M., and Silva, I. Needs for Cooperation on Field Operational Tests. *Proceedings of the 17th ITS World Congress.* Korea, Busan, 2010, ITS Asia Pacific, ITS America, ERTICO, 2010, pp. 1–9.

[12]   Barnard, Y., and Carsten O. Field operational tests: challenges and methods. In Krems, J., Petzholdt, T. and Henning, M. (eds.). *Proceeding of the European Conference on Human Centred Design for Intelligent Transport Systems*, Lyon: HUMANIST publications, 2010, pp. 323–332.

[13]   Barnard, Y. F., Gellerman, H., Koskinen, S., Chen, H., and Brizzolara, D. Anonymization of Data from Field Operational Tests. *Proceedings of the 11th ITS European Congress*, Glasgow, Scotland, June 2016, pp. 1–10.

[14]   Benmimoun, A., Benmimoun, M., Regan, M.A., van Noort, M., and Wilmink. I. Challenges in analysing data from a field operational test of advanced driver assistance systems: the euroFOT experience. *Proceedings of Transport Research Arena (TRA)*, 2010, Brussels, the Netherlands, pp. 1–9.

[15]   Goralczyk, M., Schaeufele, B., and Radusch, I. (2011). Logging design for vehicle communication field operational tests. *Proceedings of the First International Symposium on Future Active Safety Technology towards Zero-traffic Accidents (FAS-zero)*, Japan, Tokyo, Sept 2011, pp. 1–6.

[16]   Ipswich Connected Vehicle Pilot Safety Evaluation. Summary report. Department of Transport and Main Roads (Australia). 2022. Available from:

https://research.qut.edu.au/carrsq/wp-content/uploads/sites/296/2022/11/
IVCP-Safety-Evaluation-1.pdf (Accessed 2024-04-08)

[17] Pascale, M., Rodwell, D., Ho, B., Elrose, F., and Lewis, I. *Ipswich Con-nected Vehicle Pilot Safety Evaluation. Summary of the Subjective Evalua-tion Study Findings.* Centre for Accident Research and Road Safety, Queensland, 2022.

[18] Lau, C.P., Burns, P.C., and Charlebois, C. *Acceptance and experience of a vulnerable road user detection system among heavy vehicle operators: A year-long multi-city field trial.* 2020. Available from: www.researchgate.net/publication/351151942_Acceptance_and_Experience_of_a_Vulnerable_Road_User_Detection_System_among_Heavy_Vehicle_Operators_A_year-long_Multi-City_Field_Trial [Accessed 24-03-15].

[19] Libby, D.A., Morris, N.L., and Craig, C.M. *Older Driver Support System Field Operational Test. Final Report. 2019.* Center for Transportation Studies, 2019.

[20] Campbell, K.L. *The SHRP2 Naturalistic Driving Study. Addressing driver performance and behavior in traffic safety.* 2012 Available from https://insight.shrp2nds.us/documents/shrp2_background.pdf [Accessed 24-04-09].

[21] Trent, V., Bärgman, J., Gellerman, H., *et al. Swedish-Michigan Naturalistic Field Operational Test (SeMiFOT). Phase 1: Final Report.* SAFER, 2010.

[22] Barnard, Y., Innamaa, S., Koskinen, S., Gellerman, H., Svanberg, E, and Chen, H. Methodology for field operational tests of automated vehicles. *Transportation Research Procedia*, 2016; **14**: 2188–2196.

[23] Innamaa, S., Louw, T., Merat, N., Torrao, G., and Aittoniema, E. Applying the FESTA methodology to automated pilots. *Proceedings of TRS2020, the 8th Transport Research Arena: Rethinking Transport – Towards Clean and Inclusive Mobility.* Traficom Research Reports; No. 7/2020.

[24] Barnard, Y., and Koskinen, S. Methodology for Field Operational Tests: updating the FESTA methodology for connected and automated driving pilots. *Transportation Research Procedia*, 2023, **72**: 2054–2061.

[25] *FESTA Handbook Version 8.* Online. 2021 Available from www.connectedautomateddriving.eu/wp-content/uploads/2021/09/FESTA-Handbook-Version-8.pdf [Accessed 24-03-01].

[26] Innamaa, S., Barnard, Y., Geissler, T., Wilmink, I., and Harrison, G. Developing a common evaluation methodology for CCAM. *Proceedings of the 15th ITS European Congress.* Portugal, Lisbon, May 2023. Available from https://cris.vtt.fi/ws/portalfiles/portal/85140639/ITSLisbon2023Paper_CEM_FINAL.pdf [Accessed 24-04-08].

[27] Barnard, Y., Utesch, F., van Nes, Eenik, R., and Baumann, M. The study design of UDRIVE: the naturalistic driving study across Europe for cars, trucks and scooters. *European Transport Research Review*, 2016; **8**(14), pp. 1–10. Available from: https://rdcu.be/dSNYj. (Accessed 2024-04-09).

[28] Eenink, R., Barnard, Y., Baumann, M., Augros, X., and Utesch, F. UDRIVE: The European naturalistic driving study. *Proceedings of Transport Research Arena 2014*, Paris, France, 2014, pp. 1–10.

Chapter 4

Driving the future: role of artificial intelligence in road vehicles

Divya Garikapati[1], Sneha Sudhir Shetiya[2] and Chaitanya Shinde[3]

4.1 Introduction

Automated vehicles (AVs) are evolving at a much faster pace than ever starting from the past few decades. AVs now have the capability to interact with humans, and the environment, and perform the dynamic driving task (DDT) [1] as needed with very little human intervention. This transformation is possible only with the integration of artificial intelligence (AI) technologies into each aspect of the development of these vehicles. AI is needed in AVs because of the complex array of tasks and decisions required for safe and efficient navigation in dynamic and unpredictable environments. AI serves as the cognitive backbone of AVs, enabling them to perceive their surroundings, interpret sensory data, make informed decisions, and adapt to changing conditions in real-time. AI serves as a foundational component of AVs, consisting of complex techniques to mimic human cognition [2]. From recognizing traffic signs and pedestrians to predicting the behavior of other vehicles and navigating complex road networks, AI helps vehicles to operate effectively in diverse scenarios.

The vast field of AI encompasses several subfields as mentioned in [3], each with its own strengths. Machine learning (ML) forms a broad branch of AI, adept at uncovering patterns within data and adjusting to changing situations. Deep learning (DL) dives deeper within the ML realm, utilizing complex neural networks to tackle massive datasets and perform intricate calculations. Notably, DL algorithms often leverage supervised learning techniques, particularly deep neural networks, to extract knowledge from even unstructured data. Generative AI (Gen AI) further refines the DL approach as shown in Figure 4.1, specializing in mimicking human-like interactions. This is achieved through powerful models like generative adversarial networks (GANs) and generative pre-trained transformers (GPTs). GANs pit

[1]Independent Expert, USA
[2]Torc Robotics Inc., USA
[3]Independent Expert, USA

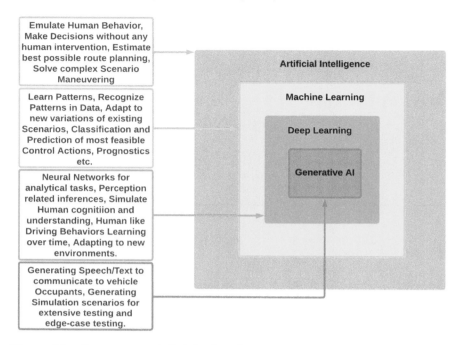

Figure 4.1   Comparison of AI, ML, deep learning, and Gen AI with some of their use cases in AVs [3,4]

a generator model against a discriminator model in a continuous learning loop, pushing both to improve until their creations become indistinguishable from real examples. GPT frameworks, on the other hand, excel at generative language modeling, enabling tasks like crafting text, generating code, and carrying on conversations.

In this section, we discuss the fundamental reasons why AI is essential for AVs, and its role in enhancing safety, improving efficiency, and enabling the full potential of self-driving technology. By leveraging the power of AI, AVs could revolutionize transportation, resulting in safer roads with zero fatalities, reduced congestion, and efficient mobility for all. There are four key areas in which AI has become critical for AVs such as perception, planning, continuous learning, and enabling personalized driving experience.

Replacing human vision [5]: AI algorithms such as computer vision and deep learning enable vehicles to "perceive or see" and understand their surroundings. They make use of the data from sensors like cameras, LiDAR, and Radar to identify objects such as pedestrians, cyclists, and other vehicles. Having multiple sensors adds to their ability to perceive even in challenging conditions like rain or fog. AI plays a crucial role in sensor fusion, as highlighted in [6]. Sensor fusion involves the integration of data from various sensors to develop a

360° understanding of the vehicle's surroundings. AI algorithms accomplish this integration by synchronizing and aligning data streams from cameras, LiDAR, Radar, Global Positioning System (GPS), and inertial sensors. Through sensor fusion, AI significantly improves perception capabilities, enabling AVs to accurately detect and recognize objects, pedestrians, traffic signs, and obstacles. It can also help with simultaneous localization and mapping (SLAM) algorithms purely perception-based or using traditional methods. Visual odometry algorithms, enhanced by AI [7], estimate the vehicle's motion and position based on visual inputs, particularly useful in GPS-challenged environments. ML models contribute to predictive localization by utilizing historical data and contextual information to refine the vehicle's position.

Making complex decisions in real time: Every driving situation calls for many decisions at the dimensions of milliseconds or microseconds. AI analyzes sensory data and traffic rules to navigate situations, plan routes, expect capacity dangers, and react. This consists of judging when to boost up, brake, or swerve to avoid barriers, all while adhering to site traffic rules and legal guidelines. The path planning algorithms using AI determine optimal routes and waypoints, considering factors like traffic conditions, traffic regulations, and the positions or velocities of the actors in the immediate scene. ML models predict the behavior or intent of other road users as well, enhancing the vehicle's ability to anticipate and plan well in advance to various scenarios effectively.

Continuously learning and adaptation: Unlike conventional programming, AI systems can learn and adapt during the operation. A huge amount of driving data is being collected by AVs for the AI models to be able to constantly improve their understanding and eventually improve decision-making capabilities, leading to safer and more efficient automated driving. This is important to move toward Level 5 autonomy where the vehicle is expected to operate without any human interaction and adapt to all new environments alike.

Personalized driving experiences [8]: AI facilitates human-machine interaction in AVs, enabling seamless communication between passengers and the vehicle. Natural language processing (NLP) systems and gesture recognition algorithms [9], powered by AI, facilitate intuitive interaction and control. AI can learn individual driving preferences, health conditions, and abilities and adjust the vehicle's behavior accordingly. The vehicle anticipates the user's preferred cruising speed or braking style, creating a smoother and more comfortable journey.

Furthermore, AI also plays a crucial role in ensuring the security and safety [10] of AVs, detecting anomalies in sensor data, and implementing robust cybersecurity measures [11]. There are a lot of other use cases for implementing the various life-cycle processes from cradle to grave for AVs like predictive maintenance, automatic detection of scenarios in collected data, automated scenario generation for simulation testing, and building the required parameter set for software-in-the-loop testing. Section 4.2 dives deeper into these use cases (Figure 4.2).

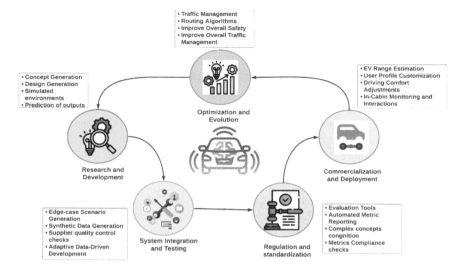

Figure 4.2   AI in automotive – use cases at various life-cycle stages

4.2   AI use cases in AVs

This section explains all the use cases [12] of AI in AVs. As identified in the introduction Section 4.1, there are a few key areas in which AI has become critical for AVs such as perception and sensing, planning and decision-making, control and actuation, learning and adaptation, and user personalization. There are also other areas such as predictive maintenance, cybersecurity, safety, human-machine interaction, and on-board diagnostics.

This section discusses a more comprehensive list of use cases [13] based on their groupings into the different life-cycle stages of AV technology development such as:

Research and development (R&D): This stage focuses on exploring the feasibility of AV technology, defining its potential applications, and identifying key challenges. It also includes the concept generation phase. Small-scale proto-types are developed as proof-of-concept (POC) and tested in controlled environments to demonstrate basic functionalities and validate core concepts. Algorithms, sensors, and hardware are developed and iteratively improved by research and refinement based on data collected from POCs and simulations.

AI Use Cases: Gen AI could be used for concept generation and for brainstorming systematically. It can also be used for design generation using AI models. For example, Volkswagen uses generative design [14] to inspire compactness in its vehicles. Gen AI rapidly iterates on permutations and combinations of simulated environments [15] and scenarios. For example, a simulated environment with rainy conditions such as modeling the road surface friction values and rainfall rate could be recreated or generated based on the specified parameter values. Also, Gen AI could be used for the prediction of outcomes using simulation to reduce experimental R&D costs.

System integration and testing: Functional prototypes are built, integrating various hardware and software components into a working system. Prototypes are rigorously tested in controlled environments like simulation and closed tracks to evaluate performance, identify bugs, and optimize algorithms. Testing expands to limited real-world roads with increased complexity and diverse driving scenarios, gathering data for further refinement.

AI use cases: Edge-case scenario generation for testing plays a key role in the safety verification and validation of the system. This is especially because most of the edge-case scenarios [16] are hard to find from existing driving datasets and are always very minimal in number, which results in AI models having a certain in-built bias to classify scenarios into existing cases or more popular scenarios. Generating more synthetic data and scenarios while training and testing the AI models can help mitigate this bias. Another way is to use historical accident data [17] to create these scenarios. From a supply chain perspective, AI could be used for quality control of supplies and finished goods with automated visual inspection and automated plug-and-play type of testing environments. Another use case is to build control systems that can pick and choose the safety limits [18] based on the immediate scenario using learning or AI techniques.

Regulation and standardization: Regulatory frameworks are drafted and implemented to ensure the safety, security, and ethical operation of AVs. Performance and safety standards are established for AV functionalities, testing procedures, and data privacy to guarantee reliable and responsible deployment. Public dialogue and education are crucial in fostering understanding, trust, and acceptance of AV technology.

AI use cases: Tools built with AI models can be trained to keep track of the performance and safety metrics for the AVs. These metrics are usually reported to the regulatory boards for approval of limited public roads or a more expanded public road operation. Some of the metrics [19] could be very complex that they could only be estimated based on existing data. ML models usually perform well for this use case of estimating the values with a pre-determined accuracy.

Commercialization and deployment: AI can be used for limited deployment or initial roll-out in restricted areas or specific use cases to gather real-world data and monitor performance. Gradual expansion based on success in limited deployments and regulatory approvals, AVs are rolled out to wider segments of the population and geographic areas. Different AV developers and manufacturers compete to offer superior performance, features, and affordability, driving innovation and consumer choice.

AI use cases: Electric vehicle range estimation, health monitoring, and prognostics algorithms for predictive maintenance capabilities are some of the major use cases. User profile customization requires learning abilities and AI models can intuitively provide this ability. It could learn user preferences for speed, and route choices, or even adjust in-cabin features like temperature or entertainment based on passenger needs. ChatGPT (Chatbot-based search by OpenAI) was introduced by

Mercedes recently to interact with the vehicle occupants [20]. There could also be scenarios in which the driving comfort related to actuation and in-cabin elements could be adjusted based on the learning of user preferences over time. AI could also be used for more entertainment-related activities like saving favorite music, and movies, booking concert tickets, and managing schedules similar to a concierge. Driver in-cabin monitoring for safety purposes would be another use case.

Optimization and evolution: Data collected from deployed AVs is used to refine algorithms, enhance performance, drive continuous improvements, and address emerging challenges. The scope of AV applications expands beyond initial use cases, exploring possibilities in different industries and transportation modes. AV applications include Traffic management, fleet management, predictive maintenance, repair guidelines, bug resolution guidelines, vehicle inspections, and end-of-life prediction for all components and subcomponents of the vehicle. Integration and collaboration with infrastructure development (e.g., smart cities) leads to enhanced communication and coordination for improved efficiency and safety.

AI use cases: For traffic management, different vehicles are treated as nodes sending and receiving information, and the routing algorithms [21] for each vehicle are determined by mapping the shortest path and the most cost-effective path to reach its destination. A more cooperative approach along with learning capabilities helps reduce congestion and improve the overall travel experience for everyone.

AI use cases – key takeaways

AI plays a key role throughout all stages of AV development, from concept design to deployment and beyond. Different stages leverage AI uniquely. R&D uses it for concept generation and simulations, Testing for edge-case scenarios, and Regulation for performance metric tracking. Deployed AVs use AI for user personalization and traffic management optimization. Continuously collected data is analyzed by AI to improve performance and explore new applications for AVs. Overall, AI drives innovation and efficiency in AV development and deployment.

4.3   AI/ML models used in AVs

Based on the key areas identified in Section 4.1, perception systems within AVs rely on AI/ML algorithms entirely to perform tasks such as detection, classification, tracking, and prediction. The following section provides an overview of the different types of algorithms used by different sensors like LiDARs, cameras and radars.

4.3.1   Sensing and perception

4.3.1.1   LiDAR

Voxel-based convolutional neural networks (CNNs) [22]: Voxelization technique can be used to convert point clouds into 3D voxel grids, which allows them to be mapped to 3D CNNs for perception.

PointNet/PointNet++ [23] *for point cloud processing:* These models are specifically developed for direct processing of raw point-cloud data (i.e., without any preprocessing using voxelization or similar).

4.3.1.2 Camera

Visible camera

Single shot multibox detector (SSD) [24] *and you only look once (YOLO)* [25]: Both of these models are designed for applications that require fast object detection, predicting both localized bounding boxes and their corresponding class probabilities on the first pass through the neural network.

Generative adversarial networks (GANs) [26] *for data augmentation:* Image-like synthetic data (be it real or imagined) produced by GANs could be used to augment training data in more realistic settings (along with data constraints) to create more robust perception models.

Thermal camera (LWIR – long wave infrared)

Semantic segmentation with fully convolutional networks (FCNs) [27]: FCNs can segment LWIR images into different classes of objects or materials based on their thermal signatures.

4.3.1.3 Radar

4D radar point cloud processing with RaTrack [28]: Radar data can also be processed as point clouds using 4D imaging radar, with the help of RaTrack network pipeline, which uses a combination of neural network called backbone network and MLP for tasks such as object detection, classification, and tracking.

4.3.1.4 Microphone

Sound source localization with deep learning [29]: Deep learning models, such as convolutional recurrent neural networks (CRNNs) or attention-based mechanisms use sound source localization, in determining the direction from which specific sounds originate.

4.3.2 Planning and controls

Another key area identified based on Section 4.1 along with efficient sensing and perception algorithms is planning and control algorithms which guide the vehicle in maneuvering safely around the objects that have been detected by perception algorithms. For this purpose, various techniques are available that try to predict the path and generate the trajectory of the Ego vehicle to the maximum accuracy. Following are some of the path planning and control algorithms that are primarily used within the AV industry:

4.3.2.1 Deep deterministic policy gradient [30]

Reinforcement learning algorithms like Deep deterministic policy gradient (DDPG) directly use the sensory inputs to learn complex control strategies which can be helpful in trajectory planning, lane keeping, and adaptive cruise control.

4.3.2.2   Recurrent neural network based MPC [31]

Model predictive control (recurrent neural network (RNN)) uses a predictive model to optimize control actions over a finite time horizon. When RNN-combined MPC models are used, they can handle tasks that require prediction and planning such as trajectory optimization

4.3.2.3   Inverse reinforcement learning [32]

Inverse reinforcement learning (IRL) relies on expert feedback to infer the underlying reward structure and therefore learns the optimal behavior. It can be used for complex tasks of decision-making or path planning.

4.3.2.4   Nonlinear model predictive control [33]

Nonlinear model predictive control (NMPC) is similar to MPC but uses nonlinear dynamics and constraints to optimize the control inputs. Tasks like precise trajectory tracking and obstacle avoidance are good examples of NMPC usage.

AI /ML models in AVs – key takeaways

Perception systems in autonomous vehicles (AVs) heavily rely on AI/ML algorithms across various sensors like LiDARs, cameras, radars, infrared cameras, and microphones. Algorithms such as Voxel-based CNNs, SSD, YOLO, FCNs, and RaTrack are employed for detection, classification, and tracking. Planning and control algorithms, crucial for maneuvering AVs safely, include techniques like DDPG for reinforcement learning-based trajectory planning, RNN-based MPC for predictive modeling, IRL for inferring reward structures, and NMPC for precise trajectory tracking and obstacle avoidance, ensuring efficient AV navigation.

4.4   Pre-processing and post-processing techniques for ML modeling in 2024

ML models rely heavily on the quality and preparation of their data. Pre-processing and post-processing techniques play a crucial role in ensuring that data is suitable for training and extracting valuable insights. Here's a glimpse into some current techniques:

4.4.1   Pre-processing

4.4.1.1   Data cleaning

Missing values: Techniques like imputation (filling in missing values) or deletion are used based on the data and context.

Outliers: Methods like capping, winsorizing, or removing extreme values address outliers that might skew the model.

Inconsistent formats: Standardizing data formats (e.g., date, currency) ensures consistency for processing.

4.4.1.2 Data integration

Merging from different sources: Techniques like feature engineering or data fusion combine data while handling schema differences and redundancy.

Normalization and scaling: Techniques like min-max scaling or z-score normalization ensure features have similar scales for model training.

4.4.1.3 Dimensionality reduction

Feature selection: Techniques like correlation analysis or feature importance ranking identify and remove irrelevant or redundant features, improving model efficiency.

Dimensionality reduction algorithms: Principal component analysis (PCA), linear discriminant analysis (LDA), or autoencoders reduce data dimensionality while preserving valuable information.

4.4.1.4 Data transformation

One-hot encoding: Categorical data is converted into numerical vectors for model processing.

Logarithmic or square root transformations: Applied to non-linear data to achieve normality for certain algorithms..

Label encoding: Categorical labels are assigned numerical values for specific algorithms.

4.4.2 Post-processing

Model explainability [34]: Local Interpretable Model-agnostic Explanations (LIME), SHapley Additive exPlanations (SHAP), or eXplainable AI (XAI) techniques: Explain model predictions, improving user trust and interpretability.

Error analysis: Confusion matrix, ROC/AUC analysis: Identify error patterns and areas for model improvement.

Calibration: Calibrating probabilities: Adjust model-predicted probabilities to align with true outcomes, enhancing reliability.

Ensemble techniques: Bagging, boosting, stacking: Combining multiple models can improve overall performance and robustness.

Hyperparameter tuning: Grid search, random search, Bayesian optimization: Optimizing model hyperparameters for better performance (Table 4.1).

4.4.3 Emerging techniques

Domain-specific pre-processing: Utilizing domain knowledge to tailor pre-processing steps for specific data types.

Active learning: Selecting the most informative data points for further labeling, improving model learning efficiency.

Federated learning: Training models on decentralized data while preserving privacy and security.

Choosing the right techniques depends on the specific data, model type, and goals. Combining several techniques can effectively optimize the ML pipeline.

Table 4.1   Summary of techniques and their key takeaways

Pre-processing technique	Takeaway	Popular techniques
Scaling	Features are brought down to values which are comparable with one another, so the optimization function doesn't have to take major leaps to reach the optimal point	Min-Max Scaler Standard Scaler Robust Scaler Max-Abs Scaler
Outlier treatment	Outliers can be detected and treated with the help of box-plots. Box plots are used to identify the median, interquartile ranges and outliers.	
Feature encoding	Sometimes, data is in a format that can't be processed by machines. For instance, a column with string values, like names, will mean nothing to a model that depends only on numbers. So, we need to process the data to help the model interpret it	Classical Encoders: Ordinal Encoding, One hot encoding, Binary Encoding, BaseN encoding, Hashing Bayesian Encoders: Target Encoding, Weight of Evidence Encoding, Leave One Out Encoding, James-Stein Encoding
Feature creation and aggregation	Features can be appropriately aggregated to reduce data bulk, and also to create relevant information.	
Dimensionality reduction	Reduced processing time Improved accuracy Reduced overfitting	
Univariate selection	Variance, correlation, mutual information	Pearson Correlation Spearman Rank Correlation Kendall Rank Correlation Chi square
Multivariate selection	Also referred to as Wrapper methods, multivariate selection techniques take a group of features at a time and test the group's competence in predicting the target variable.	Forward Selection Backwards Elimination Recursive Feature Elimination Linear Discriminant Analysis (LDA) ANOVA Embedded methods

4.5   Datasets

4.5.1   *Vehicle class imbalance in automated driving datasets*

The research paper titled "Cityscapes 3D: Dataset and Benchmark for 9 DoF Vehicle Detection" [35] provides annotations for all vehicle types. Popular datasets like Karlsruhe Institute of Technology and Toyota Technological Institute at Chicago dataset (KITTI) [36] and Waymo open are biased toward cars in their 3D annotation distribution, with a higher average number of annotations for cars compared to other vehicle classes such as bicycles and motorbikes. This bias is further observed that it is toward car detection performance.

Similarly, the research on Honda 3-dimensional (H3D) datasets for 3D multi-object detection [37] reveals a significant class imbalance favoring cars over pedestrians. The training, validation, and test sets contain 157k, 85k, and 228k samples for cars, respectively, compared to 148k, 65k, and 242k for pedestrians in the corresponding sets. This imbalance highlights there is a need for more balanced datasets to improve model performance on diverse object classes. The study majorly utilizes the VoxelNet neural network architecture with the KITTI dataset.

In comparison, the paper "nuScenes: A Multimodal Dataset for Autonomous Driving" [38] emphasizes datasets originating from China, including ApolloScape, AS-Lidar, and D2-City. This adds on to the existing US-centric repository that includes prominent datasets like Argoverse, Lyft Level 5 (Lyft L5) [39], and Waymo Open [40].

4.5.2   Sensor suite selection for adverse weather conditions

For automated driving in challenging winter weather conditions, particularly prevalent in Canada [41], the selection of sensor suites becomes crucial. GNSS (Global Navigation Satellite System) and IMU (Inertial Measurement Unit) data accuracy plays a pivotal role in fine-tuning ML models for robust performance in such scenarios. Datasets collected under various snowfall intensities (low, medium, heavy, and extreme) are invaluable for training and evaluating these models.

One noteworthy dataset was captured using an Audi Q7 e-tron [42], providing comprehensive 360° sensor coverage of the vehicle's environment. This dataset offers annotations for approximately 10,000 images. While Ford follows suit with a multi-sensor, seasonal dataset facilitating improved training on diverse data, it additionally provides valuable ROS (Robot Operating System) based tools for visualization.

4.5.3   Dataset size comparison

The paper "PandaSet: Advanced Sensor Suite Dataset for Autonomous Driving" [43] also analyzes class imbalance, revealing a similar trend toward a higher percentage of car frames compared to cyclists and pedestrians. However, the ONe million sCEnes (ONCE) dataset surpasses all other datasets mentioned in this review in terms of the captured scenes. With a staggering 1 million scenes, it significantly exceeds the scene counts of KITTI (15k), Lyft L5 (30k), Waymo Open dataset (230k), and nuScenes (400k) (Figure 4.3).

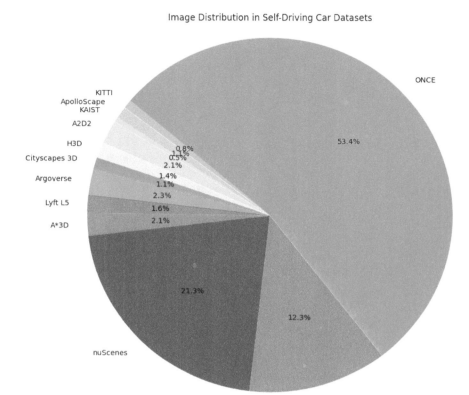

Figure 4.3   Image distribution in self-driving car datasets

Datasets – key takeaways

Datasets favor cars: Existing datasets like KITTI and Honda 3-Dimensional Dataset (H3D) show bias toward cars in annotations and metrics, neglecting other vehicles and pedestrians. Need for diverse data: Studies highlight the importance of including datasets from various regions and with a better balance between different vehicle classes (e.g., ONCE dataset with its massive size). Sensor selection matters: GNSS and IMU accuracy are crucial for winter driving, with dedicated datasets collected under different snowfall intensities being valuable for training robust models.

4.6   Truck driver shortage and the case for driverless trucks

Industry experts like the American Trucking Associations warn of a potential shortage of over 100,000 truck drivers by 2030, potentially doubling by 2050.

While the Bureau of Labor Statistics doesn't predict a crisis, their slower-than-average growth projection indicates difficulty meeting future demand. This paints a complex picture, especially considering the aging workforce, demanding job conditions, and regulatory hurdles contributing to the issue. AI in AVs is not trying to replace any truckers. This scenario in fact might create a surprising business case for driverless trucks, contrasting with concerns about their impact on driver jobs. Research suggests that self-driving trucks can significantly reduce driver costs, improve truck utilization, and enhance safety, as highlighted in [44]. However, there's a crucial gap in research compared to driverless cars, highlighting the need for studies tailored to road freight transport and its unique value chain.

Eliminating the driver cabin for mere value discount may not be the answer. Most groups developing self-driving trucks appear to recognize "hub-to-hub" transport, in contrast to the "door-to-door" model of passenger vehicles. This implies drivers might still be wanted for preliminary and final legs of trips, refocusing the narrative on their endured relevance at the same time as addressing long-haul demanding situations. This situation indicates that as opposed to displacing drivers altogether, the self-using generation may want to cope with modern-day industry-demanding situations and enhance performance. Truck drivers with specialized capabilities could collaborate with self-reliant structures, that specialize in price-introduced tasks like preliminary/final mile shipping and navigating complex city environments.

There is a potential for driverless trucks to address industry challenges while ensuring the continued value of human expertise in the future of road freight transport [45].

4.6.1   Future of trucks: self-driving needs more attention

Industry leader Dieter Zetsche laid out a vision for "CASE – Connected, Autonomous, Shared, and Electric" vehicles: connected, automated, shared, and electric. While he acknowledged the potential disruption each facet brings, the focus seems to be primarily on passenger cars (Zetsche, 2018) [46]. However, research suggests we're fast approaching Level 4 autonomy by 2030 and even Level 5 by 2040. Combining this prediction with studies and expert interviews, it's clear the impact of automation on trucks demands serious consideration.

Currently, the research landscape heavily favors passenger cars, with numerous players vying for market dominance. Yet, data on self-driving trucks remains scarce. This imbalance is concerning, given the crucial role trucks play in our economy and the unique challenges automated trucking presents compared to passenger vehicles.

4.6.2   Understanding the road ahead: key differences in automated technology for cars vs. trucks

Developing self-driving technology presents unique challenges for both cars and trucks. Let's explore the key parameters that differentiate their automated journeys (Table 4.2).

Table 4.2   Parameter set for differences in using AI in automated trucks and cars

Parameter	Trucks	Cars
Environment and dynamics	Operate primarily on predictable highways, prioritizing stability and visibility due to wind concerns and heavier cargos. Weather conditions: trucks may prioritize stability and visibility for cargo safety.	Tackle diverse environments, navigating congested urban streets and adapting to varying road conditions and weather. Weather conditions: cars may prioritize maneuverability for passenger comfort.
Sensor suite and data acquisition	Rely on long-range LiDAR and fine-tuned radars for blind-spot detection, focusing on highway lane markings, traffic flow, and weather data.	Require 360° perception with high-resolution cameras for near-field obstacles, gathering data on object detection, pedestrian recognition, and detailed lane information.
Maneuverability and control	Demand larger turning radii and slower acceleration/deceleration, emphasizing smooth lane changes, safe distances, and stable cargo handling.	Need agility and responsiveness for quick maneuvers, traffic avoidance, and emergency braking, focusing on precise steering control, collision prediction, and pedestrian safety.
Regulations and legal considerations	Subject to stricter regulations regarding driver fatigue, cargo securing, and safety protocols for automated operation, requiring AI models to comply and ensure responsible driving.	Navigate evolving regulations for automated features, with AI models needing adaptability and prioritizing passenger safety in all scenarios.
Development and training data	Require specialized datasets and algorithms encompassing highway scenarios, weather variations, and cargo dynamics simulations.	Leverage large publicly available datasets for initial training, focusing on diverse traffic situations, pedestrian interactions, and various road types.
Use case differences	Need additional load considerations due to varying payloads and trailer types, potentially utilizing platooning for fuel efficiency.	Primarily focus on individual ownership or ride-sharing models, with shorter, dynamic routes.

4.6.2.1   Additional factors

Public perception: Trust in self-driving vans is probably slower to build because of size and capacity cargo dangers.

Economic and business models: Truck automation might contain fleet control and logistics optimization, at the same time as vehicle automation could put consciousness on trip-sharing and individual possession.

Vehicle characteristics: Trucks may additionally rely upon devoted infrastructure for communique (platooning, V2X) at the same time as vehicles often use present mobile networks.

Infrastructure-related differences: Lane markings and signage: Lane widths and markings would possibly need modifications to deal with trucks' wider turns and longer stopping distances. Signage for truck-particular statistics or restrictions might be necessary.

Dedicated infrastructure for trucks: Designated truck lanes, charging stations, and parking regions is probably required for freight motion efficient.

Understanding those key differences is crucial for growing safe, efficient, and ethical automated cars for each trucks and motors. By addressing their particular demanding situations and possibilities, we need to be capable of have a smoother and greater sustainable future on the street.

Driverless trucks v/s cars – key takeaways

A potential truck driver shortage looms, and self-driving technology could offer solutions. Driverless trucks can bring benefits beyond cost reduction, including improved safety and utilization. Research should focus on the specific needs of road freight transport compared to passenger cars. Truck drivers might not be replaced, but their roles could evolve alongside automated technology. The industry anticipates major disruption from "CASE" vehicles, particularly automated functionality. Predictions point toward a rapid advancement in self-driving technology, reaching Level 4 by 2030 and potentially Level 5 by 2040. Despite the impact on the trucking industry, research and market focus lean heavily toward passenger cars, leaving a critical gap in knowledge about self-driving trucks.

4.7   Software package size, processing power

The burgeoning length of AV software applications presents a big roadblock for developers like Nvidia and Qualcomm as mentioned in [3]. This increase brings numerous technical hurdles. Larger programs translate to multiplied demands on processing power and memory. This necessitates expensive, excessive-overall performance hardware, probably leading to bulkier, much less green systems. Download and installation instances emerge as protracted with great programs, specifically in areas with limited internet bandwidth. This can seriously impact consumer enjoyment and deployment efficiency. The multiplied assault surface brought by large programs amplifies safety worries, particularly cybersecurity vulnerabilities. More code and data equate to more ability entry factors for malicious actors, raising protection dangers. Platforms like Nvidia's Drive Orin utilize bendy scalable hardware architectures capable of coping with large software footprints at the same time as retaining performance.

Software optimization techniques like code compression, hardware-particular optimizations, and other strategies can significantly reduce software program length without compromising functionality. Cloud-based totally solutions play a crucial function. Offloading processing and statistics storage to the cloud can alleviate the burden on onboard hardware, taking into consideration smaller software program programs at the tool itself.

4.7.1   *Qualcomm's Snapdragon Ride: tackling AV software*

Similar to Nvidia's Orin platform [47], Qualcomm's Snapdragon Ride adopts a scalable architecture tailored for efficient processing of bulky AV software packages. By leveraging heterogeneous computing, Ride strategically employs diverse processing units like central processing units (CPUs), graphics processing units (GPUs), and neural processing units (NPUs). This specialization optimizes performance for specific tasks, potentially reducing overall software size by minimizing redundant computations. Furthermore, Ride emphasizes a modular software architecture, segmenting the software into smaller, independent components. This modularity not only simplifies updates and maintenance but also reduces the cumulative footprint of the software package. These technical approaches contribute to streamlined performance, minimized software size, and potentially more efficient use of hardware resources within the Snapdragon Ride platform [48].

Standardized software across the industry can eliminate duplicate code and inconsistencies, leading to leaner software packages. Additionally, cutting-edge compression techniques can drastically shrink data and code size without sacrificing functionality. Machine learning can further optimize performance and resource allocation, minimizing the overall software footprint. The race to create smaller AV software packages is a top priority, with industry leaders like Nvidia and Qualcomm spearheading innovative solutions. As technology matures and these approaches become more refined, we can anticipate a future with smaller, more efficient AV software, paving the way for broader adoption of self-driving vehicles.

Software package size – key takeaways

Large AV software packages are problematic: They require powerful hardware, take longer to install, and increase security risks. Solutions are being developed: scalable hardware, software optimization, cloud-based processing, and specialized hardware are all being explored. Nvidia and Qualcomm are leading the charge: Both companies offer platforms with specific solutions like scalable architecture and modular software.

4.8   Unveiling the correlation between autonomy and package size in AVs

The insightful observation that a direct correlation exists between an AV's level of autonomy and its software package size necessitates a deeper exploration.

4.8.1 Going down the pyramid

Visualize a pyramid, with Level 0 (no automation) representing the base vehicle, which has common power train functionalities. We estimate its size at a mere few mega bytes (MB). As we ascend, functionality and complexity increase.

Level 1 has adaptive cruise control and lane departure warnings. Sensor inputs and processing power requirements increase the package size to tens/hundreds of MB.

Level 2 is a hands-free highway driving and automated parking feature set. LiDAR, Radar, and advanced algorithms contribute to the SW package size, inflating the package to several hundred MB or a few giga bytes (GB).

Level 3 can handle specific scenarios such as driver intervention which is a remarkable feature addition. Robust perception, planning, and decision-making algorithms become key factors in contributing the size to several GB or even tens of GB.

Level 4 is the highest point that allows the vehicle to manage most driving tasks in different environments urban and rural and different weather conditions as well. Complex AI and ML algorithms thus end up contributing to tens to hundreds of GB or even more.

Level 5 is the tallest mountain – the vehicle tackles all driving tasks across any environment. Software complexity reaches complex levels, with package sizes potentially soaring to hundreds of GB or even Tera Bytes (TB) in the future.

4.8.2 Software package size based on the level of autonomy (navigating the maze of challenges) [49]

4.8.2.1 Limited hardware capacity

Current onboard storage and processing might struggle to accommodate the large chunks of Level 4 and 5 software. Updating these extends the download times and potential operational disruptions. Increased complexity exposes vulnerabilities to cyberattacks and other cybersecurity concerns necessitating robust security measures. While higher autonomy levels entice convenience and potential safety benefits, managing the ballooning software size presents a formidable challenge. Overcoming these hurdles in storage, processing, download times, and security is crucial for paving the path toward a future where larger-than-life AV software packages don't become roadblocks to truly automated driving.

Autonomy and package size – key takeaways

Software size grows exponentially with autonomy level: As AVs become more sophisticated, their software packages balloon in size from megabytes at Level 0 to potentially terabytes at Level 5. Massive software presents challenges: Limited hardware capacity, lengthy updates, and increased cybersecurity risks pose significant hurdles for higher autonomy levels. Overcoming these hurdles is crucial: Addressing storage, processing, download times, and security is essential for the successful implementation of fully AVs.

4.9   AI challenges in automotive

This section discusses the challenges of using AI in automotive and provides an overview of a few key areas as follows:

Uncertainty: Uncertainty and variability in real-world driving scenarios include varying lighting conditions, weather, and road surface conditions. These coupled with the interaction with several road actors pose challenges to AI/ML models used for sensing and perception. Model robustness can be improved for the AI/ML model by employing techniques like data augmentation and domain adaptation [50].

Real-time processing: Real-time computation by running ML models for scene understanding and/or decision-making is a very challenging task in Automated driving. The AD systems require real-time sense, thinking, and acting capabilities in complex and dynamic environments. Some of the solutions/mitigations for this problem are using lightweight neural network architectures, or use of hardware accelerators (e.g., GPUs, field programmable gate arrays (FPGAs)), and also algorithmic optimizations (e.g., data pruning, quantization) [25].

Safety-critical decision-making: AI/ML models in Automated Driving that are relevant to safety-critical decision-making like emergency braking or collision avoidance should perform sufficiently safely and reliably. Robust safety mechanisms and fail-safe strategies, such as model-based uncertainty estimation, safe reinforcement learning, and formal verification techniques can be useful to address the safety-critical issues in AI/ML [51].

Generalization across diverse scenarios: One of the biggest challenges in the AV use case of ML models is: not able to generalize well across diverse driving scenarios, especially if they are not in the training set. A typical AV drives through various scenarios like urban, suburban, and rural environments, day and night conditions, and varying traffic densities. Techniques like meta-learning, domain adaptation, and simulation-to-real transfer learning can help enhance model generalization by exposing the model to a wide range of scenarios during training [52]. Additionally, incorporating diversity-aware training objectives and data augmentation strategies can improve generalization.

Interpretability and explainability: AI/ML models lack interpretability and explainability. This problem becomes particularly challenging in AVs and other safety-critical applications of ML due to a lack of reason and inference. Techniques like attention mechanisms, saliency maps, and model-agnostic explanation methods help improve model interpretation and explanation. (e.g., LIME, SHAP) [53]. The use of interpretable architectures and incorporating human-in-the-loop approaches can also enhance model transparency.

Cybersecurity and privacy: Hacking, data tampering, and spoofing attacks on sensor inputs are some of the cybersecurity threats that AVs may have to face in operating in the real world. To prevent malicious attacks and increase system integrity, AI/ML models and their data need to be protected. Cybersecurity measures, like cryptographic encryption, authentication, intrusion detection, and secure communication protocols, are some of the ways to protect AI/Ml models.

Additionally, cybersecurity threat and vulnerability assessments and penetration testing can help identify and mitigate potential security vulnerabilities in the system [54].

Regulatory and legal compliance: Ensuring that the AVs to be deployed on public roads and the AI/ML models used in their development meet safety, homologation, and legal requirements established by governmental agencies and industry organizations is a big challenge for the regulatory bodies to provide approval and overall public acceptance. Active collaboration between regulatory bodies, policymakers, and industry stakeholders to establish clear standards and guidelines for AI/ML-based automated driving systems is an important step toward regulating AVs. Developing certification processes, safety frameworks, and compliance testing methodologies to evaluate the safety and performance of AI/ML models will help societies with these emerging technologies [55].

Ethical decision-making: AI/ML models in automated driving similar to humans face ethical dilemmas, such as how to prioritize the safety of AV passengers versus other road users, and how to navigate moral conflicts in complex driving scenarios. Ensuring that AI/ML models adhere to ethical principles and societal values is important for responsible deployment, especially in safety-critical applications like Automated driving [56]. So of the potential solutions are adhering to ethical frameworks, such as utilitarianism, deontology, and virtue ethics, and incorporating them into the design and development of AI/ML models. Engaging relevant stakeholders, like regulators, policymakers, ethicists, and the general public, in ethical decision-making processes for AV and emerging technologies. Establishing effective mechanisms for transparency, accountability, and oversight in automated driving systems could be some of the ways to ensure AI/ML ethics.

AI challenges – key takeaways

Integrating AI into automotive systems faces challenges like real-world uncertainties and the need for real-time processing. Safety-critical decision-making demands robust mechanisms while generalizing AI models remains a challenge. Ensuring interpretability and addressing cybersecurity threats are crucial. Ethical decision-making is paramount, requiring adherence to ethical frameworks for responsible deployment.

4.10   Standards review

This section delves into the current landscape of standards development for AVs. Standards serve as a crucial foundation for the automotive industry, establishing a set of guidelines, specifications, and criteria. These ensure uniformity, safety, quality, and interoperability across various components and systems within AVs which addresses some of the challenges mentioned in Section 4.9. Given the multifaceted nature of AV technology, numerous standards are under development by various international Standards Development Organizations (SDOs) like the International Organization for Standardization (ISO), Society of Automotive Engineers (SAE), Institute of Electrical and Electronics Engineers (IEEE), European Telecommunications Standards Institute (ETSI), British Standards Institution (BSI), and others. These standards encompass a broad spectrum of AV technology, including safety, interoperability, communication protocols, cybersecurity, and ethical considerations. By establishing these common frameworks and guidelines, Standards Development Organizations aim to achieve compatibility, reliability, and safety.

In the case of AI being used within AV technologies, standards development is currently evolving at a faster pace. There is a need to be able to provide guidelines and to keep pace with the research and innovation that is ongoing in the areas mentioned above. ISO 8800 [57] provides industry-specific guidance on using AI and ML-based development in safety systems of road vehicle safety functions. It establishes a framework aligned with ISO 26262 [58] and ISO 21448 [59] standards, addressing functional safety risks and performance limitations of AI systems. By proposing a causal model, it derives safety requirements and risk reduction measures, offering guidance for project-specific assurance arguments. It also classifies AI technology classes and aligns with other AI-related standards, enhancing its applicability in ensuring AI system safety in road vehicles. ISO/TC 204/WG 20 is currently focusing on Big Data and AI supporting ITS and it is a standard under progress.

SAE's AI Committee for Ground Vehicles addresses a lot of AI-related topics like "SAE J3298 - Data for AI of Ground Vehicles" [60], "SAE J3313 – AI – terms & definitions taxonomy" [61], "SAE J3312 – Artifical Intelligence Use Cases in Ground vehicles" [62], SAE J3329 AI Regulations, Standards & Applications Challenges and SAE J3321 Verification & Validation of AI/ML based Components & Systems in Ground Vehicles. There is also ISO/IEC DIS 5259-2 [63] standard which covers the data quality measures topic. All of these topics address the new and existing topics specific to AI and AI in ground vehicles. Following is a list of other standards in the AI area that could also apply to AV applications and domains:

ISO/IEC CD TR 5469:2024 Artificial intelligence – functional safety and AI systems [64].

ISO/IEC TS 25058:2024 Systems and software engineering – Systems and software Quality Requirements and Evaluation (SQuaRE) – guidance for quality evaluation of artificial intelligence (AI) systems [65].

NIST AI 100-1 Artificial Intelligence Risk Management Framework (AI RMF 1.0) [66].

UL4600 – Standard for Safety for the Evaluation of Autonomous Products [67].

ISO/IEC 38507:2022 Information technology – governance of IT – governance implications of the use of artificial intelligence by organizations [68].

ISO/IEC TR 24029:2021 Artificial Intelligence (AI) – assessment of the robustness of neural networks – Part 1: Overview [69].

ISO/IEC TR 24028:2020 Information technology – artificial intelligence – overview of trustworthiness in artificial intelligence [70].

ISO/IEC FDIS 5259-1:2024 Artificial intelligence – data quality for analytics and machine learning (ML) – Part 1: Overview, terminology, and examples [71].

ISO/IEC DIS 5259-2:2023 Artificial intelligence – data quality for analytics and machine learning (ML) – Part 2: Data quality measures [72].

ISO/IEC DIS 5259-3:2024 Artificial intelligence – data quality for analytics and machine learning (ML) – Part 3: Data quality management requirements and guidelines [73].

ISO/IE 23053:2022 Framework for Artificial Intelligence (AI) Systems Using Machine Learning (ML) [74].

ISO/IEC 22989:2022 Information technology – artificial intelligence – artificial intelligence concepts and terminology [75].

DIN SPEC 92001-1:2019 Artificial intelligence – life cycle processes and quality requirements – Part 1: Quality Meta Model [76].

DIN SPEC 92001-2:2019 Artificial intelligence – life cycle processes and quality requirements – Part 2: robustness [77].

CAN/CIOSC 101:2019 Ethical design and use of automated decision systems [78].

Furthermore, government bodies like the National Highway Traffic Safety Administration (NHTSA), Department of Transportation (DOT), and Euro NCAP, along with various regional organizations, are actively developing regulations for AVs. These regulations aim to cover similar topics as the standards but from a compliance perspective. By implementing comprehensive regulatory frameworks, these organizations strive to promote the responsible development and deployment of AVs. This ensures public safety, upholds ethical considerations, and maintains existing road safety standards while paving the way for the widespread adoption of AVs.

Standards review – key takeaways

Importance of standards: standards are crucial for ensuring uniformity, safety, quality, and compatibility across various components of AVs. Focus areas of standards: Key areas and challenges addressed by standards include safety (fail-safe mechanisms), interoperability (communication between AVs and infrastructure), communication protocols, cybersecurity, and ethical considerations. Due to the rapid development of AI, standards for its use in AVs are evolving quickly. ISO 8800 provides specific guidance for AI and ML safety in road vehicles. Standardization efforts by different organizations: Several organizations are actively developing standards for AVs. This includes ISO (focusing on safety and AI), SAE (focusing on AI terminology and use cases), and others. Regulations for AVs: Government bodies are developing regulations to ensure the safe, secure, and ethical deployment of

AVs. These regulations cover areas similar to those addressed by standards. Overall Goal: Establishing robust standards and regulations aims to promote the responsible development and deployment of AVs, prioritizing public safety and ethical considerations.

4.11   Conclusion

This chapter provides an overview of the transformative role of AI in the development and deployment of AVs. It identifies multiple use cases of AI including concept creation, simulations, testing edge cases, performance tracking, user personalization, and traffic management optimization. It also points out that continuous data collection and analysis by AI algorithms helps with performance improvements and drive innovation within the AV industry. Another important aspect being covered is the AI/ML algorithms being used by the perception systems and how do they process data from diverse sensors, enabling precise object detection, classification, and tracking. Similarly, for planning and control tasks, to generate feasible and safe trajectories with reinforcement learning and other methods. A short review of the different types of datasets that are available currently has been provided as most of the AI models and systems rely on data-driven methodologies, inferences and training.

A review of the driverless trucks use case has been provided as it proves a more promising opportunity for AI technology with the ongoing truck driver shortage in the trucking industry. In spite of the remaining challenges, the AV industry anticipates significant advancements and predictions point toward Level 4 autonomy becoming a reality by 2030, with Level 5 potentially achievable by 2040. Another important challenge that this chapter talks about is the large software packages needed for AVs . Scalable hardware architectures, software optimization, cloud-based processing, and specialized hardware are proposed as potential solutions under development. There is also a comprehensive review of AI challenges related to real-world uncertainties, real-time processing demands, interpretability, cybersecurity, and ethical decision-making. Lastly, the chapter clearly states that standardization and regulations are essential for ensuring uniformity, safety, quality, and compatibility across AV components. In conclusion, robust standards and regulations are foundational for the responsible development and deployment of AVs. By harnessing the power of AI and addressing existing challenges, the future of transportation promises to be safer, more efficient, and increasingly autonomous.

References

[1]   Taxonomy and Definitions for Terms Related to Driving Automation Systems for On-Road Motor Vehicles. SAEJ3016:202104: On Road Automated Driving Committee (ORAD), Society of Automotive Engineers (SAE); 2021-04-30.

[2]   Ruijten PA, Terken JM, and Chandramouli SN. Enhancing trust in autonomous vehicles through intelligent user interfaces that mimic human behavior. *Multimodal Technologies and Interaction*. 2018;2(4):62.

[3]   Garikapati D, and Shetiya SS. Autonomous vehicles: evolution of artificial intelligence and the current industry landscape. *Big Data and Cognitive Computing*. 2024;8:42.

[4]   Zhuhadar LP, and Lytras M. The application of AutoML techniques in diabetes diagnosis: current approaches, performance, and future directions. *Sustainability*. 2023 9(15):13484.

[5]   Rosique F, Navarro PJ, Fernández C, *et al.* A systematic review of perception system and simulators for autonomous vehicles research. *Sensors*. 2019; 19(3):648.

[6]   Fayyad J, Jaradat MA, Gruyer D, *et al.* Deep learning sensor fusion for autonomous vehicle perception and localization: a review. *Sensors*. 2020; 20(15):4220.

[7]   Ahrens S, Levine D, Andrews G, *et al.* Vision-based guidance and control of a hovering vehicle in unknown, GPS-denied environments. In: *2009 IEEE International Conference on Robotics and Automation*. IEEE; 2009. pp. 2643–2648.

[8]   Sundar SM. Evaluating the impact of AI-based personalization of automotive user interfaces in fully autonomous vehicles, MS thesis, Eindhoven University of Technology. 2021.

[9]   Zheng L, Bai J, Zhu X, *et al.* Dynamic hand gesture recognition in in-vehicle environment based on FMCW radar and transformer. *Sensors*. 2021;21 (19):6368.

[10] Garikapati D, and Liu Y. Dynamic control limits application strategy for safety-critical autonomy features. In: *2022 IEEE 25th International Conference on Intelligent Transportation Systems (ITSC)*; 2022. pp. 695–702.

[11] Kaja N. Artificial intelligence and cybersecurity: building an automotive cybersecurity framework using machine learning algorithms; PhD diss., 2019.

[12] Ma Y, Wang Z, Yang H, *et al.* Artificial intelligence applications in the development of autonomous vehicles: a survey. *IEEE/CAA Journal of Automatica Sinica*. 2020;7(2):315–329.

[13] UnfoldLabs. AI & Automotive — 8 Disruptive Use-Cases. Medium. 2020 03; Accessed 28 March 2024, https://unfoldlabs.medium.com/ai-auto-motive-8-disruptive-use-cases-fd079926aea9.

[14] Halvorson B. Volkswagen tests AI-informed design process to cut weight – and flaunt it. Green Car Reports. July 10, 2019; Accessed 28 March 2024. www.greencarreports.com/news/1123945_volkswagen-tests-ai-informed-design-process-to-cut-weight-and-flaunt-it.

[15] Xu M, Niyato D, Chen J, *et al.* Generative AI-empowered simulation for autonomous driving in vehicular mixed reality metaverses. *IEEE Journal of Selected Topics in Signal Processing*. 2023;17(5):1064–1079.

[16] Drayson G, Panagiotaki E, Omeiza D, *et al.* CC-SGG: corner case scenario generation using learned scene graphs. arXiv preprint arXiv:230909844. 2023.

[17]   Holland JC, and Sargolzaei A. Verification of autonomous vehicles: scenario generation based on real world accidents. In: *2020 SoutheastCon*. vol. 2. IEEE; 2020. pp. 1–7.

[18]   Garikapati D, Liu Y, and Huo Z. Systematic selective limits application using decision making engines to enhance safety in highly automated vehicles. *SAE Int. J. CAV* 8(1), 2025, https://doi.org/10.4271/12-08-01-0005.

[19]   Wishart J, Como S, Elli M, *et al.* Driving safety performance assessment metrics for ads-equipped vehicles. *SAE International Journal of Advances and Current Practices in Mobility*. 2020;2(2020-01-1206):2881–2899.

[20]   Mercedes-Benz. Mercedes-Benz takes in-car voice control to a new level with ChatGPT. Mercedes Media Newsroom USA. June 15 2023; Accessed: 28 March 2024. https://media.mbusa.com/releases/release-b35d2af89e06f55 6bbd8fe420412e9c2-mercedes-benz-takes-in-car-voice-control-to-a-new-level-with-chatgpt.

[21]   Bang H, Chalaki B, and Malikopoulos AA. Combined optimal routing and coordination of connected and automated vehicles. *IEEE Control Systems Letters*. 2022;6:2749–2754.

[22]   Zhou Y, and Tuzel O. Voxelnet: end-to-end learning for point cloud based 3d object detection. In: *Proceedings of the IEEE Conference on Computer Vision and Pattern Recognition*; 2018. pp. 4490–4499.

[23]   Qi CR, Su H, Mo K, *et al.* Pointnet: deep learning on point sets for 3d classification and segmentation. In: *Proceedings of the IEEE Conference on Computer Vision and Pattern Recognition*; 2017. p. 652–660.

[24]   Liu W, Anguelov D, Erhan D, *et al.* Ssd: single shot multibox detector. In: *14th European Conference on Computer Vision–ECCV 2016*, Amsterdam, The Netherlands, 11–14 October 2016, Proceedings, Part I 14. Springer; 2016. pp. 21–37.

[25]   Redmon J, Divvala S, Girshick R, *et al.* You only look once: unified, real-time object detection. In: *Proceedings of the IEEE Conference on Computer Vision and Pattern Recognition*; 2016. pp. 779–788.

[26]   Goodfellow I, Pouget-Abadie J, Mirza M, *et al.* Generative adversarial nets. *Advances in Neural Information Processing Systems*. 2014;27, *Curran Associates, Inc.* https://proceedings.neurips.cc/paper_files/paper/2014/file/ 5ca3e9b122f61f8f06494c97b1afccf3-Paper.pdf.

[27]   Long J, Shelhamer E, and Darrell T. Fully convolutional networks for semantic segmentation. In: *Proceedings of the IEEE Conference on Computer Vision and Pattern Recognition*; 2015. pp. 3431–3440.

[28]   Pan Z, Ding F, Zhong H, *et al.* Moving object detection and tracking with 4D radar point cloud. *2024 IEEE International Conference on Robotics and Automation (ICRA)*, Yokohama, Japan, 2024, pp. 4480–4487, doi: 10.1109/ ICRA57147.2024.10610368.

[29]   Krause D, Politis A, and Kowalczyk K. Comparison of convolution types in CNN-based feature extraction for sound source localization. In: *2020 28th European Signal Processing Conference (EUSIPCO)*. IEEE; 2021. pp. 820–824.

[30] Lillicrap TP, Hunt JJ, Pritzel A, *et al.* DeepMind Technologies Ltd, 2020. *Continuous control with deep reinforcement learning.* U.S. Patent 10,776,692

[31] Hochreiter S, and Schmidhuber J. Long short-term memory. *Neural Computation.* 1997;9(8):1735–1780.

[32] Abbeel P, and Ng AY. Apprenticeship learning via inverse reinforcement learning. In: *Proceedings of the 21st International Conference on Machine Learning*; 2004. p. 1.

[33] Findeisen R, and Allgöwer F. An introduction to nonlinear model predictive control. In: *21st Benelux Meeting on Systems and Control.* vol. 11. Veldhoven; 2002. pp. 119–141.

[34] Mankodiya H, Obaidat MS, Gupta R, *et al.* XAI-AV: Explainable artificial intelligence for trust management in autonomous vehicles. In: *2021 International Conference on Communications, Computing, Cybersecurity, and Informatics (CCCI).* IEEE; 2021. pp. 1–5.

[35] Gählert N, Jourdan N, Cordts M, *et al.* Cityscapes 3d: dataset and benchmark for 9 dof vehicle detection. arXiv preprint arXiv:200607864. 2020.

[36] Liao Y, Xie J, and Geiger A. Kitti-360: A novel dataset and benchmarks for urban scene understanding in 2d and 3d. *IEEE Transactions on Pattern Analysis and Machine Intelligence.* 2022;45(3):3292–3310.

[37] Patil A, Malla S, Gang H, *et al.* The h3d dataset for full-surround 3d multi-object detection and tracking in crowded urban scenes. In: *2019 International Conference on Robotics and Automation (ICRA).* IEEE; 2019. pp. 9552–9557.

[38] Caesar H, Bankiti V, Lang AH, *et al.* nuscenes: a multimodal dataset for autonomous driving. In: *Proceedings of the IEEE/CVF conference on computer vision and pattern recognition*; 2020. pp. 11621–11631.

[39] Perez A, Scott CC, and Cherry J Lyft SDV Trajectory Prediction. 2021, 10. 13140/RG.2.2.36322.91848.

[40] Ettinger S, Cheng S, Caine B, *et al.* Large scale interactive motion forecasting for autonomous driving: the Waymo open motion dataset. In: *Proceedings of the IEEE/CVF International Conference on Computer Vision*; 2021. pp. 9710–9719.

[41] Pitropov M, Garcia DE, Rebello J, *et al.* Canadian adverse driving conditions dataset. *The International Journal of Robotics Research.* 2021;40(4–5):681–690.

[42] Geyer J, Kassahun Y, Mahmudi M, *et al.* A2d2: Audi autonomous driving dataset. arXiv preprint arXiv:200406320. 2020. www.a2d2.audi/content/dam/a2d2/dataset/a2d2-audi-autonomous-driving-dataset.pdf.

[43] Xiao P, Shao Z, Hao S, *et al.* Pandaset: advanced sensor suite dataset for autonomous driving. In: *2021 IEEE International Intelligent Transportation Systems Conference (ITSC).* IEEE; 2021. pp. 3095–3101.

[44] Fritschy C, and Spinler S. The impact of autonomous trucks on business models in the automotive and logistics industry – a Delphi-based scenario study. *Technological Forecasting and Social Change.* 2019;148:119736.

[45] Engholm A, Björkman A, Joelsson Y, *et al.* The emerging technological innovation system of driverless trucks. *Transportation Research Procedia.* 2020;49:145–159.

[46]   Parekh D, Poddar N, Rajpurkar A, *et al.* A review on autonomous vehicles: progress, methods and challenges. *Electronics*. 2022;11(14):2162.

[47]   Marie L. NVIDIA enters production with DRIVE Orin, announces BYD and Lucid Group as new EV customers, unveils next-gen DRIVE Hyperion AV platform [Press Release]. Nvidia Newsroom; 2022. Accessed 28 March 2024. https://nvidianews.nvidia.com/news/nvidia-enters-production-with-drive-orin-announces-byd-and-lucid-group-as-new-ev-customers-unveils-next-gen-drive-hyperion-av-platform.

[48]   Tharakram K. Snapdragon Ride SDK: a premium platform for developing customizable ADAS applications. Qualcomm OnQ Blog; 2022. Accessed 28 March 2024. Available from: https://www.qualcomm.com/news/onq/2022/01/snapdragon-ride-sdk-premium-solution-developing-customizable-adas-and-autonomous.

[49]   Garikapati D, and Sudhir Shetiya S. Autonomous vehicles: evolution of artificial intelligence and learning algorithms. vol. arXiv:2402.17690; 2024.

[50]   Schneider S, Rusak E, Eck L, *et al.* Improving robustness against common corruptions by covariate shift adaptation. *Advances in Neural Information Processing Systems*. 2020;33:11539–11551.

[51]   Berkenkamp F, Turchetta M, Schoellig A, *et al.* Safe model-based reinforcement learning with stability guarantees. *Advances in Neural Information Processing Systems*. 2017;30:1–8.

[52]   Huang W, Zhang T, Rong Y, *et al.* Adaptive sampling towards fast graph representation learning. *Advances in Neural Information Processing Systems*. 2018;31:4–8.

[53]   Ribeiro MT, Singh S, and Guestrin C. "Why should I trust you?" Explaining the predictions of any classifier. In: *Proceedings of the 22nd ACM SIGKDD International Conference on Knowledge Discovery and Data Mining*; 2016. pp. 1135–1144.

[54]   Tian Y, Pei K, Jana S, *et al.* DeepTest: automated testing of deep-neural-network-driven autonomous cars. In: *Proceedings of the 40th International Conference on Software Engineering*; 2018. pp. 303–314.

[55]   Montemerlo M, Thrun S, Dahlkamp H, *et al.* Winning the DARPA grand challenge with an AI robot. In: *AAAI*; 2006. pp. 982–987.

[56]   Geisslinger M, Poszler F, Betz J, *et al.* Autonomous driving ethics: from trolley problem to ethics of risk. *Philosophy and Technology*. 2021;34(4): 1033–1055.

[57]   Road vehicles – safety and artificial intelligence. ISO/CD PAS 8800: ISO/TC 22/SC 32, International Organization for Standardization (ISO); 2024.

[58]   Road vehicles – functional safety. ISO26262:2018: ISO/TC 22/SC 32, International Organization for Standardization (ISO); 2018-12.

[59]   Road vehicles – safety of the intended functionality. ISO21448:2022: ISO/TC 22/SC 32, International Organization for Standardization (ISO); 2022-06.

[60]   Ground vehicle – Artificial Intelligence Data Information Report. SAE J3298, Artificial Intelligence: Society of Automotive Engineers (SAE); 2023-09-28.

[61] Artificial intelligence – terms & definitions taxonomy. SAE J3313, Artificial Intelligence: Society of Automotive Engineers (SAE); 2023-12-15.

[62] Artificial intelligence (AI) – use cases in ground vehicle applications. SAE J3312, Artificial Intelligence: Society of Automotive Engineers (SAE); 2023-12-15.

[63] Data quality for analytics and machine learning (ML) Part 2: Data quality measures. IISO/IEC DIS 5259-2: ISO/IEC JTC 1/SC 42, International Organization for Standardization (ISO); 2023-10-27.

[64] Artificial intelligence – functional safety and AI systems. ISO/IEC TR 5469:2024: ISO/IEC JTC 1/SC 42, International Organization for Standardization (ISO); 2024-01-08.

[65] Systems and software engineering – systems and software quality requirements and evaluation (SQuaRE) – guidance for quality evaluation of artificial intelligence (AI) systems. ISO/IEC TS 25058:2024: ISO/IEC JTC 1/SC 42, International Organization for Standardization (ISO); 2024-01-24.

[66] Artificial intelligence risk management framework (AI RMF 1.0). NIST AI 100-1: National Institute of Standards and Technology (NIST) U.S Department of Commerce; 2023-01.

[67] Standard for safety for the evaluation of autonomous products. UL4600, ANSI/UL: UL Standards & Engagement partners; 2022-12.

[68] Information technology – governance of IT – governance implications of the use of artificial intelligence by organizations. ISO/IEC 38507:2022: ISO/IEC JTC 1/SC 42, International Organization for Standardization (ISO); 2022-04.

[69] Artificial Intelligence (AI) – assessment of the robustness of neural networks – Part 1: Overview. ISO/IEC TR 24029:2021: ISO/IEC JTC 1/SC 42, International Organization for Standardization (ISO); 2021-03.

[70] Information technology – artificial intelligence – Overview of trustworthiness in artificial intelligence. ISO/IEC TR 24028:2020: ISO/IEC JTC 1/SC 42, International Organization for Standardization (ISO); 2020-05.

[71] Artificial intelligence – data quality for analytics and machine learning (ML) Part 1: Overview, terminology, and examples. ISO/IEC FDIS 5259-1:2024: ISO/IEC JTC 1/SC 42, International Organization for Standardization (ISO); 2024-01-19.

[72] Artificial intelligence – data quality for analytics and machine learning (ML) Part 2: Data quality measures. ISO/IEC FDIS 5259-1:2024: ISO/IEC JTC 1/SC 42, International Organization for Standardization (ISO); 2023-10-27.

[73] Artificial intelligence – data quality for analytics and machine learning (ML) Part 3: Data quality management requirements and guidelines. ISO/IEC FDIS 5259-1:2024: ISO/IEC JTC 1/SC 42, International Organization for Standardization (ISO); 2024-01-19.

[74] Framework for artificial intelligence (AI) systems using machine learning (ML). ISO/IEC 23053:2022: ISO/IEC JTC 1/SC 42, International Organization for Standardization (ISO); 2022-06-20.

[75]   Information technology – artificial intelligence – artificial intelligence concepts and terminology. ISO/IEC 222989:2022: ISO/IEC JTC 1/SC 42, International Organization for Standardization (ISO); 2022-07-19.

[76]   Artificial intelligence – life cycle processes and quality requirements – Part 1: quality meta model. DIN SPEC 92001-1:2019: ANSI Webstore, American National Standards Institute (ANSI); 2019-04.

[77]   Artificial intelligence – life cycle processes and quality requirements – Part 2: robustness. DIN SPEC 92001-1:2019: European Standards; 2020-12-01.

[78]   Ethical design and use of automated decision systems. DIN SPEC 92001-1:2019: CIO Strategy Council, Standards Council of Canada; 2021-15-10.

Chapter 5

Assessment method for prioritising transport measures and infrastructure development

Henk Taale[1,2] and Jan Kiel[3]

5.1 Introduction

The interplay between infrastructural development and transport policy measures requires comprehensive assessment methodologies to guide decision-making processes. Existing methodologies often fall short of encompassing the multifaceted nature of transport systems, the variety of stakeholder interests, and the broader social and environmental impacts. This chapter introduces the Assessment Method for Policy Options (AMPO), a framework designed to address this complexity by integrating cost-benefit analysis (CBA) and multi-criteria analysis (MCA) while promoting stakeholder engagement.

The background to this initiative stems from an identified gap within existing transport policy assessment frameworks. Traditional methods such as CBA, while methodologically sound, often overlook the socio-political dimensions inherent in infrastructure development and maintenance. Moreover, the dynamic nature of societal needs and environmental considerations calls for a more adaptive and inclusive approach to transport planning and assessment. In response to these challenges, the AMPO emerges as a comprehensive tool that aims not only to bridge the gap between quantitative and qualitative analysis but also to harmonise the myriad perspectives and objectives of stakeholders.

This chapter defines the structure and application of the AMPO and highlights its potential as a tool for transport policy and infrastructure planning. The sections in this chapter are methodically structured to provide insights into the AMPO. Section 5.2 describes the scope and other aspects of the AMPO and provides clarity on its application and relevance. Section 5.3 explains the ten steps of the AMPO and provides a detailed guide to its implementation. Section 5.4 presents a case study that demonstrates the practical application and effectiveness of the AMPO in a real-life scenario. Section 5.5 introduces the AMPO tool, a digital tool designed to facilitate the application of the AMPO framework. Finally, Section 5.6 provides a synthesis of the findings and provides recommendations for the future application of the AMPO to ensure its continuous development and relevance for transport policy and infrastructure development.

[1]Rijkswaterstaat, Department Water, Traffic and Environment, The Netherlands
[2]Delft University of Technology, Transport & Planning, The Netherlands
[3]Panteia, Zoetermeer, The Netherlands

5.2   Scope and other aspects

5.2.1   Scope

The AMPO has a broad coverage and is designed to address the complexities and versatility of transport systems by covering a comprehensive set of transport measures that span urban, regional, national, and international scales. This comprehensive coverage ensures that the applications of the AMPO are diverse and relevant to different types of infrastructure projects, including both passenger and freight transport. Because the AMPO can accommodate such a diverse range of transport measures, it addresses the challenges and opportunities that characterise transport policy and infrastructure planning, making it an indispensable tool in this field.

The methodological basis of the AMPO is characterised by its holistic approach to transport policy assessment. Instead of focusing on isolated aspects of transport systems, different actions and policy goals are considered, which is essential for understanding the interactions within transport systems and their related infrastructures. This holistic perspective is essential for conducting assessments that reflect the nature of transport systems, promoting informed decision-making and more effective planning.

Recognising the crucial role of stakeholders in the successful planning and implementation of transport policy measures, the AMPO emphasises inclusiveness in its design. It is accessible and relevant to a diverse group of users, including policymakers, practitioners, and regional stakeholders, and ensures that the assessment process is not only comprehensive but also collaborative. By integrating the insights, needs, and preferences of all stakeholders, the AMPO ensures a balanced and representative assessment of transport measures, contributing to sustainable results that are widely accepted and supported.

The scope of the AMPO attests to its comprehensive and adaptive nature. By covering a broad spectrum of transport measures, adopting a holistic approach to assessment, and prioritising stakeholder inclusiveness, the AMPO sets a new standard for transport planning and development. It provides a robust and multi-faceted framework that is equipped to navigate the complexity of transport systems and ensures a more comprehensive, informed, and collaborative approach to transport policy and infrastructure planning.

5.2.2   Relation with other methods

The AMPO plays an innovative role in advancing methodologies for transport policy and infrastructure planning. A comprehensive framework integrating CBAs, multi-criteria analyses (MCAs), and discussion tools, the AMPO has proven itself as a tool for evaluating transport policy measures. As a follow up on the AMPO, the Spatial Planning and Development Evaluation (SPADE) project is emerging as a next development, extending the application of AMPO principles at the European level, and to spatial development. The transition from AMPO to SPADE marks an important evolution in methodologies for transport and infrastructure assessment. SPADE not only builds on the foundation laid by the AMPO but also underpins both tools with an extensive literature analysis, reinforcing and validating its methodologies with a robust scientific basis and ensuring that the tool is based on the latest research and best practices in the field [1,2]. But it also ensures that the tool reflects contemporary challenges and opportunities in transport planning and infrastructure development.

The AMPO's engagement with other methodologies is crucial and underlines its ability to synergise with a broad spectrum of evaluation frameworks and practices. This integration of methodologies facilitates a holistic approach to transport planning, allowing assessment of both quantitative economic impacts and qualitative stakeholder concerns. The versatility of the AMPO is further illustrated by its alignment with collaborative planning and stakeholder engagement models, which reflects a commitment to inclusive and participatory decision-making processes.

The AMPO's relationship with other (Dutch) methodologies [3–7], including those that emerged in the SPADE literature analysis, shows the potential for methodological innovation and synthesis and illustrates a comprehensive and evolving evaluative ecosystem. By integrating insights from a wide range of evaluation practices, the AMPO enhances its analytical capabilities and offers new perspectives on the environmental, social, and spatial implications of transport policy. This methodological enrichment is important for developing more sophisticated, dynamic, and responsive assessments of transport initiatives and for meeting the complex needs of contemporary transport policy and infrastructure planning.

5.2.3   Basic principles

The basic principles underlying the AMPO embody a comprehensive, inclusive, and forward-looking approach to transport policy and infrastructure planning. This section provides the basic principles that guide this methodology, focussing on assessment, stakeholder engagement, and flexibility in application.

Central to the AMPO is the principle of conducting thorough and methodologically sound evaluations of transport policies. This involves a balanced integration of cost-benefit analysis (CBA) and multi-criteria analysis (MCA), which ensures that both quantitative economic impacts and qualitative aspects, such as environmental sustainability and social equity, are comprehensively considered [8]. This approach ensures that assessments are not only methodologically robust but also reflect the multifaceted nature of transport systems and their impacts on society and the environment.

The AMPO prioritises the inclusion of a wide range of stakeholders in the planning and assessment process. This principle recognises the diverse interests and concerns of various parties involved in transport and infrastructure projects, from policymakers and practitioners to community members and environmentalists. By promoting an environment of collaborative planning and dialogue, these methodologies ensure that assessments are participatory, transparent, and responsive to stakeholder input. This inclusive approach is important for building consensus, facilitating equitable decision-making, and improving the social acceptability of transport initiatives.

Recognising the dynamic nature of transport systems and the societal needs and policy objectives, the AMPO is designed to be adaptable and responsive. This principle ensures that the methodology can be applied across a range of contexts, scales, and types of transport interventions, ensuring its relevance and effectiveness in addressing contemporary challenges. Moreover, the flexibility of the framework supports the integration of emerging trends, technologies, and analytical tools, enabling continuous improvement and innovation in transport policy assessment.

AMPO integrates a process with a digital tool designed to facilitate collaboration between stakeholders with divergent backgrounds and interests, with an emphasis on fair participation and transparency. The process advocates collaborative planning, with stakeholders with different resources and objectives working together so that both resource-rich and resource-poor groups have an equal voice.

The core tool combines a digital workshop with an evaluation tool (see section 5.5). This combination allows for nuanced evaluation through multi-criteria and cost-benefit analyses. The digital workshop, moderated to ensure balanced participation, allows stakeholders to discuss and deliberate on policy measures using electronic support such as e-participation tools or the Delphi method. This setup ensures that every participant contributes to the discussion, bypassing the traditional dynamics of meetings where voices can dominate or silence. The innovative use of electronic means to support structured discussions and the ability to make quick, accurate assessments without extensive modelling represents a significant advance, especially for early planning and decision-making in both small- and large-scale projects. AMPO therefore not only improves stakeholder engagement, but also streamlines and speeds up the planning process and enables iterative use as more detailed information becomes available, thus anchoring decisions more firmly with stakeholders (Figure 5.1).

The AMPO's basic principles reflect a holistic and integrated approach to the assessment of transport policy and infrastructure planning. These principles – accurate assessment, stakeholder involvement, and flexibility in application – serve as the foundation for the AMPO, which is methodologically sound, inclusive, and

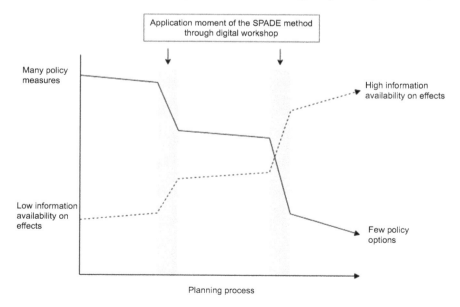

Figure 5.1   AMPO method in the entire planning process

adaptable to changing contexts and needs. By adhering to these principles, the AMPO contributes to the advancement of transport planning practices and ensures that it is equipped to navigate the complexities of modern transport systems and contribute to the sustainable development of societies.

5.3   The method in ten steps

5.3.1   Introduction

The AMPO is aimed at determining the impacts and the costs and benefits of (packages of) measures but is not intended to select these measures. Selecting measures is covered by other methods, such as the 'General Guidance for CBA' [9] or the ones mentioned in previous sections. A good CBA can obviously contribute to the realisation of a good package of measures.

To realise a good set of measures, the first step is to determine the traffic and transport problems and challenges in a region. Then, the bottlenecks and solution directions for these bottlenecks can be determined. The solutions can be tested and discussed with stakeholders and can be converted into specific measures. This set or sets of measures form the starting point for the AMPO. Stakeholders in a region often include various parties, such as local authorities, public transport companies, the regional business community and citizens. The AMPO aims at a proper assessment of the proposed measures. Involving all stakeholders in the selection and assessment process ensures that the final result will be supported.

A CBA uses a baseline scenario and project alternatives. The impact of a project is defined as the difference between the developments with the project (project alternatives) and without the project (baseline). The baseline scenario is not simply doing nothing, but the most likely development that would take place without new policies. It may consist of continuation of the existing policies but may also include other choices. In the AMPO, several measures combined in a package form a 'project alternative'. The method aims to weigh the resulting packages of measures against each other. The method also allows – in an interactive setting – to vary the composition of the package and choose the best package, based on the impacts and costs, but also based on underlying choices such as available budget, objectives and the importance that stakeholders attach to these objectives.

5.3.2   The ten steps

5.3.2.1   Step 1 – stakeholder engagement

The beginning of the AMPO depends on the critical first step of stakeholder engagement. This stage prioritises identifying and engaging a broad spectrum of stakeholders so that the assessment process is based on inclusiveness and diversity from the outset. It includes a process to identify all entities that could be affected by or have an interest in the transport policies, ranging from government agencies and the private sector to community groups and the general public. This inclusive approach not only enriches the assessment with a multitude of perspectives, but also lays the foundation for a transparent, trust-building process that recognises and addresses the concerns and expectations of all stakeholders.

Engaging stakeholders is an effort that goes beyond initial identification. Ongoing contact and dialogue are needed to maintain engagement and trust throughout the assessment and implementation phase. Challenges such as reconciling divergent interests and countering engagement fatigue are addressed with strategies such as facilitated discussions and regular updates to keep stakeholders engaged and informed. By ensuring that the different views and interests of stakeholders are integrated into the planning and decision-making process, this step lays a solid foundation for the next stages of the AMPO. This leads to the development of transport policies that are not only more sustainable and equitable, but also enjoy a higher degree of social acceptance and support.

5.3.2.2   Step 2 – determining the interaction between measures

The second step examines the extent to which policy measures influence each other. Practice shows that measures can reinforce or weaken each other. Measures that reinforce each other are preferable to those that weaken each other. Therefore, the interaction of measures should first be identified so that it can be taken into account in the next steps. For example, a measure that introduces a green wave on a route and another measure that expands capacity on the same route both have a positive effect on travel time. Travel time will decrease. Especially, if it is a busy route, the interaction between the two measures will be high.

In this step, all proposed measures are compared. It involves the full list of measures, without having already been combined into packages. The interaction between measures is determined by stakeholders by assigning scores for example from −3 to +3, where the score '=3' represents a strong negative correlation and '+3' a strong positive correlation. Table 5.1 shows an example. The scores are subjective, but if all stakeholders complete the form, a relatively clear, 'intersubjective' picture should emerge. The result of this step is taken into account in step 6 and is reflected in the following steps.

5.3.2.3   Step 3 – assessment of costs and benefits

For each measure within a policy package, it is essential to determine both costs and benefits. This stage often requires considerable effort, as extensive information has to be collected or calculated. A standardised assessment requires monetising the benefits where possible, depending on the quantifiability of the effects of the measures. Quantification becomes feasible when impacts on primary indicators such as distance, time and transport costs can be identified, along with derived indicators such as speed, time loss or emissions. For monetisation, a time factor is usually used; in freight transport this can be applied in a similar way, or alternatively using fixed and variable costs (e.g. salaries and fuel), supplemented by additional transport costs such as tolls or parking fees. To monetise safety and environmental impacts, conversion factors for emissions and traffic-related deaths and injuries are used. If monetising environmental or safety variables is deemed inappropriate, they should instead be quantified, a process described in the next step.

Table 5.1 *Example of interaction matrix*

	Bundle goods	Eco-transport	Green wave	Information	Measure X	Measure Y	Measure Z	Sum
Bundle goods		2	0	1	1	0	0	4
Eco-transport	3		1	-1	0	0	0	3
Green wave	0	1		0	0	0	0	1
Information	1	-1	0		0	0	0	0
Measure X	0	0	0	0		-1	0	-3
Measure Y	0	0	0	0	0		0	0
Measure Z	0	0	0	0	1	-3		0
Sum	4	2	1	0	2	-4	0	5
Relative score	10	8	7	3	2	1	3	

Determining the costs associated with the measures is equally crucial and includes all expenditures, including future maintenance. A certain time horizon and discount rate must also be applied to ensure the accuracy of the assessment. The culmination of this step is a comprehensive overview of the costs and benefits of each measure, as illustrated in Table 5.2. Early on in the policy process this step can be skipped and replaced entirely by a multi-criteria analysis as detailed information that serves as input for a cost-benefit analysis probably is not available.

5.3.2.4   Step 4 – investigation of other quantifiable impacts

If monetisation of the impacts on aspects such as safety and environment is not desirable or possible, it can be chosen to quantify these effects. For safety, the number of accidents, deaths and injuries are quantified. For the environmental impacts all types of emissions and noise can be quantified. For safety and the environment, the partial effects of each measure are put together in absolute terms. Table 5.3 shows an example where Measure X reduces the number of accidents by 30, the number of casualties by 14 and the number of deaths by 3. For the measures it is clear what the absolute effects are and how they relate to each other. Based on the sum of the absolute effects the measures can be ranked on safety and environment. Apart from summing them also other operators can be used to determine a score for safety.

5.3.2.5   Step 5 – determine qualitative impacts

While assessing impacts on accessibility, safety and the environment often yields quantitative results, qualitative impacts require a separate analysis, complementing steps 3 and 4. Instead of monetary values or absolute numbers, these impacts are assessed using a scoring system, for example ranging from '−5' to '+5'. In this ranking, '0' stands for 'no effect', '+5' for a 'significant positive effect' and '−5' for a 'significant negative effect'. This stage is inherently more subjective and can lead to discussions about the magnitude or direction of an effect. Asking for the scores of different stakeholders helps build a collective, intersubjective consensus. Discrepancies in scores may prompt a review to understand divergent perceptions.

The scoring process involves stakeholders completing a questionnaire to evaluate impacts, ensuring that each perspective is captured. The scores collected from all stakeholders are then aggregated to calculate an average impact for each measure. This approach culminates in a comprehensive summary of the average estimated impact for each measure, which provides a balanced view of the qualitative impacts. An illustration of this process can be found in Table 5.4, where the aggregated results clearly reflect the perceived impact of each measure, facilitating informed comparison and discussion among stakeholders.

5.3.2.6   Step 6 – determine the priority order of impacts

From steps 3 to 5, the ranking of measures for individual impacts can be determined. For each measure, for each aspect (accessibility, environment, safety,

Table 5.2 Example of matrix with monetised impacts

	Bundle goods	Eco-transport	Green wave	Information	Measure X	Measure Y	Measure Z
Costs	1.000	500	200	200	300	250	0
Benefits in travel costs (k€)	2.000	0	1.000	500	25	15	10
Benefits in emissions (k€)	15	20	25	2	3	2	1
Benefits in safety (k€)	0	1.300	900	0	300	0	0
Balance of costs and benefits	1.015	820	1.725	302	28	−233	11
Relative score accessibility	10	1	5	3	1	1	1
Relative score environment	6	8	10	1	2	1	1
Relative score safety	1	10	7	1	3	1	1
Relative score balance	7	6	10	3	2	1	2

Table 5.3 *Example of matrix with quantified impacts*

	Bundle goods	Eco-transport	Green wave	Information	Measure X	Measure Y	Measure Z
Reduction emissions (tonnes)	234	345	445	25	30	15	10
Reduction accidents		−20	30		30		
Reduction casualties		−10	16		14		
Reduction deaths		−1	2		3		
Score on environment	234	345	445	25	30	15	10
Score on safety	0	−31	48	0	47	0	0
Relative score environment	6	8	10	1	1	1	0
Relative score safety	5	1	10	5	10	5	5

Table 5.4   Example of matrix with qualitative impacts

	Bundle goods	Eco-transport	Green wave	Information	Measure X	Measure Y	Measure Z
Comfort	1		2		2	2	
Image		3	3	2	1	1	1
Barriers							
Landscape quality	−3				−2		
Score on perception	−2	3	5	2	1	3	1
Relative score per-ception	1	7	10	6	5	7	5

quality aspects and interaction) and impact (e.g. costs, number of accidents, emissions, comfort, reliability), it is possible to examine how the scores of the measures compare with other measures. Within the aspects, the scores are added together so that each aspect shows one score. The choice for rankings now is that they scale proportionally and so are continuous ranges of 1–10. But other choices for ranking are possible and the choice of proportional scale is something to test in a case study. Table 5.5 gives an example of the effects that can be monetised.

The measures all have some impact on the previously mentioned aspects of accessibility, safety, environment, quality aspects, and interaction. Some comments on these:

For accessibility, this involves an impact on the key variables distance, volume and capacity, travel time, travel cost and revenue. More than these variables are not needed to determine the impact on accessibility, as almost all other indicators can be derived from them. These impacts can be monetised.

Environmental impacts involve effects on emissions and noise. These too can be monetised, but this is not necessary. A quantification of the volume of emissions (such as NOx and CO_2) may be sufficient.

For safety, we look at the impact of measures on the number of accidents, deaths and injuries. In principle, these effects can be monetised, but not necessarily. There may be ethical reasons that prevent it. If that is the case, it is enough to quantify the number of accidents, injuries and deaths.

Qualitative aspects lead to effects that cannot be quantified, such as emotion, comfort, image or landscape quality. Sometimes the number of relevant aspects can be large.

The importance of interaction between measures (mutual reinforcement and weakening of measures) is also included in the assessment of the effects of measures. Mutual reinforcement earns extra points.

Table 5.5 *Example of matrix comparing monetised impacts*

	Bundle goods	Eco-transport	Green wave	Information	Measure X	Measure Y	Measure Z
Costs per measure	1.000	500	200	200	300	250	0
Changes in travel costs (k€)	2.000	0	1.000	500	25	15	10
Changes in emissions (k€)	15	20	25	2	3	2	1
Changes in accidents (k€)		200	300		100		
Changes in casualties (k€)		1.000	2.000				
Changes in deaths (k€)		100	200		200		
Balance	1.015	820	3.325	302	28	−233	
Relative score balance	7	6	10	3	2	1	2

5.3.2.7   Step 7 – determine the weights for the different aspects

Within a region or sector, different objectives and stakeholder interests may result in certain aspects – such as accessibility, safety, environment, qualitative factors or interactions – being given higher priority than others. This variability is addressed in this step, where it becomes possible to assign specific weights to these different aspects, facilitating the formulation of a balanced set of measures. This process not only helps to gain a full understanding of how each aspect affects the overall outcome, but also ensures that the package with measures matches regional or sectoral priorities and needs. Weights can be assigned on a scale of 1 to 100, with the total sum of the weights across all aspects equal to 100.

In some cases, stakeholders may hesitate to assign weights, fearing that this could politicise the assessment process. In such cases, this step is omitted so that the assessment can be carried out without this prioritisation layer. It must be clear that this method allows for a tailored approach to prioritising measures and reflects the unique context and objectives of each region or sector.

5.3.2.8   Step 8 – perform a multi-criteria analysis

Once the weights are determined, a multi-criteria analysis (MCA) can be carried out, taking all aspects into account. This exercise reveals which measure is the best, based on a weighted average across all aspects, and which gives the least result. In Table 5.6, the safety and environmental aspects have not been monetised, but their relative scores have been used. This shows, in this example, that before weighting, green wave is the best, but after weighting, bundling of goods is slightly better.

If the aspects accessibility, safety and environment are monetised, we can use the relative scores by balance (see Table 5.5). The outcome is shown in Table 5.7. In this case, the green wave is the best measure because of its lower cost and better safety score.

5.3.2.9   Step 9 – perform a sensitivity analysis

With the available budget and cost per measure, the sensitivity of several different packages of measures can be determined. This is done using the results of steps 3 and 6. The cost of a measure can be extracted from step 3 and the average ranking across all aspects and effects from step 6 can be used. Different composite measure packages are possible within the available budget. Normally the budget is exhausted, but of course there is a possibility that the budget is exceeded or under-utilised. There is also the possibility that the optimal measure package in terms of benefits requires additional funding compared to the original budget. In that case, the discussion will focus on where additional financial sources can be found.

5.3.2.10   Step 10 – discuss the results and determine the measures

The last step in the process is a discussion among stakeholders about the composite packages from step 9 and the weights from step 7. In the beginning, basically every

Table 5.6 Example of the rankings without and with weighing (without monetisation)

	Bundle goods	Eco-transport	Green wave	Information	Measure X	Measure Y	Measure Z
Accessibility	10.0	1.0	5.5	3.2	1.1	1.0	1.0
Safety	4.5	1.0	10.0	4.5	9.9	4.5	4.5
Environment	5.6	7.9	10.0	1.3	1.4	1.1	1.0
Perception	1.0	7.4	10.0	6.1	4.9	7.4	4.9
Interaction	10.0	7.5	6.7	3.5	1.8	1.0	3.5
Total score	**31.2**	**24.9**	**42.2**	**18.7**	**19.0**	**15.1**	**14.8**

	Weight	Bundle goods	Eco-transport	Green wave	Information	Measure X	Measure Y	Measure Z
Accessibility	50	5.0	0.5	2.7	1.6	0.5	0.5	0.5
Safety	15	0.7	0.2	1.5	0.7	1.5	0.7	0.7
Environment	15	0.8	1.2	1.5	0.2	0.2	0.2	0.2
Perception	10	0.1	0.7	1.0	0.6	0.5	0.7	0.5
Interaction	10	1.0	0.8	0.7	0.3	0.2	0.1	0.3
Total score		**7.6**	**3.3**	**7.4**	**3.4**	**2.9**	**2.2**	**2.2**

Table 5.7 Example of the rankings without and with weighing (with monetisation)

	Bundle goods	Eco-transport	Green wave	Information	Measure X	Measure Y	Measure Z
Balance	6.7	5.8	10.0	3.5	2.2	1.0	2.1
Perception	1.0	7.4	10.0	6.1	4.9	7.4	4.9
Interaction	10.0	7.5	6.7	3.5	1.8	1.0	3.5
Total score	**17.7**	**20.8**	**26.7**	**13.1**	**8.9**	**9.4**	**10.4**

	Weight	Bundle goods	Eco-transport	Green wave	Information	Measure X	Measure Y	Measure Z
Balance	80	5.4	4.7	8.0	2.8	1.8	0.8	1.7
Perception	10	0.1	0.7	1.0	0.6	0.5	0.7	0.5
Interaction	10	1.0	0.8	0.7	0.3	0.2	0.1	0.3
Total score		**6.5**	**6.2**	**9.7**	**3.7**	**2.4**	**1.6**	**2.5**

aspect gets the same weight. In practice, however, it will turn out that stakeholders attach more value to certain aspects. In step 7, the weights can be adjusted, and it is therefore possible that some measures get a different ranking if the valuation of individual aspects changes. Optionally, a second iteration can be done from step 6 to step 9 to change the composition of the packages based on the changed weights.

The composite packages show, given the cost and available budget, which measures can be taken together in one package. Discussion among stakeholders should lead to agreement on which package of measures is best and should be chosen.

5.4   Case study A15 'River land'

5.4.1   Description of the case

The project 'A15 River land' was carried out with the objective to make an analysis of the current and future problems on the A15 motorway from Rotterdam to Germany (more specifically between Gorinchem and Valburg), as well as the underlying road network in this region. But also, to provide a broad package of policy measures to overcome these problems.

Different policy measures have been drafted within this project. A selection of these policy measures has served as input for the case study. The AMPO could not be used to a full extent as for most policy measures the costs and benefits of implementation were unknown. Therefore, only an MCA could be used. Input for the MCA was collected by means of a workshop with different stakeholders, consisting of road authorities in the region, Rijkswaterstaat (motorway operator) and logistic organisations.

5.4.2   Measures and interactions

Some 22 measures were considered. Table 5.8 provides an overview of the measures and their assumed interaction scores.

5.4.3   Costs and benefits

The costs are usually given in Euros, but at this stage of the project, most cost data were not available. By means of a discussion and rating tool, the costs were scored, ranging from 'very low' to 'very high'. After that, relative scores were calculated (a high relative score means low costs). Table 5.10 shows the costs. As can be seen, the costs for extra capacity on the motorway were assessed as very high. All other costs were lower and due to the large difference in costs, all other measures do not score very different on this aspect.

5.4.4   Unquantifiable impacts

As benefits could not be expressed in quantitative or monetary terms, all benefits were assessed by the stakeholders in a qualitative way. Table 5.9 shows the scores on accessibility, but also on safety, environment and other qualitative aspects. As can be seen the extra capacity on the A15 scores highest on accessibility, while

Table 5.8   Measures and interaction for the case A15 'River land'

No.	Measure	Total score	Relative score
1	Accessibility Kerkdriel-North	0	1
2	Better access logistic hotspot Medel	4	4
3	Promotion of the usage of e-bikes	4	4
4	High-speed bicycle path between Geldermalsen and Tiel	2	3
5	Bundling freight	4	5
6	Rescheduling freight transport	2	3
7	Improvement traffic flow N322/N323	1	2
8	Increase capacity of the N323 (Kesteren – Rhenen)	1	2
9	Extra lane on the A15 motorway	6	4
10	Peak-hour lane A15 between Tiel and motorway junction Deil	5	8
11	Automated driving	1	2
12	Mobility management Medel	4	5
13	Traffic safety N320 (Culemborg – Kesteren)	2	3
14	Safety on intersection Maas-Waal road	1	2
15	Remove billboards A15	0	2
16	Awareness driving behaviour	5	8
17	Better design of curves in the A15	1	2
18	Ramp metering on-ramps to the A15	8	10
19	Usage of the A15 in off-peak hours	3	5
20	Control traffic on the A15 using real-time data	3	7
21	Longer acceleration lanes on slip roads to the A15	4	6
22	Motorway traffic management system on the A15 (queue tail warning)	5	7

measures for the bike score the highest on environment. For quality, automated driving scores the highest, probably due to the comfort involved.

Summing the scores (unweighted) shows that Motorway traffic management scores best with 35.2 points. Bundling freight flows scores second best with 34.2 points, followed by Controlling A15 traffic using real-time data and longer acceleration lanes. So, the best measures are related with the motorway. If we involve costs, we get Table 5.10. Because the costs of an extra lane on the A15 is very high compared to the other measures, the costs don't differentiate the measures further. That is a critical point to consider: if the costs of certain measures have a different magnitude, then scaling them from 1 to 10 decreases the difference between measures and the real difference is not reflected in the scores.

5.4.5   Multi-criteria analysis

The weights for the different measures have been determined by the stakeholders, with and without costs. They are shown in Table 5.11.

Applying the weights without costs leads to the situation in which the extra capacity of the motorway becomes most attractive. The peak hour lane has the highest score and also the extra lane scores high, because both score well on accessibility,

Table 5.9 Scores for all aspects without costs

No.	Measure	Costs	Accessibility	Environment	Safety	Quality	Interaction	Total Score
1	Accessibility Kerkdriel-North		7.4	4.5	5.0	1.9	1.0	19.8
2	Better access logistic hotspot Medel		7.4	6.0	5.0	1.0	4.4	23.8
3	Promotion of the usage of e-bikes		3.6	10.0	1.0	6.8	4,4	25.8
4	High-speed bicycle path Geldermalsen – Tiel		4.2	10.0	4.0	9.2	3.3	30.7
5	Bundling freight		5.8	9.5	6.0	7.6	5.3	34.2
6	Rescheduling freight transport		3.6	7.0	4.5	1.9	3.3	20.3
7	Improvement traffic flow N322/N323		7.9	5.5	4.5	4.7	1.6	24.2
8	Increase capacity N323 (Kesteren – Rhenen)		8.4	2.0	3.5	2.8	1.6	18.3
9	Extra lane on the A15 motorway		10.0	1.0	5.5	5.1	3.8	25.4
10	Peak-hour lane A15 between Tiel and Deil		10.0	3.0	4.0	6.0	7.8	30.8
11	Automated driving		3.9	2.5	6.5	10.0	2.1	25.0
12	Mobility management Medel		4.7	8.5	4.0	5.1	4.9	27.2
13	Traffic safety N320 (Culemborg – Kesteren)		1.0	5,5	9.5	5.6	2.7	24.3
14	Safety on intersection Maas-Waal road		1.0	5.5	10.0	5.6	2.1	24.2
15	Remove billboards A15		2.1	5.5	6.5	4.7	1.6	20.4
16	Awareness driving behaviour		2.1	7.2	7.5	2.7	8.3	27.8
17	Better design of curves in the A15		2.6	5.5	7.5	1.9	2.1	19.6
18	Ramp metering on-ramps to the A15		5.2	6.6	6.9	1.9	10.0	30.6
19	Usage of the A15 in off-peak hours		4.7	8.0	6.0	4.4	4.9	28.0
20	Control traffic on the A15 using real-time data		5.8	7.5	6.0	6.8	6.6	32.7
21	Longer acceleration lanes A15		5.8	6.6	7.5	6.8	5.5	32.2
22	Motorway traffic management system A15		5.8	7.5	8.5	6.8	6.6	35.2

Table 5.10 *Scores for all aspects with costs*

No.	Measure	Costs	Accessibility	Environment	Safety	Quality	Interaction	Total Score
1	Accessibility Kerkdriel-North	9.5	7.4	4.5	5.0	1.9	1.0	29.3
2	Better access logistic hotspot Medel	10.0	7.4	6.0	5.0	1.0	4.4	33.8
3	Promotion of the usage of e-bikes	10.0	3.6	10.0	1.0	6.8	4,4	35.8
4	High-speed bicycle path Geldermalsen – Tiel	10.0	4.2	10.0	4.0	9.2	3.3	40.7
5	Bundling freight	10.0	5.8	9.5	6.0	7.6	5.3	42.2
6	Rescheduling freight transport	10.0	3.6	7.0	4.5	1.9	3.3	30.3
7	Improvement traffic flow N322/N323	10.0	7.9	5.5	4.5	4.7	1.6	34.2
8	Increase capacity N323 (Kesteren – Rhenen)	9.9	8.4	2.0	3.5	2.8	1.6	28.2
9	Extra lane on the A15 motorway	1.0	10.0	1.0	5.5	5.1	3.8	26.4
10	Peak-hour lane A15 between Tiel and Deil	6.0	10.0	3.0	4.0	6.0	7.8	36.8
11	Automated driving	9.8	3.9	2.5	6.5	10.0	2.1	34.9
12	Mobility management Medel	10.0	4.7	8.5	4.0	5.1	4.9	37.2
13	Traffic safety N320 (Culemborg – Kesteren)	9.9	1.0	5,5	9.5	5.6	2.7	34.2
14	Safety on intersection Maas-Waal road	10.0	1.0	5.5	10.0	5.6	2.1	34.2
15	Remove billboards A15	10.0	2.1	5.5	6.5	4.7	1.6	30.4
16	Awareness driving behaviour	10.0	2.1	7.2	7.5	2.7	8.3	37.8
17	Better design of curves in the A15	10.0	2.6	5.5	7.5	1.9	2.1	29.6
18	Ramp metering on-ramps to the A15	10.0	5.2	6.6	6.9	1.9	10.0	40.6
19	Usage of the A15 in off-peak hours	10.0	4.7	8.0	6.0	4.4	4.9	38.0
20	Control traffic on the A15 using real-time data	10.0	5.8	7.5	6.0	6.8	6.6	42.7
21	Longer acceleration lanes A15	10.0	5.8	6.6	7.5	6.8	5.5	42.2
22	Motorway traffic management system A15	9.9	5.8	7.5	8.5	6.8	6.6	45.1

Table 5.11   Two different set of weights

	Weight		Weight
		Costs	44
Accessibility	47	Accessibility	27
Safety	18	Safety	10
Environment	20	Environment	11
Qualitative aspects	6	Qualitative aspects	5
Interaction	6	Interaction	4
Total	**100**	**Total**	**100**

which has a high weight. Measures which score low on accessibility (for example, Promotion of e-bike) have a low total score. All scores are shown in Table 5.12.

When the costs are taken into account, measures to add extra capacity to the motorway drop in the list and now the motorway traffic management scores best. The relative low weight of accessibility and the high weight for the costs are the reasons for this. The relative scores for all measures are given in Table 5.13. It shows that taking into account the costs at this stage of the project has significant implications for the ranking of the measures.

5.4.6   Conclusions from the case

During the meeting with the stakeholders the discussions were determined by the costs of the measures and the scale that was used (5-step scale) and finally who will pay for the measures. The discussion partially influenced the scoring procedure as well as making choices between measures. The assessment method has been developed to weigh the different measures equally. Therefore, a focus should be put on the impacts, then ranking the measures based upon their scores and finally look at the scores based upon their (financial) feasibility, support and available information. Scoring the measures may lead to quick wins of some measures which can be realised on short term. For other more expensive measures the costs need to be determined first. This asks for more information. The amount of investment needs further insight in the financial feasibility of the policy measures before they can be implemented. Finally, this case shows that it is difficult to take into account transport measures that differ much, such as a parking facility for bikes and extra capacity on a motorway. We conclude that the transport measures should have some relation to each other concerning size and costs.

5.5   AMPO tool

The AMPO, described in this chapter, has been translated into a web-based tool. This tool makes gives the user an interface to the steps involved and makes it easy to involve stakeholders into the process and to do the calculations and sensitivity analysis.

The measures and aspects can be defined freely and for the aspects it can be chosen to make them quantitative or qualitative, using a scale. Also, the levels for the scales can be chosen, e.g. from -3 to $+3$, or from 1 to 10.

Table 5.12 Results of MCA, without costs

No.	Measure	Costs	Accessibility	Environment	Safety	Quality	Interaction	Total Score
1	Accessibility Kerkdriel-North		3.5	0.9	0.9	0.2	0.1	5.5
2	Better access logistic hotspot Medel		3.5	1.2	0.9	0.1	0.3	5.9
3	Promotion of the usage of e-bikes		1.7	2.0	0.2	0.6	0.3	4.7
4	High-speed bicycle path Geldermalsen – Tiel		2.0	2.0	0.7	0.8	0.2	5.7
5	Bundling freight		2.7	1.9	1.1	0.7	0.2	6.6
6	Rescheduling freight transport		1.7	1.4	0.8	0.2	0.2	4.3
7	Improvement traffic flow N322/N323		3.7	1.1	0.8	0.4	0.1	6.1
8	Increase capacity N323 (Kesteren – Rhenen)		3.9	0.4	0.6	0.3	0.1	5.3
9	Extra lane on the A15 motorway		3.7	0.2	1.0	0.5	0.2	6.6
10	Peak-hour lane A15 between Tiel and Deil		4.7	0.6	0.7	0.5	0.5	7.0
11	Automated driving		1.8	1.5	1.2	0.9	0.5	5.9
12	Mobility management Medel		2.2	1.7	0.7	0.5	0.3	5.4
13	Traffic safety N320 (Culemborg – Kesteren)		0.5	1.1	1.7	0.5	0.2	3.9
14	Safety on intersection Maas-Waal road		0.5	1.1	1.8	0.5	0.1	4.0
15	Remove billboards A15		1.0	1.1	1.2	0.4	0.1	3.8
16	Awareness driving behaviour		1.0	1.4	1.4	0.2	0.5	4.5
17	Better design of curves in the A15		1.2	1.1	1.4	0.2	0.1	4.0
18	Ramp metering on-ramps to the A15		2.4	1.3	1.2	0.2	0.6	5.8
19	Usage of the A15 in off-peak hours		2.2	1.6	1.1	0.4	0.3	5.6
20	Control traffic on the A15 using real-time data		2.7	1.5	1.1	0.6	0.4	6.3
21	Longer acceleration lanes A15		2.7	1.3	1.4	0.6	0.3	6.3
22	Motorway traffic management system A15		2.7	1.5	1.5	0.6	0.4	6.8

Table 5.13 Results of MCA, with costs

No.	Measure	Costs	Accessibility	Environment	Safety	Quality	Interaction	Total Score
1	Accessibility Kerkdriel-North	4.2	2.0	0.5	0.5	0.1	0.0	7.3
2	Better access logistic hotspot Medel	4.4	2.0	0.7	0.5	0.1	0.2	7.8
3	Promotion of the usage of e-bikes	4.4	1.0	1.1	0.1	0.3	0.2	7.0
4	High-speed bicycle path Geldermalsen – Tiel	4.4	1.1	1.1	0.4	0.4	0.1	7.5
5	Bundling freight	4.4	1.6	1.0	0.6	0.3	0.1	8.0
6	Rescheduling freight transport	4.4	1.0	0.8	0.5	0.1	0.1	6.8
7	Improvement traffic flow N322/N323	4.4	2.1	0.6	0.5	0.2	0.1	7.8
8	Increase capacity N323 (Kesteren – Rhenen)	4.4	2.3	0.2	0.4	0.1	0.1	7.4
9	Extra lane on the A15 motorway	0.4	2.7	0.1	0.6	0.2	0.2	4.2
10	Peak-hour lane A15 between Tiel and Deil	2.6	2.7	0.3	0.4	0.2	0.3	6.6
11	Automated driving	4.3	1.1	0.8	0.7	0.4	0.3	7.6
12	Mobility management Medel	4.4	1.3	0.9	0.4	0.2	0.2	7.4
13	Traffic safety N320 (Culemborg – Kesteren)	4.4	0.3	0.6	1.0	0.2	0.1	6.5
14	Safety on intersection Maas-Waal road	4.4	0.3	0.6	1.0	0.2	0.1	6.6
15	Remove billboards A15	4.4	0.6	0.6	0.7	0.2	0.1	6.5
16	Awareness driving behaviour	4.4	0.6	0.8	0.8	0.1	0.3	6.9
17	Better design of curves in the A15	4.4	0.7	0.6	0.8	0.1	0.1	6.6
18	Ramp metering on-ramps to the A15	4.4	1.4	0.7	0.7	0.1	0.4	7.7
19	Usage of the A15 in off-peak hours	4.4	1.3	0.9	0.6	0.2	0.2	7.5
20	Control traffic on the A15 using real-time data	4.4	1.6	0.8	0.6	0.3	0.3	7.9
21	Longer acceleration lanes A15	4.4	1.6	0.7	0.8	0.3	0.2	7.9
22	Motorway traffic management system A15	4.4	1.6	0.8	0.9	0.3	0.3	8.1

After the definitions, a scoring form is compiled, and this can be sent to the stakeholders for the scoring process. The stakeholders can fill in the form and send it back to the tool. The forms are managed, and results are calculated (see Figure 5.2)

The outcome, which includes average scores, but also standard deviations, can be input for a discussion with the stakeholders. Why does a certain measures scores higher than another on this aspect? Why is the standard deviation for that score so high? That means that people have valued that aspect very differently for a certain measure and that could be used for a good discussion and eventually could lead to a better understanding of the viewpoints and interests of stakeholders.

Finally, the tool provides a priority order for the measures. This does not mean that one a certain measure is better than the other, but simply that this measure has the best overall score (see Figure 5.3).

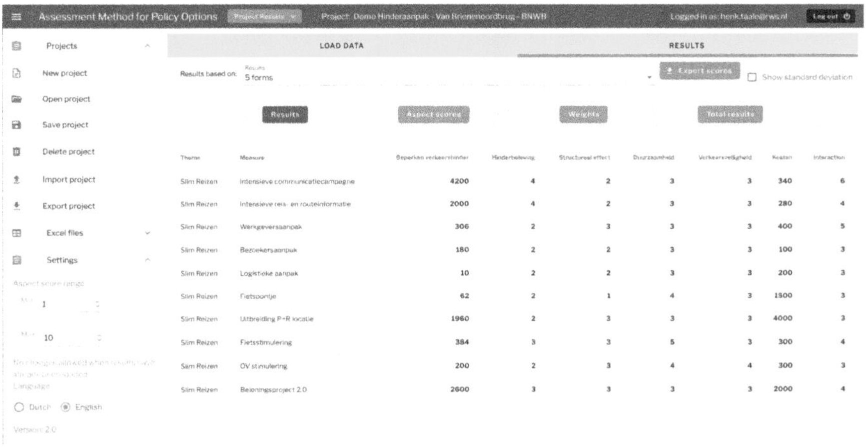

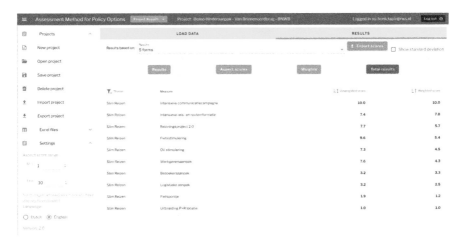

Figure 5.2   AMPO tool – results

Figure 5.3   AMPO tool – final score

5.6   Conclusions

Based on the use cases we have done so far, we can conclude that the assessment method can be applied for the selection of different packages of policy measures, both similar and divergent measures, such as infrastructural projects and traffic management projects. The AMPO has added value in an early stage of policy-making. It provides a feeling of the relative impacts of the policy measures. In a later stage the AMPO can also be applied when more detailed cost–benefit information becomes available.

When real costs are not available, discussions may arise on the question 'Who will pay?' and this may disturb the assignment of scores or the choice between measures. Policy measures with high benefits may be rejected in an early stage due to high costs, even though it is not yet clear what the available budget is. Therefore, we feel it would be better to put benefits in front of the process instead of costs: a Benefit–Cost analysis in other words.

The case of A15 River land shows that the policy measures should not diverge too much if only an MCA is carried out. The differences between small and large policy measures, such as parking bikes versus development of a motorway, may hamper a good evaluation of the projects. The projects should be more or less in line with each other. A possible solution is to do the assessment in two steps: one assessment for a package of smaller measures and an assessment for a package of big measures.

Currently, the allocation of budgets for infrastructure and other policy measures has become more dynamic than in the past, largely owing to the opportunities for co-financing from public and private organisations. This flexibility allows for the consideration of more costly measures that are financially viable due to their significant benefits. The prospect of co-financing not only makes measures more attainable but also enhances public support for them, leveraging shared financial responsibility to broaden the base of endorsement.

The AMPO maintains an expansive view, encouraging the exploration of innovative solutions for policy packages. This holistic approach fosters creativity and avoids the limitations of a cost-centric analysis by means of a social cost-benefit analysis. By including benefits such as accessibility, safety, environment, and qualitative improvements a more balanced evaluation emerges, guiding the selection process to consider the overall value of measures. This strategy suggests that the emphasis should be on assessing the benefits, deferring cost considerations to later discussions. This is instrumental in the assessment, where the costs are reintegrated into the dialogue alongside financing strategies, ensuring a comprehensive review of each package's feasibility.

Adopting this approach illuminates the most impactful measures, beyond their cost, facilitating informed decisions about which initiatives are practical. It also aids in strategically phasing policy measures, distinguishing between immediate gains and more complex, long-term projects. This transparent methodology not only clarifies the potential return on investment for each measure but also aligns the selection process with broader strategic objectives, ensuring that chosen initiatives are both impactful and financially sustainable.

Acknowledgement

The authors would like to thank former colleagues Michèle Coëmet and Arnoud Muizer for their valuable input in developing and applying the AMPO.

References

[1] Kiel, J., Hindriks, I., Sollitto, F., *et al.* Assessment methods for spatial development and infrastructure planning. *Paper Presented at the European Transport Conference* 2019, Dublin, Ireland, 2019.

[2] Kiel, J., Biesinger, B., Galonske, B., *et al.* SPADE – Assessment of the collaborative planning of infrastructure and spatial development. *Paper Presented at the Transport Research Arena* 2020, Helsinki, Finland, 2020.

[3] Rijkswaterstaat. *Manual for the Accessibility Solutions Method.* RWS Water, Traffic and Environment, Rijswijk, the Netherlands, 2014.

[4] Ecorys. Assistance for Cost-Effectiveness Analysis. Report for Rijkswaterstaat, Rotterdam, the Netherlands, 2014.

[5] MuConsult. Procedure Ex ante CBA Tool Toekan. Report for Rijkswaterstaat, Amersfoort, the Netherlands, 2011.

[6] CROW. Wikken & Wegen [Online]. 2014. Available at www.wikken-wegen.nl, accessed 13 February 2024.

[7] HU and Ministry of Infrastructure and the Environment. CORT & Krachtig. 2014.

[8] Sijtsma, F.J. Project Evaluation, Sustainability and Accountability. PhD Thesis, University of Groningen, 2006.

[9] Romijn, G., Renes, G. General Guidance for Cost-Benefit Analysis. CPB Netherlands Bureau for Economic Policy Analysis, PBL Netherlands Environmental Assessment Agency, The Hague, 2013.

Chapter 6

Evaluation of ITS: opportunities and challenges in the era of new pervasive technology

Susan Grant-Muller[1] and Frances Hodgson[1]

6.1 Introduction

One of the features of intelligent transport systems (ITS) is the large number of possible configurations that arise in practice, due to the various combinations of connected fixed-based technologies and equipment such as ramp metering, VMS (Variable Message Signs) and lane marshalling. The advent of web2.0-enabled pervasive technologies (e.g. smartphone and wearables) has broadened the definition of schemes that can be considered as falling within the family of ITS. These personal technologies and others such as Bluetooth and connected sensors, Internet of Things (IoT), effectively form part of the ITS when they record (or are linked with) individuals' transport choices and/or contribute data to the operation of the transport system. The definition of ITS now includes a legacy system of fixed-based infrastructure, such as regional traffic control centres with responsibility for a network of managed motorways, plus a network of connected and pervasive technologies such as personal devices, IoT and individuals connected through web-enabled social media (Figure 6.1).

The addition of pervasive and connected technology has added to the variety and granularity of data potentially available for the tactical and strategic management of the transport system, and to support more informed choices for the travelling public. However, they also create two different challenges for evaluation methodology:

1. How new data forms can contribute to the evaluation of a variety of established ITS-infrastructure schemes?
2. How the impacts of ITS schemes based on new technology can be evaluated, and whether the approach needed is different to that used to evaluate established ITS infrastructure schemes?

The first of these challenges arises as new connected technologies provide opportunities to collect data with a different granularity, quality and scope

[1]Institute for Transport Studies, University of Leeds, UK

Fixed-base ITS Infrastructure

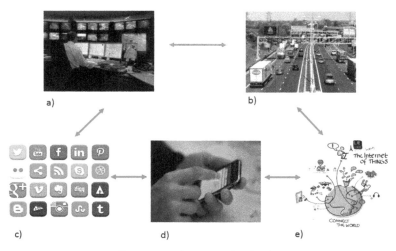

a)   b)

c)   d)   e)

Pervasive and connected technology

Figure 6.1   Extended scope of intelligent transport systems with pervasive and connected technology. (a) Traffic control, (b) managed motorways, (c) social media, (d) personal devices and (e) Internet of Things.

concerning transport choices, traffic conditions and system performance alongside data streams from fixed-based technologies. Examples include social media posts by either vehicle passengers (not drivers), local residents or pedestrians nearby that can rapidly disseminate user-contributed information on incidents and traffic conditions to supplement data from embedded loops and video cameras. This field has had some attention in the literature based on early research. Understanding the strengths, weaknesses and potential of new data forms is key, see for example [1] and [2].

The second challenge is the main focus of this chapter however and is a field for which relatively little research has been published to date. A new tranche of transport initiatives and interventions has emerged in recent years. Some of these are based on the ability to detect individuals' location within the transport system and to engage in two-way data exchange and information flow directly with the traveller. This has led to the development of sophisticated and tailored location-based services and persuasive incentivisation to encourage individuals towards behavioural change. Other initiatives have been concerned specifically with the one-way provision of near-real time, dynamic travel information to individuals allowing better informed choices about travel options before and during a journey. A further set of initiatives have been focused on web-enabled social networks and the opportunities for individuals to share their transport related resources (e.g. knowledge of the transport system and vehicles) with other travellers. A better understanding of the impacts of all these new types of schemes is important to being able to justify investment and expenditure on the technology, training staff in key skills areas and on-going "maintenance" costs (such as the cost of provision of incentives for behavioural change). Identifying and enumerating the impacts is also important to policy-makers. Due to the relative novelty and granularity of these types of schemes, there is still little in the way of established evaluation methodology to demonstrate their impacts or the success of the scheme overall. The goal of this chapter is therefore to summarise the issues in evaluating such schemes, to propose outline evaluation approaches for two main scheme typologies and to provide an overview comparison with established evaluation practice for fixed-based ITS.

The remainder of this chapter is organised as follows. Section 6.2 describes the main characteristics of the new wave of technology driven pervasive ITS, with two particular types (new mobility schemes and social innovation schemes) being selected for further elaboration in Sections 6.3 and 6.4. In these sections, we describe the respective features of these schemes in more detail, highlighting the challenges for evaluation methodology and proposing outline evaluation approaches. In Section 6.5, we provide a comparative overview across evaluation of fixed-based established ITS, new mobility schemes and social innovation schemes. The chapter concludes with some issues for future consideration regarding the commodification of data arising from new technologies.

6.2   Pervasive technology in the transport sector

ITS is described in the literature (e.g. [3–5]) as an "umbrella" term covering a variety of combinations of Information Communications Technology (ICT) systems and various transport infrastructures to provide "intelligent" services. The intelligence element of ITS is essentially the ability to assist in operational and tactical decision making concerning parts of the transport system in response to current conditions, typically the highways, although parts of the urban network are also ITS rich. Deakin *et al.* [6] refer to some ITS as discreet invisible agents (services which are not obvious or undetected by the public) due to their role in supplying data for traffic control centres that may be remotely located and therefore not visible to the travelling public. The infrastructures connected by ICT include overhead gantries, roadside variable message signs (VMS), speed indicators, traffic signals, inductive loops, automatic number plate recognition systems, cameras, central control rooms, roadside monitoring stations and many more. ITS also includes future connected and automated vehicles (CAV), though implementations of CAV are still rare and therefore beyond the scope of this chapter. Typical examples of ITS schemes are as follows:

1.   VMS, where messages are displayed on overhead or roadside gantries providing general traffic information such as estimated travel times to major destinations, incident information or congestion related information. They may also be used to advise on alternative routes to avoid incidents or congestion ahead.

2. Toll collection via an electronic tag installed within the car to pay for road use and parking (see e.g. [7,8]).
3. Highway lane marshalling using overhead gantries to separate and direct vehicles into appropriate lanes to avoid last minute lane changing or vehicles performing dangerous manoeuvres around junctions.

These may be categorised as established, fixed-based, ITS as the main channels of detection and communication are at fixed locations on the transport system, whilst vehicles moving through the system generate data that are detected by the infrastructure. These examples have different operational goals in the transport system (informed travellers, congestion charging and operational efficiency) and as a result, the evaluation of their impacts should be expected to cover different impact areas with particular indicators and data needed to capture those impacts.

These fixed-based ITS form the backdrop infrastructure against which a new set of technologies have emerged and which offer the potential to either work in complement to the existing ITS or to offer alternative transport services and functions. These new technologies may be web-enabled, are typically small scale, pervasive and may not have a primary function as part of the transport infrastructure. Examples include (but are not limited to) smartphone, tablet, laptop, Bluetooth, smartcards, electronic tagging (barcode, radio-frequency identification (RFID)) wearables and accelerometers. Large volumes of micro-level data are typically collected by the technologies, reflecting the choices, preferences and activity of individuals before, during and after (or not-related to) travel. A distinction can be drawn between those technologies that involve one-directional, passive data flow, such as tap-in-tap-out smartcards and those that are web-2.0 enabled. The latter may include established social media (e.g. X, Facebook and Instagram) or bespoke software applications ("Apps") that may have been specifically designed with transport related functions (such as traveller information) or be unrelated but still reflect individuals' activity patterns (e.g. health/exercise Apps). The contribution that data and information harvested from these new technologies can make towards the tactical, operational and strategic direction of the transport system and towards the evaluation of the success of fixed-based ITS is outside the scope of this chapter, but the reader is referred to [1] for more discussion.

Aside from the serendipitous contribution of data for ITS evaluation, a new wave of initiatives and interventions has emerged that harness the new technologies and are specifically transport and mobility focused. As with fixed-based ITS, the numerous technology configurations and functionalities available mean that there is considerable variability in the focus and scope of these initiatives. A broad categorisation is as follows:

1. Smarter choices: the provision of travel information, personalised travel planning (e.g. the WAZE app).
2. Purchasing, financial discounts and transactions – packages for Public Transport discount cards, parking and toll charging (e.g. M-ticket, First Bus, UK).

3. New mobility schemes – incentivisation to persuade behavioural change – rewards for "peak avoidance" [9], feedback and self-monitoring, rewards [10]). Where incentivisation is accompanied by personal mobility profiling (track and trace functions which indicate the location of the individual as they move on their journey), these are referred to as new mobility schemes.

4. Social innovation schemes: walking buses, car share initiatives, peer-to-peer feedback via social networks, information sharing via social networks or dedicated websites (public participation schemes) [11].

In practice, there is considerable overlap between these categories with many Apps offering several functions and a soft distinction between types – for example social sharing schemes can also act as positive incentives for behavioural change as individuals are positively encouraged to try a new mode with peer support. The provision of dynamic travel information may also be viewed as a behavioural change initiative. For the purposes of considering the evaluation methodology for the impacts of these schemes, we focus in the remainder of this chapter on two examples: new mobility schemes and social innovation schemes.

Within the transport sector, new mobility schemes and social innovation-based schemes are becoming more widespread; however, there is no established practice in terms of evaluating the impacts of such schemes. This knowledge would be beneficial for practitioners, policy-makers and academics involved in transport sector scheme evaluation and design. Accurate evaluation of these types of new technology schemes would enable practitioners and policy-makers to develop and implement such schemes within transport and across policy sectors to promote policy aims for the improvement of economic development, public health/well-being [12], congestion impacts, carbon and other pollutant emissions.

6.3 Evaluation of technology-enabled new mobility schemes

In this section, we focus on the scope of the evaluation framework that is needed for an ICT enabled new mobility scheme. We first outline some of the main features and components of a new mobility scheme, identifying illustrative examples. We subsequently discuss the evaluation challenges they create and propose the main elements of an evaluation approach.

6.3.1 Components of a new mobility scheme

The aim of a new mobility scheme is to harness the opportunity offered by pervasive personal devices and two-way communication software applications to encourage individuals to re-think their transport and travel choices. The overall goal may be as broad as the long term sustainability of the transport system, local network efficiency, improving traveller experience or tactical road/route management. A new mobility scheme essentially comprises a combination of ICT enabled technology components, databases and information sources, plus a consortium of transport and other stakeholders who may variously be responsible for providing transport services, incentives, travel timetables and more. The basic technical components of a new mobility scheme are as follows:

1. Software application (App) based scheme, whereby individuals download software onto a personal device such as a smartphone or tablet. The software may be themed as a travel information/assistant, rewards scheme, game, purchasing facility, activity-challenge, for example. It may also include a social media or social network function and thereby support social innovation schemes (see Section 6.4)
2. Other software held by the transport supplier or transport authority may include an "operators dashboard" which allows transport operators or city authorities to design and issue transport and travel related messages, challenges, incentives and/or rewards directly to individuals
3. A mobility monitoring facility, whereby (with permission) individuals' travel choices and movements can be detected using the sensing functionality of the personal device. These travel choices can be converted into whole journeys and mobility profiles, which are anonymised, stored and accessed by the operator of the dashboard in order to personalise communications and influence future transport choices towards more sustainable modes or to achieve the broader goals mentioned above.

The consortium of stakeholders responsible for providing transport services, incentives, travel timetables and other services is a key element to the scheme, as the technology and software alone will not deliver the ITS. The distinctive feature of these services is that they are "location based" and often tailored to the individuals profile e.g. of past travel choices, home location or other characteristics. An operator is needed to periodically issue communications – though some may be automated, a "helpdesk" may be needed to deal with individuals queries and a sustainable business model for the provision of dynamic travel information, incentives, and rewards is also needed. In the same way that a traffic control centre is needed for central co-ordination of managed motorways for example, central co-ordination is needed for the delivery and effectiveness of a new mobility scheme.

6.3.2   Challenges in evaluating a new mobility scheme

In order to understand the challenges in evaluating the success of a new mobility scheme it is important to note that the impacts for this type of scheme are the result of an accumulation of micro-level choices and changes at the level of the individual. Assessing impacts by observing the transport network at macro level over a period of time (i.e. the traditional approach to assessing many transport infrastructure schemes) is unlikely to detect changes and impacts. The reason for this is that the noise or variation in aggregate traffic indicators such as link travel time, is in general, too large to detect changes which have occurred at the level of the individual traveller.

Impact assessment is therefore based on detecting and monitoring changes in individuals' choices concerning travel time, mode, and route following their agreement to participate in the scheme. These individual changes may be "scaled up" to population segment sizes or used in estimates (through modelling) of the total scheme impacts in terms of e.g. reduced vehicle km, increased public transport patronage, route changes or others as appropriate to the scheme.

Further challenges exist in capturing the success of the scheme as responses to particular offers, challenges or rewards may be delayed in time from the original time of offering or may result in impacts on parts of the network or transport system that are some geo-distance from the area of focus. For example, an offer of a discounted bus or train ticket may not be responded to immediately where an individual has strong travel habits by private car or responsibility to transport family members. Informational incentives to encourage individuals to avoid a particular route or road segment due to congestion problems, an incident or even a pollution hotspot, may result in choices to divert across a number of other routes. The evaluation challenge is therefore complicated in space and time.

Evaluating the success of the new mobility scheme may also entail capturing the flow of costs and benefits to a different set of stakeholders than those traditionally involved in a transport infrastructure scheme and well outside the range of common transport indicators. Third-party providers of incentives such as discounted tickets for local entertainment or refreshments, for example, may benefit from longer term and potentially undetected increases in patronage from travellers revisiting their services outside the scheme boundaries.

According to [13], the evaluation framework for a new mobility scheme should ideally fulfil the high-level requirements.

High-level requirements that should be fulfilled by evaluation framework for a new mobility scheme
Allow comparability of the scheme with "traditional" ITS or other transport schemes
 Captures performance against the specific objectives for the technology configuration
 Ability to reflect the dynamic nature of impacts
 Ability to reflect long term costs and benefits
 Flexibility for different schemes/contexts, particularly given the numerous features possible
 Ability to monetise some or all of impacts (in order to allow comparability with standard evaluation methods)
 Disaggregate outputs by stakeholder (considering the different sets of stakeholders involved in this type of scheme)
 Ability to reflect "intangibles" and broader socio-economic impacts
 Practical with respect to measurability and data demands

The final requirement may act as a determining one on the evaluation approach – whilst the potential presence of a wide range of impacts can be envisaged with a new mobility scheme, the ability to measure these in practice and access the necessary data may remain challenging.

6.3.3   Proposal for an evaluation approach for an ICT enabled new mobility scheme

As a general proposal (and reflecting the basic high-level framework outlined in Section 3.2), we suggest the following evaluation approach for a new mobility scheme. The approach reflects that adopted in the EU-funded SUNSET project [13], as implemented in The Netherlands, Sweden and UK.

An evaluation approach for a new mobility scheme
The general structure and main categories may follow those used in the basic evaluation of a fixed-based ITS scheme and which is also foundational to the evaluation of other types of transport infrastructure schemes. This involves the following categories: costs and financial flows, efficiency, safety, environmental, social, equity and other impacts. Following this structure allows some consistency with the evaluation of other transport schemes.

However the components in each category will be very different in the case of the new mobility scheme. "Costs" will need to include ICT integration costs for the scheme operator for example, in order to capture, store and process large volumes of individual mobility data. Incentives will also be a cost component, as may be the provision and training of staff with skills in social media, staff to design and implement incentives packages. Marketing costs may also be accrued as this type of scheme needs proactive individual participation (by downloading an app for example) and does not have the same physical "presence" as a gantry based lane marshalling scheme.

Stakeholder groups, by which both costs and benefits are usually disaggregated, should have an expanded and alternative membership. They should include any third parties involved in the provision of location based information or other services, providers of incentives or rewards, providers of data or ICT services.

Some impact types and indicators that are often included in evaluation may be omitted. The set of tactical highway management indicators related to specific lane use for example, as the new mobility scheme may not be able to accurately detect the data needed and the scheme is not directed towards that objective.

Further impact categories are needed that are specific to this type of scheme, including the impacts generated by any social media or social network function in the application (these are discussed further in Section 6.4). Similarly, impacts concerning changes in perception or attitude to transport choices that have not yet resulted in a tangible outcome, but which have predisposed the individual to future changes should be captured. These represent positive moves along the classic behavioural-change and attitudinal axis.

The evaluation will need to consider the whole journey taken by the individual and aggregate impacts over different stages as appropriate, for example where an individual walks to a bus terminus then catches a bus and walks the final stage. This is a different approach to capturing scheme impacts than is the case for fixed-based ITS where only the instrumented sections of the network are captured, though usually supplemented by social surveys in a broader area.

In summary, the evaluation of a new mobility scheme stretches (and in part, extends well beyond) the traditional bounds of the ITS evaluation paradigm. However that paradigm is foundational to the approach advocated here and we perceive a number of benefits in retaining some common impact categories and structure. Whilst new mobility schemes have become far more widely accepted and adopted by cities across Europe and internationally, the ability to determine benefits clearly and in a way that is consistent with the way in which alternative (more traditional) schemes are evaluated is key for public acceptability. A comparison between the evaluation approaches is outlined in Section 6.5.

6.4   Evaluation of technology-enabled social innovation schemes

The aim of this section is to outline an evaluation framework that will overcome some of the challenges inherent in evaluating social innovation based schemes within the transport sector. We first provide a definition of social innovation schemes and identify a number of examples. Subsequently, we identify the challenges in evaluating these types of schemes, proposing an evaluation approach.

6.4.1   Definitions of social innovation in the transport sector

Definitions of social innovation vary across different fields of study and a number of terms are used interchangeably including, for example, "social entrepreneurship", social enterprise, "shared value creation", and "sharing economy" resulting in a disjointed and under-researched area [14]. Discussions around social innovation generally highlight three elements: the first is that it refers to the "finding of solutions", the second element is that the benefit is to society or meeting a social need and the third element refers to social organisation. Mulgan [15] defined social innovation as "innovation activities and services that are motivated by the goal of meeting a social need" and similarly in 2008 the Stanford Centre for Social Innovation coined the following definition: "A novel solution to a social problem that is more effective, efficient, sustainable, or just than existing solutions and for which the value created accrues primarily to society as a whole rather than private individuals" [16].

Social innovation schemes within the transport sector necessarily means a focus on "shared economy", or "sharing economy" schemes. Largely because these forms of "collaboration consumption" which has been defined as an economic model that emphasizes "access" or "sharing" instead of "ownership" [17]. Within the transport sector there are a number of examples of these types of schemes including, e.g.

- Car sharing: Zipcar, COMMUNAUTO, eGO CarShare
- Peer to peer carsharing: Relay rides, GoMore
- Ridesharing: Carma, Carticipate, Zimride, Uber
- BikeSharing or micromobility: Publibike, Citibikes, Lime
- Other information and performance sharing schemes: Waze, Strava, mapmyrun

These schemes differ in that some are owned and managed by large, sometimes multi-national businesses and others are more small-scale businesses or not for profit businesses. Other, less formal schemes which are much more likely to be small scale or community stimulated and run include: Bike Buddy systems, travel buddy schemes for more vulnerable travellers and Walking School Buses. This range and diversity of organisational forms is evolving rapidly as new pervasive technologies enable new forms of communication and social organisation e.g. UBER resulting in an increased interest in innovative business models.

6.4.2   *Challenges in evaluating social innovation schemes in transport*

The evaluation of social innovation schemes in pervasive technology presents a number of challenges. The first challenge is one of resolving how to deal with "time", that is finding the "right" time to evaluate a social innovation based initiative. This problem can be partially resolved depending on how the social innovation scheme has been initiated. If it started with a local authority or similar structured organisation then there may well be a clearly identified start date. This would enable a date for evaluation to be identified. Alternatively if the social innovation scheme has been more organic in development and it is less easy to define the start date then it is more problematic to set an appropriate date for evaluation.

   Similarly with social innovation based schemes there is the expectation and possibility that they will change and evolve as they grow and mature. Once again this makes the date of the evaluation less obvious to determine. Obviously it can be argued that the appropriate date for evaluation could be based on "exposure". Using the exposure of the individuals involved in the social innovation based scheme would require decisions about the definition of exposure and decisions about the level of involvement or participation. In a social innovation based scheme an individual could be a long standing registered participant but also passive, not taking an active part in the scheme.

   This brings us to the second set of challenges associated with social innovation based schemes, which is that of "capturing" the type of involvement each individual has and how the sum of the individual parts creates social value. As alluded to above, some participants may get "value" simply from observing the participation of others for example altruistic feelings towards bike sharing schemes, alternatively they may be able to use information that others generate for example congestion estimates in Waze, without participation or contributing of information themselves. So the challenge in evaluation is to capture the "value created" and the involvement of the participants.

A further challenge is to capture the role of participants in social innovation based schemes. These can be roles such as influencers, enablers, early adopters, and passive participants. This is because the actual activities of individual participants and the relations they have between each other may be a determinant of the success or failure of social innovation based schemes. The challenge is to identify those participants who are most influential, who by their actions or engagement with social media influence others to join a scheme or to engage in particular activities thus generating "snowball" effects and for example, making ridesharing seem a viable alternative to driving into work or the use of shared e-scooter a fun way to take the final leg of the journey. A further challenge is to determine the extent of the influence of specific combinations of individual participants and specific actions at influential times. Added to that a challenge is to establish what would constitute a "critical mass" for any social innovation based scheme particularly if the scheme is part of the "sharing economy". For example, the New York bike share scheme seems to have underestimated the number of casual, one-off users such as visitors to the city.

The technological developments enabling the growth and diversity in social innovation based schemes in the transport sector that are of most interest are: location based services, social media use and strategies for building trust. This latter development being an integral key to schemes within the "sharing economy" and policy developments such as "Mobility as a Service". One of the greatest challenges in evaluation of social innovation based schemes is the measurement of "trust".

6.4.3 *Evaluation proposal for social innovation schemes in transport*

It follows that an evaluation proposal for social innovation schemes could decide to focus entirely on either outcomes or process indicators or some combination of the two.

1. Outcome indicators would have to focus on travel behaviour and transport related choices. Changes on a before-and-after basis in travel choices (e.g. mode, route and distance by mode) which can then be converted into energy, carbon, equity, health, safety and other impacts for the scheme
2. A focus on process would be a (productive) focus on identifying those factors that have created success or failure in the schemes. The success of the scheme may be determined by whether the post-scheme' travel choices involve the sharing of modes or use of communal assets (such as shared bikes) or involvement in buddy/peer support related travel.

Any evaluation framework for social innovation based schemes within the transport sector will need to exploit location-based services (LBS) in pervasive technology in order to estimate travel behaviour, e.g., walking kilometres, passenger kilometres or cycling/riding kilometres.

The specific challenge with social innovation schemes is to determine when someone is travelling by car but is also ridesharing, or using another car sharing scheme. It is possible to identify a number of ways to refine LBS to identify if car travel is in a carsharing or ridesharing car scheme. The first is to exploit the unique identifiers in Bluetooth and mobile phone technologies to coordinate between cars, and individuals within a car sharing or ridesharing scheme. The aim being to enable the automatic "track and trace" of cars and individuals within a car and ride sharing scheme. This will enable coordination and identification of use of ride or car share and vehicle occupancy. The collection of this type of data collection requires a corresponding "social" element in that individuals need to agree to their movements being recorded and used in subsequent analysis and outputs.

An alternative is more interactive (and potentially invasive) and involves engaging with the social media accounts individual's may have, e.g. X or Instagram or Facebook, according to their permission settings, to ask the individual how they are travelling. Once again "track and trace" has to be used, but coordinated and automatic functionality should be explored and utilised so that questions using social media technology are generated by "track and trace" thresholds and indicators and sent automatically.

"Track and trace" functionality goes some way to solving the issue of the right "time" to evaluate in that it allows the generation of data over a period of time with relatively little cost impact or effort burden for the respondent. However, "track and trace" applications using LBS must be downloaded by the participant and there has to be active consent to be traced. Nevertheless with sufficient knowledge and foresight it should be possible to sign up a community of scheme users and vehicles, bikes or e-scooters (as appropriate) for the collection of data over the individual's scheme membership lifetime. Such an initiative would have the potential to provide a substantial dataset for evaluation.

There are additional aspects of complexity to consider when using "track and trace" functionality, particularly whether route-based information and time stamps are required. There are still errors associated with the geo-location function though, which need to be considered in the interpretation of the individual traces produced and which can cause errors in estimating mode choice.

Capturing data on participant's actions and roles requires other tools and techniques. For this we suggest that future evaluation frameworks explore the potential of social media analytic software such as Hootsuite. The software focuses on content analysis of textual data from posts on Facebook or X, including sentiment analysis. Sentiment analysis allows some understanding of the "feeling" behind the content that the individual posts. However, there are fundamental issues with this type of data, including, (a) it is not necessarily contemporaneous, and (b) the required understanding of the lexicon including shortened phrases is onerous. In addition the analytical software available is not yet sufficiently evolved to precisely analyse images and pictures, rather than textual data, which is a drawback given that an increasing volume of social media content is in the form of images and video. However, the principal problem is that analysis of the social media presence and content, the 'digital footprint', of an individual does not necessarily reveal their *actions* within a social innovation scheme unless the actions are specifically reported. Nor does it reveal much about the link between the individual, the actions and the importance or serendipitous nature of timing – unless an individual proactively posts a narrative to reflect this.

Sophisticated data harvesting and analysis techniques are required to reveal an individual's actions within a social innovation scheme and underpin an evaluation of impacts from the scheme. Existing techniques such as social network analysis could be used to uncover an individual's influence within a social network. Social network analysis would be able to demonstrate such factors as density and centrality within a social network highlighting the importance of individual characters. Combinations of "track and trace", social media content analysis, sentiment analysis and social network analysis could all be undertaken without additional burden on the participant, but would require informed consent and could be quite revealing about the day-to-day activities of any one individual and therefore a sensitive data stream. However, social network analysis is not an exact way to explore or uncover the role and actions of individual's and it is quite possible that additional social survey methods would be needed.

The final element in the proposal for the evaluation of social innovation based schemes is to develop new techniques for the gathering of dataset on "trust". One approach would be to begin by exploring the potential for using existing data such as the "rate my driver" or "rate my passenger" functions, examples being those used in UBER, using existing social media analytic software, e.g. GoogleAnalytics or HootSuite.

In summary, then the principal suggestions for the evaluation proposal starts with the identification and exploitation of existing datasets closely followed by a principle to use existing "off-the-shelf" analytical software packages and to combine those with bespoke surveys that fill the gaps. The primary gaps are around the lack of "social" data: activities, social roles and "trust". It follows that an implicit principle is to use a mix of methods allowing quantitative and qualitative dataset generation.

6.5   Overview and summary remarks

The advent of new pervasive technology has enabled a new tranche of ITS schemes to develop, based on personal devices and often employing social functionality alongside functions that are more clearly transport focused. As with established, fixed-based ITS, the possible permutations of technology configurations and functionality is vast. In this chapter we have focused on two particular types to illustrate how evaluation methodology should now evolve – new mobility schemes and social innovation schemes.

In Table 6.1, a summary comparison between these two new types of scheme and established ITS is provided, drawing out the main features and areas of difference.

Table 6.1 Comparison of main impact categories for fixed-based ITS, new mobility scheme and social innovation scheme

Examples of fixed-based ITS (FB-ITS) impacts	Corresponding examples for new mobility scheme (NMS)	Corresponding examples for social innovation (SI) scheme	Summary comparison
Costs: Highways Authority/ stakeholder Investment costs, Finance, Operating and maintenance costs	Costs: Individual traveller investment in technology, Transport authority/third party costs and benefits (e.g. provision of incentives), investment and maintenance of "city dashboard", staff resource for support/CRM	Costs: investment to establish and maintain a social forum to facilitate sharing and peer support. Costs of owning and operating particular modes now distributed between individuals or communal owners/ consortia	Shift in costs from centralised investment in FB-ITS to individual and local/regional costs for NMS. New business models needed to evaluate the value chain and impacts for third parties, e.g. in data supply, incentives, establishing a community/social forum.
Efficiency: travel time, travel time reliability	Efficiency: Individual travel time and travel time reliability. Particular NMS are focused on route change and departure change, implying the need to monitor both over time and in a wide spatial dimension.	Efficiency: Individual travel time and travel time reliability. Particular SI schemes are focused at reducing total travel demand through sharing, reflected by reduced congestion related indicators with large scale uptake	Individuals' observed departure and arrival time for whole journey possible through mobility profiling with NMS and SI. Efficiency impacts may be distributed/re-distributed over a wider spatial area than directly observed. Observed FB ITS efficiency impacts for instrumented portion of journey and instrumented spatial area only.
Safety: accidents, speeding/ compliance	Safety: exposure to accidents through mode choice and route choice. Personal security impacts through exposure.	Safety: exposure to accidents through mode choice and route choice. Personal security impacts through both travel related exposure and sharing of information in community forums.	Bidirectional information flow in NMS allows data collection on contextual perceived safety and low level/unreported security incidents. Exposure through posting personal information on social network forums or through connecting with peers/buddies difficult to assess other than through self-reporting and perceived risk.

Environmental: local environment (emissions) + global (climate change), habitat disruption	Environmental: individual carbon equivalent, energy + ICT carbon costs/benefits through mode switch	Environmental: individual carbon equivalent, energy + ICT carbon costs/benefits through mode switch and sharing. Need to assess the level of sharing to enumerate.	FB ITS employs a mixture of observed roadside emissions and modelling from traffic conditions. NMS and SI uses modelled/ estimated emissions only based on mode choice and either vehicle/ mode characteristic assumptions or user input vehicle information.
System/operation: throughput, lane behaviour (weaving and swooping), junction hopping, unauthorised hard shoulder running	System/Operation: unlikely to generate or effectively detect these impacts from NMS	System/Operation: unlikely to generate or effectively detect these impacts from NMS	Speed impacts may be detectable through location tracking with NMS, however micro lane movements best detected through FB ITS
Social and others: liveability (residents), driver burden/stress	Social and others: Equity, health/ wellbeing, comfort, peer/community standing, community inclusion	Social and others: Equity, health/ wellbeing, comfort, peer/community standing, community inclusion	A wide range of additional impacts are possible for NMS and SI depending on the scheme definition and geo-context

This comparison is also intended to highlight the interface with the standard evaluation paradigm (typically a cost–benefit approach supplemented by an environmental and safety assessment), the shift in impacts and the ways in which the traditional evaluation approach should be extended.

A final point to consider is that this is a time of growing understanding of the usefulness and commodification of data. Simultaneously, individuals are becoming more aware of the "value" in their own data and increasingly sensitive to the "sharing" of data. Increasingly it is important to include this understanding in any consideration of ethical practice in evaluation research. Any evaluation framework wishing to use datasets generated by recording individuals in everyday activities, such as mobility "track and trace" Apps or social media input, has an obligation to question the extent to which the individual can "own" or use and have access to the dataset generated from their own activities and actions. In addition, there is an obligation to explore the extent to which this "sharing" of the "trace" can form the basis for individuals to engage and participate in research.

References

[1]   Grant-Muller, S.M., Gal-Tzur, A., Minkov, E., Kuflik, T., Nocera, S., and Shoor, I. (2014) Enhancing Transport Data Collection through Social Media Sources: Methods, Challenges and Opportunities for Textual Data, *IET Intelligent Transport Systems*, 9(4): 407–417. doi:10.1049/iet-its.2013.0214

[2]   Harrison, G., Grant-Muller, S.M., and Hodgson, F.C. (2020). New and Emerging Data Forms in Transportation Planning and Policy: Opportunities and Challenges for "Track and Trace" Data. *Transportation Research Part C: Emerging Technologies*, 117: 102672. https://doi.org/10.1016/j.trc.2020.102672

[3]   Cottrill, C.D. (2009). "Approaches to Privacy Preservation in Intelligent Transportation Systems and Vehicle-Infrastructure Integration Initiative" *Transportation Research Record* (2129): 9–15.

[4]   Žilina, U. (2009). Present and Future Challenges of ICT for Intelligent Transportation Technologies and Services. *2009 1ST International Conference on Wireless Communication, Vehicular Technology, Information Theory and Aerospace & Electronic Systems Technology*. 1(2): 112–115.

[5]   Lee, W.-H., Tseng, S.-S., and Shieh, W.-Y. (2010). Collaborative Real-Time Traffic Information Generation and Sharing Framework for the Intelligent Transportation System. *Information Sciences* 180(1): 62–70.

[6]   Deakin, E., Frick, K., and Skabardonis, A. (2009). Intelligent Transport Systems – ACCESS 34 (Spring 2009), 1(34): 29–34.

[7]   Levinson, D., and Chang, E. (2003). A Model for Optimizing Electronic Toll Collection Systems. *Transportation Research Part A: Policy and Practice* 37 (4): 293–314.

[8]   Boyles, S. D., Kockelman, K.M., and Travis Waller, S. (2010). Congestion Pricing under Operational, Supply-Side Uncertainty. *Transportation Research Part C: Emerging Technologies* 18(4): 519–535.

[9] Li, T., Chen, P., and Tian, Y. (2021) Personalized Incentive-Based Peak Avoidance and Drivers' Travel Time-Savings. *Transport Policy*, 100: 68–80.

[10] Huang, B., Thomas, T., Groenewolt, B., Claasen, Y., and van Berkum, E. (2021). Effectiveness of Incentives Offered by Mobile Phone App to Encourage Cycling: A Long-Term Study. *IET Intelligent Transport Systems*, 15(3): 406–422.

[11] Shaheen, S., Cohen, A., Chan, N., and Bansal, A. (2020). Sharing Strategies: Carsharing, Shared Micromobility (Bikesharing and Scooter Sharing), Transportation Network Companies, Microtransit, and Other Innovative Mobility Modes. *Transportation, Land Use, and Environmental Planning (Elsevier)*. 237–262.

[12] Grant-Muller, S., Yang, Y., Panter, J., and Woodcock, J. (2023). Does the Use of E-Scooters Bring Well-Being Outcomes for the User? A Study Based on UK Shared E-Scooter Trials. *Active Travel Studies*, 3(1): 1–21.

[13] EU SUNSET: Accessed 5/9/24. https://cordis.europa.eu/project/id/270228

[14] Phillips W., Lee, H., Ghobadian, A., O'Regan, N., and James, P. (2015), Social Innovation and Social Entrepreneurship: A Systematic Review, *Group and Organization Management*, 40(3) 428–461.

[15] Mulgan G. (2006) The Process of Social Innovation. *Innovations: Technology, Governance, Globalisation*, 1: 145–162.

[16] James, A., Deiglmeier, P.Jr.K., and Miller, D.T. (2008). *Stanford Social Innovation Review*, Fall 2008. 6(4): 34–43. http://ssir.org/articles/entry/rediscovering_social_innovation#sthash.5JWJ5aQc.dpuf

[17] Shaheen S.A. (2013) Introduction Shared-Use Vehicle Services for Sustainable Transportation: Carsharing, Bikesharing, and Personal Vehicle Sharing across the Globe, *International Journal of Sustainable Transportation*, 7(1): 1–4. DOI:10.1080/15568318.2012.660095

Part III

Intelligent transport systems evaluation results

Chapter 7

The potential benefits of heavy goods vehicle (HGV) platooning

Richard Cuerden[1], Chris Lodge[1] and Matthias Seidl[1]

7.1 Introduction

The UK's first real-world trial of Heavy Goods Vehicle (HGV) platooning ran between 2017 and 2022 and was called HelmUK [1]. The program was funded by National Highways and the UK's Department for Transport. HelmUK used vehicle to vehicle (V2V) communications and Advanced Driver Assistance Systems (ADAS) to enable three HGVs to platoon and safely minimise the distance between each other to as low as 19.6 m, when travelling at 89 km/h (55 mph) on motorways. The aim of the project was to conduct a comprehensive evaluation of the benefits and risks of the platooning concept using information gathered from real-world operational trials.

HelmUK found that platooning saved small amounts of fuel compared to a like by like comparison with the same three vehicles using Adaptive Cruise Control (ACC) and transporting the same freight loads on the same motorway routes. Further analysis found that on a road network optimised for platooning, fuel savings could increase to between 2.5% and 4.1%. HelmUK also found that platooning was as safe as ACC operation, when risks associated with merging vehicles at junctions are managed.

This chapter describes the purpose, approach, design, and results of the HelmUK trials covering road safety, fuel savings, emissions, effects on the road network, and economic benefits. It also discusses the future of platooning for road operators, governments, and the freight industry. Finally, the chapter makes a series of recommendations for platooning development and deployment. These include consideration of regulation of low headways at junctions, and a strong recommendation to deploy the underlying platooning systems, at more typical ACC and larger headway driving conditions, where they offer safety benefits.

7.2 What is platooning?

Platooning uses wireless communication technology and ADAS to allow vehicles to safely travel much closer to one another than would be safe with human drivers

---

[1]TRL Ltd. (Transport Research Laboratory), Wokingham, UK

alone. The aims are to reduce fuel consumption and emissions by making use of the slipstream effect, to improve safety at close following distances thanks to coordinated automatic braking between the vehicles, to reduce congestion by improved road space usage, and to reduce the workload of the non-lead drivers by automated acceleration and braking (longitudinal control).

Platooning is not driverless technology, although these vehicles are called cooperative and automated vehicles (CAV). The driver of the lead vehicle controls the speed, acceleration and braking of the whole platoon. The drivers of the following vehicles are required to steer their vehicles manually (lateral control) and must be always ready to take over, and if necessary, leave the platoon. In contrast to a single tractor pulling more than one trailer, platooning allows vehicles to connect and disconnect at any time during the journey, enabling vehicles to reach their own delivery destinations.

The concept of HGV platooning has been researched for over two decades and in 2022, seven HGV manufacturers and the European Commission collaborated on the ENSEMBLE (ENabling SafE Multi-Brand pLatooning for Europe) project [2], to pave the way for interoperable platooning. The project defined two different ways of multi-brand platooning, of which one, "platooning as a support function", was demonstrated, whilst reflecting the full diversity of trucks.

Test-track studies have shown fuel benefits as high as 7% for a lead vehicle and 16% for a following vehicle demonstrated by the SARTRE (Safe Road Trains for the Environment) project [3]. However, these were achieved in highly controlled conditions, driving at short headways without any interruptions. A real-world road trial was required to measure any benefits in real operational use and to investigate whether they can be realised without negative consequences.

The road freight industry is critical to the UK economy. England's Strategic Road Network (SRN) accounts for 2% of all roads by length and carries 66% of all freight. This makes the combination of road freight and the SRN an important enabler for economic growth. HelmUK assessed the potential benefits of HGV platooning, especially focusing on decarbonisation and improving safety.

7.3   Trial design

The HelmUK trials were funded by National Highways and the Department for Transport (DfT) to thoroughly evaluate the real-world benefits and impacts of platooning and what this means for the future of the technology in the UK. The project was carried out by the HelmUK consortium, formed of program lead, TRL and the partners: Apollo Vehicle Safety, Connected Places Catapult, Costain, DAF Trucks, DHL, fka, Fusion Processing, Ricardo, UTAC, TNO, TransportPR, VisionTrack, and ZF. The project consisted of three phases (see Table 7.1)

The road trials were conducted using three articulated HGVs fitted with a prototype DAF platooning system, which was released for a field trial after tests by

Table 7.1 Three phases in HelmUK

Phase No.	Period	Description
Phase 1	August 2017 to March 2020	Detailed planning, design and build: The development and validation of the DAF platooning system and TRL safety case including driver training and track trials
Phase 2	October 2020 to August 2021	Network familiarisation: On-road operation and system validation on the approved route
Phase 3	September 2021 to March 2022	Road trials: Real-world trials and data collection followed by data analysis and final reporting

Note: In Phase 1, the project was paused from March to October 2020 due to the global COVID-19 pandemic and UK government restrictions.

DAF and TRL. The key features of this platooning system, which allowed safe operation at close following distances, are:

1. **Cooperative adaptative cruise control (CACC)**, which keeps the distance between the vehicles constant using short-range V2V communication.
2. **Brake performance estimator**, which adjusts the distance between the vehicles based on vehicle weight to ensure safety no matter the load or order of the trucks.
3. **Cooperative collision avoidance**, which ensures coordinated automatic emergency braking.
4. **Lane keeping assistance (LKA)**, which supports the driver's steering to prevent vehicles from veering out of lane.

The key consideration from the start of planning for a real-world platooning trial was the necessity to carefully manage the significant safety risk associated with permitting three HGVs to travel in close proximity at speeds of up to 89 km/h. The safety of all road users was paramount and the study design to monitor and evaluate changes in fuel efficiency, sustainability, and safety was governed by a comprehensive safety case, which followed National Highways' GG104 [4] risk assessment framework, and the associated internal safety review boards.

In addition, TRL arranged an independently chaired ethics committee (Chaired by Professor Paul Jennings from Warwick University), and an independent specialist Safety Reviewer (Professor Graham Braithwaite from Cranfield University). For such complex trials of new technology, a large part of the preparation for on-road trials is the development of a safety case. The project benefited from invaluable safety expertise and advice afforded by many individuals from across the project consortium and Advisory Committee.

Potential safety risks covered a broad range of areas including the platooning technology, operating procedures, driver selection and training, infrastructure and characteristics of the route, operating conditions, the behaviour of other road users, and emergency response and more.

This safety case ensured that HelmUK met its objectives, while pro-actively managing risk to keep the trials safe for all. The comprehensive safety case also enabled HelmUK to record a wide range of safety indicators and apply a full system approach and to confidently conduct the trial, with the acceptance that new knowledge and learnings were constantly added to the Risk Register and assessed. The Risk Register was a living document that evolved throughout the project as new potential hazards were identified and evaluated.

The development of the safety case for the HelmUK platooning system guided the selection of the trial route, the assessment of junctions which were safe for platooning (about one in every ten junctions), and the operational procedures which supported the trial. The safety case took inputs from simulator studies and driver training to ensure a best-in-class approach to risk management. This included daily safety monitoring and a "GO or NO-GO" decision at the end of each day on the road. As part of this daily safety monitoring, over 4000 separate events were reviewed, and four incidents were escalated through the safety management procedure for review with the client.

When on the road, vehicles would engage platooning where it was safe to do so and only on the approved route. Based on the safety case, some junctions were considered suitable for platooning operation while others were not. When platooning, the trucks travelled in lane 1 only, 0.8 seconds apart (ca. 19.6 metres at up to 89 km/h).

The trial route was on the M5 and M6 motorways between Avonmouth and Stafford (and then back again) in England. The total round trip was 351 km. The route was chosen to be as representative as practicable of UK motorway freight journeys. A key road feature of the route was the number and density of junctions. For motorways in England the average distance between junctions is 5.56 km, and for the HelmUK route, the average distance between junctions was very close to this, at 5.59 km.

7.4   Phase 3: Road trial

The design of the road trials needed to balance the requirements of a wide range of research questions and data sources to achieve those objectives. The key considerations when designing the trials were: (1) enabling a comparison between platooning and non-platooning operation; (2) maximising data consistency in real world conditions; and (3) ensuring the trials remained safe.

7.4.1   Comparison between platooning and non-platooning

The main element which was deliberately changed during trials was whether platooning functionality was used for a journey. This was varied per day, so a trial day was either "platooning" or "non-platooning". On a "platooning" day, platooning operation was permitted where the drivers deemed it safe, and where it was permitted by the safety case, i.e. the vehicles did not spend all the time on a platooning day in platooning operation.

This was compared with non-platooning operation. On these days the drivers were encouraged to drive as they normally would. The vehicles travelled together and made use of all the driver assistance systems on the vehicle such as ACC.

It was considered in the trial design if use of ACC should be discouraged. It was decided that this would provide an unrealistic comparison between platooning and non-platooning and limit the insights for the future of platooning in the UK. Therefore, the non-platooning baseline was set using new HGVs equipped with ACC and drivers were encouraged to drive as they normally would.

Figure 7.1 illustrates the three trial vehicles travelling in platoon and non-platoon formation on the motorway. For platooning, the acceleration and braking of Helm02 and Helm03 were automated to maintain a constant time-headway to the preceding vehicle; steering was manually controlled with support from a LKA system.

7.4.2   Fixed parameters

A number of variables were controlled during the trials in order to give consistency to the data. The variables kept constant are summarised in Table 7.2.

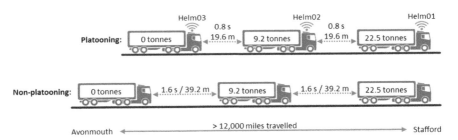

Figure 7.1   HelmUK trial formation, payloads and vehicle spacing for platooning and non-platooning operation at 89 km/h

Table 7.2   Variables constantly used in HelmUK trials

Variable	Description
Number of trial vehicles	There were three tractor units outfitted with platooning equipment and there were always three vehicles in the platoon
Position of the vehicles in the platoon	Helm01 always led followed by Helm02 and Helm03
Vehicle load	Each vehicle was loaded so that the combined gross combination weight of the entire platoon represented the average UK load carried by comparable vehicles on domestic freight journeys. The loads were kept constant throughout Phase 3: • Helm01: 22.5 tonnes/37.2 tonnes (payload/gross combination weight) • Helm02: 9.2 tonnes/23.9 tonnes • Helm03: 0 tonnes (unladen)/14.7 tonnes

7.4.3   Variable parameters

HelmUK's core aim was to conduct trials in a real-world setting. Therefore, there were a number of variables which could not be controlled for during the trials, but which can be controlled when analysing the data. These included, weather, wind direction, speed, acceleration, gradient, and driver.

The route was largely oriented in north-south direction. The prevailing wind direction during the road trials was from the south-southwest at Staverton weather station (nearer the south of the route), and from the west at Birmingham weather station (nearer the north of the route). The average wind speed was between 3 and 4 metres per second (approximately 8 miles per hour).

7.4.4   Duration and statistical significance

Phase 3 aimed to collect around 30 days of on-road data; the number of days planned on-road was greater than 30 to provide contingency. To compare the modes of operation the trials targeted an equal number of platooning and non-platooning days. A break point was agreed approximately halfway into the trials where the statistical significance of the fuel benefit data was examined to determine if any change to the duration of the trials was needed.

The interim statistical test on the fuel consumption data showed small differences between platooning and non-platooning days at the overall journey level. This small difference meant that to achieve 95% confidence in the difference between platooning and non-platooning would have required thousands of journeys, which was beyond the scope of the HelmUK project. The original target of 30 days remained the focus for data collection and a larger emphasis on statistical modelling was required for data analysis.

7.4.5   Data sources

HelmUK's research relied on data captured by several systems during the trials. The project team used the VisionTrack integrated camera and vehicle telematics platform for day-to-day trial management. VisionTrack captured all of the video data for HelmUK and its principal use was in daily safety monitoring activities, where any flagged incidents could be reviewed using this video footage. The key data from VisionTrack was:

1. Video data for all three vehicles – from five cameras per vehicle with forward facing, rear facing, left side, right side and driver facing views. This data was used for incident review and safety monitoring as well as research data.
2. Event marker button presses – these allowed the drivers to mark events, while driving, by pressing a button on the dashboard. These events were then reviewed daily by the safety team.
3. Driver behaviour flags – these are generated by VisionTrack using the driver facing camera. The flags which were reviewed each day for this project were smoking, phone use, and lane departure warning.

TRL worked with DAF to identify the key Controller Area Network (CAN bus) signals that would be needed for HelmUK. These ranged from platooning status to vehicle speed. Fusion Processing data loggers were used to collect and store the CAN signals both for research and safety assurance activities. The key data from the data loggers were:

1. CAN signals – all relevant CAN signals were captured from the trucks at 10 Hz, this hugely rich data source enabled both research and daily safety monitoring (this process is further described in [1], Section 2.7.1).
2. Fuel data – the data loggers also captured the data from the fuel meter fitted to the vehicles as part of the trials (see [1], Section 2.6.3).
3. Calculated fields – a series of more complex data fields were calculated by the data logger based on the raw CAN signals or fuel input. These fields captured elements such as culminative fuel consumption and braking magnitude.
4. Safety critical signals – within the data above (primarily the CAN signals) there was a subset of data which was critical for our safety review process; these flags were reviewed daily along with associated video footage to capture any near misses or other events.

To provide accurate fuel consumption measurements, each of the three trial vehicles was equipped with a mechanical precision diesel consumption flow meter, type AIC-4008 Veritas. To measure the true diesel consumption, the fuel return flow from the injectors was switched from the tank directly to the fuel supply line. This measuring system provided 800 pulses per litre to determine instantaneous and cumulative fuel consumption values, which were captured by the Fusion Processing data loggers.

In the HelmUK project there were two distinct on-road activities, Phase 2 and Phase 3. The on-road aspects of the HelmUK project were the focus of the program and this section captures the key achievements and challenges of the project during the on-road activities. Across the project, there were 58 days on-road covering over 19,000 km (12,000 miles) and capturing 12 million lines of data and 4000 hours of video footage.

7.5   Trial findings

7.5.1   Safety

HelmUK evaluated a range of road safety factors associated with platooning as part of the trial. Surrogate Safety Measures, which are proxies for safety, were calculated or measured from vehicle data. A simple example of a Surrogate Safety Measure is "harsh braking" over a certain threshold. The research considered: (1) the safety of the vehicles themselves using data from the HelmUK trucks to monitor Surrogate Safety Measures; (2) the impact on other road users (focusing on junctions) using vehicle data, video footage, and traffic flow data; and (3) the impact on driver workload from platooning operation.

Overall, HelmUK found that platooning is at least as safe as ACC when risks at junctions are controlled. The safety findings are summarised in Table 7.3.

Table 7.3   Summary of safety findings

	The DAF platooning system was fail-safe in road trials as demonstrated in Phase 1 of the project.
	Platooning kept good control over the position of the vehicles relative to one another; no time was spent closer than 0.8 s which did occur in non-platooning operation.
	Surrogate Safety Measures taken from vehicle data evidenced that, overall, the vehicles were safer in platooning mode.
	Cooperative Collision Avoidance functionality reduced collision risk on a few occasions; however, there were some activations which could have posed a risk to following vehicles.
	Platooning is not expected to introduce new collision types and is predicted to have a small beneficial effect on common, existing HGV collision types overall.
	Driver workload did increase slightly due to platooning, but only by a very small amount – overall, driver workload was not substantially impacted.
	For junctions which were platooned through, platooning did not increase risk to other road users or disrupt traffic flow.
	The residual risk for platooning systems is conflict with merging vehicles at junctions, which is caused by close following distances.

7.5.2   Fuel efficiency

Comparisons were made to calculate platooning fuel savings that can be realised over existing driving technology. HelmUK found that the measured fuel savings across three vehicles were 0.5% in the trial. Statistical modelling showed no evidence that fuel consumption was significantly different to trucks using ACC. Analysis confirmed that the number of junctions which could not be safely platooned through was the main reason for the low fuel savings. The vehicles could spend only 53.5% of their driving time in a platoon.

Junctions which cannot be platooned through required the vehicles to break the platoon. When the platoon is then reformed, the vehicles accelerate which costs fuel. The more often vehicles do this, the lower the fuel benefits overall.

To further explore the potential of platooning, two optimised scenarios were created from the HelmUK data to represent possible future scenarios for platooning deployment. The first looked at a route where all junctions were suitable for platooning, and the second looked at areas of the route where a high degree of platooning occurred. The measured fuel savings across three vehicles for these two scenarios rose to 1.7% (at 74.7% of time in platoon) and 1.8% (at 85.7% of time in platoon), respectively. Further statistical modelling performed on these two scenarios, which accounted for other variables on fuel consumption (such as traffic flow, journey direction, platooning states, weather, etc.), showed that the fuel savings of platooning alone were marginally greater than the fuel savings directly measured at 2.5% and 2.6%, respectively.

Finally, it was investigated what platooning could achieve in perfect conditions, where there was completely uninterrupted platooning. In this scenario, fuel measurements from HelmUK showed that platooning could produce fuel savings of 4.1% across three vehicles.

7.5.3   Regulated emissions

Based on the fuel savings identified in HelmUK, the following reductions of nitrogen oxide and particulate matter emissions can be expected for a three-vehicle platoon compared to ACC operation, depending on vehicle load, driving speed and associated base fuel consumption:

(1) Current road network:	NOx: 1.0% to 1.1%; PM: 0.7% to 0.8%
(2) Optimised road network:	NOx: 5.3% to 5.5%; PM: 3.6% to 4.0%
(3) Uninterrupted platooning:	NOx: 8.4% to 8.7%; PM: 5.7% to 6.8%

Adding more than three vehicles to a platoon could increase the average reductions beyond these levels. The "optimised road network" value reflects a situation where high proportions of platooning could be achieved on the network. The "uninterrupted platooning" figure reflects a hypothetical upper boundary of what could be achieved if uninterrupted steady-state platooning could be realised.

7.5.4   Freight industry readiness for platooning technology

Engagement with the freight industry found that while operators would be willing in principle to consider making operational changes to facilitate platooning, they identified concerns about the operational costs and commercial risks arising from rescheduling trips to enable platooning.

The fuel savings found in HelmUK are below the level which most operators considered worthwhile and are lower than other fuel saving interventions that the industry does invest in (which are operationally less complex and deliver more predictable results).

The very low savings for the lead vehicle could discourage different operators from platooning together. This aspect, and low fleet penetration, is likely to rule out any "open access" platooning model in the short term. In this model, vehicles platoon opportunistically with nearby vehicles in the course of a trip, rather than pre-planning platooning within a single operator's own fleet, or close partners.

Operators should be aware that the order of the vehicles by weight is likely to affect the minimum achievable headway so some coordination may be required to optimise fuel savings – but this is impractical for an ad-hoc platoon.

Platoons of only two vehicles may not be viable, as only one of a pair might actually be saving fuel, so operators would have the greater complexity of needing to find at least three vehicles whose trips can be combined. Consideration has been given to a payment structure to distribute benefits evenly between operators participating in a platoon, but this is believed to be too complex to justify the level of benefits to be redistributed.

Economic analysis showed that a positive business case is likely to be made only for a very small number of operators with quite specific favourable circumstances. It would be necessary for such an operator to undertake regular long-distance trunk haulage trips between fixed locations located close to a motorway, and where vehicles can be relatively easily grouped without significant rescheduling being required.

7.6   What does this mean for the future of platooning?

Overall, the business case for platooning is currently weak – the fuel benefits over ACC operation are negligible in a real-world setting at 0.5%. HelmUK's research has shown that the primary reason for this is the safety requirement to disband the platoon at 9 out of every 10 junctions along the route. With these fuel benefits, meaningful uptake from the freight industry is considered unlikely because HelmUK's stakeholder engagement found that a minimum fuel saving of 2% would be required for operators with large fleets to invest in platooning technology.

Therefore, unless more junctions on motorways (and other motorway-standard roads) are compatible with platooning operation, the fuel benefits from platooning technology will be hard to realise with current generation systems. Most traditional civil engineering-based ways of achieving this, such as changes to junction design to accommodate platooning or introducing dedicated lanes for platooning, are likely to require significant investment, but they could produce fuel savings in the 2.6–4.1% region across three vehicles.

The business case for this level of investment to support platooning is weak in isolation. However, as English Motorways and the wider SRN are made more "CAV-ready", merge junctions are likely to be a key pinch point for safe CAV deployment. Platooning should be considered as a use case when assessing potential changes to junctions and road layouts. Fuel savings could be further improved by:

1. **Reducing the distance between vehicles** – HelmUK initially targeted a headway of 0.5 seconds, but the variability of brake performance in real-world conditions meant this had to be increased to 0.8 seconds. Future vehicles might be able to reduce this headway, but this will be challenging with a mixed-age fleet.
2. **Increasing the number of vehicles in a platoon** – HelmUK highlighted risks at junctions from a three HGV platoon, and subsequent research has found that the longest platoon possible for the most suitable junction on the HelmUK route was likely to be no greater than four HGVs.

In contrast to fuel savings, it is plausible that safety benefits of the systems which enable platooning would also be realised at larger headways. These technologies could be introduced without following at close distances regardless of the future success of the platooning concept. The two main recommendations from HelmUK based on 5 years of research and more than 12,000 miles travelled are given below.

7.6.1   Benefits of platooning technology without low headways

As well as being the first on-road trial of platooning in the UK, HelmUK is believed to be the first on-road trial of cooperative adaptive cruise control and cooperative collision avoidance, these systems enable safe platooning operation using vehicle-to-vehicle communication. These systems offer safety benefits over existing systems. HelmUK has proven this functionality using DAF vehicles and the European ENSEMBLE project has proven this functionality across multiple manufacturers.

These systems could be deployed in the near term when sufficiently developed. If implemented without allowing operation closer than the lowest ACC settings (1.6–1.4 s), they require no additional regulation and do not increase risk at junctions. However, benefits realisation will be dependent on the market penetration of these systems.

7.6.2   Potential for future ITS applications to enhance platooning performance

A key characteristic of the trial was the requirement for drivers to manually disband their platoon at "non-platoonable" junctions. This was a reasonable approach because HelmUK used specially trained drivers, in three dedicated HGVs, on the same route each day. The drivers also had inter-vehicle voice communication which permitted them to coordinate their actions. The comprehensive safety management of the trials ensured disengagement was achieved.

Assuming that significant infrastructure changes are not made to junction designs and/or dedicated lanes are not made available for platooning vehicle fleets, it will be necessary for a potential future commercial deployment of platooning to incorporate an automated headway increase functionality. HelmUK has identified three ways, or via a combination of methods, that this could be achieved. This would be applicable for the successful deployment in the UK and in countries with similar road design and traffic conditions.

Junction database: HelmUK captured the characteristics of the junctions on the approved trial route, assessing their type, slip-lane length, and other factors to determine risk. The same static feature assessment approach could be applied for the network suitable for platooning. This would provide a database of "platoonable" and/or "non-platoonable" junctions for vehicle and system manufacturers and operators, which would naturally need to be maintained and updated.

On-board systems: Cameras and other sensors could be used to inform AI systems to recognise road features and signage, which would initiate an increase in headway prior to vehicles approaching a junction. However, this approach without additional data input and associated algorithms would not be capable of safely differentiating between platooning and non-platooning junctions. This would mean it would be required to disband the platoon at all junctions. However, this system would not be reliant on outside data sources.

Vehicle to Infrastructure (V2I) communication: This ITS approach would permit decisions about platooning operation and whether to disband at junctions to be made based on real time traffic information. If there was no traffic present on the slip road, either naturally or because of additional traffic signals such as ramp metering temporarily stopping traffic to allow the platoon to safely proceed through the junction, and this information was shared with the platoon, it would be reasonable to not disband and to maximise the efficiency of the cooperative journey. The costs of V2I systems would be significantly less and more practicable than physical infrastructure changes to junction design. However, it would require road authorities to install V2I equipment at junctions with data feeds from existing sensors or potentially with additional sensors added. A standard international V2I communication protocol applicable for platooning would be essential.

Consideration would need to be given as to whether it is possible to sense individual vehicles on individual slip roads and to communicate this to a platoon in time, it is likely that this will vary based on the nature of the local design. Therefore, it is logically to seek to strengthen the V2I approach by combining it with the junction database and the on-board systems methods. It is conceivable that a future vehicle based platooning management system would be able to distinguish between junctions that could (a) "normally be platooned", subject to road works or other transient features such as adverse weather, others (b) that should "never be platooned", and finally some (c) that are "variable" based on V2I communication.

V2X communication has significant potential to enable more consistent and uninterrupted platooning journeys, especially at nighttime when there is typically less traffic joining and leaving motorway via slip roads. Further work is required to fully understand the benefit to cost ratio, but as we look to decarbonise transport there remains considerable merit in considering how V2I and Vehicle-to-Vehicle (V2V) communications can safely enable more efficient journeys, including assisting vehicle platoons. There is the potential in the future to support longer platoons comprised of multi-vehicle types and to design solutions which interface with the advances in ADAS and automated driving technologies.

7.6.3   *Managing platooning deployment in the UK*

A fundamental feature of platooning operation is safely enabling close following distances to save fuel. HelmUK has found that junctions constitute the key risk from platooning, but that disbanding the platoon at most junctions has a major impact on fuel savings. Based on these findings, there are three possible options for the future of platooning in the UK in the medium term (before 2030).

A. Maintaining the "status quo"

No regulation or control of platooning. This approach would leave risks at junctions down to either driver judgement or to systems which might be implemented by the freight industry to increase headways at junctions to alleviate the risk to merging traffic. HelmUK's safety case judged that by the time the platooning drivers were able to decide to increase headway at junctions, it was likely to be too late for the vehicles to reach safe headways for most junctions.

B. Regulate platooning

Do not permit vehicles traveling at low headways to travel past junctions, low headways being below typical ACC minimum settings of 1.4 seconds. This will mitigate the main risk from platooning technology. Also consider restricting platooning to suitable roads such as motorways.

HelmUK has shown that for current platooning systems there is a limited business case if no junctions can be platooned through as the fuel savings are negligible. This option would allow platooning's deployment as a driver support system while preserving road safety. It would also keep the door open to operation on routes with lower junction density and for future platooning developments by manufacturers to provide fuel benefits despite this restriction.

C. Regulate and support platooning development

To further support platooning development, more junctions on the network would need to be safe for platooning. This could be done through ITS solutions or signal control (ramp metering) but the business case for such interventions is weak based on HelmUK's fuel results. Wider changes to English motorways (and the wider SRN) to enable Connected and Automated Vehicles should consider platooning as a use case for these changes. In the short term a more realistic way of supporting platooning may be a framework for identifying and managing risk at junctions, which would allow a case to be made for platooning past certain junction types or in certain conditions.

A practical step would be to apply HelmUK's junction assessment criteria to English motorways (or roads built to these standards) to determine which junctions are likely to be safe for platooning. HelmUK saw no increase in risk at junctions that were identified to be safe for platooning through the safety case development process. This could be developed further to build a case for platooning through junctions under other conditions such as night-time operation, which is a plausible low-risk scenario for platooning deployment.

However, this framework for managing platooning risk would require either UK Government, road operators, the freight industry, or a combination thereof to take ownership of implementing and maintaining this framework.

7.7   Accelerating innovation and ITS adoption

HelmUK is an example of a successful R&D partnership between government agencies, research organisations and industrial partners, including technology providers and operators. The next challenge is to understand how the findings can be adopted and integrated, where appropriate, into a holistic approach to tackle today's macro transport challenges? HelmUK has demonstrated that platooning technology can improve road safety for all. The trial also highlighted the potential, albeit with other interventions around junctions, to reduce fuel consumption. The barriers to adoption include relevant and appropriate regulation to authorise safe and viable commercial solutions. Also, there must be an acceptable business case with an associated return on investment that would be realised by the stakeholders, including the HGV manufacturers, the fleet operators, and road authorities.

ITS solutions require independent monitoring and evaluation to assess the benefits and costs (financial, environmental, and societal). If we assess the results in isolation, the costs often outweigh the benefits, because each ITS solution is deemed to bear all the costs. However, when system-based models are applied, we can, for example take the costs of V2X equipment, and share these for multiple ITS applications. This requires policy-based coordination and a technology agnostic approach and is an important stage to overcome to achieve change.

The justification behind the adoption of ITS solutions is often complex, because a range of interrelated things need to be coordinated, and cooperation and agreement reached by different organisations. For example, HelmUK has highlighted the potential environmental benefits that could be achieved if multi-brand platooning systems were introduced and supported by dynamic V2X communication at junctions to maximise journey efficiency (minimise disbanding of the platoon). This would require investment by public and private stakeholders and the benefits are challenging to calculate based on today's policy impact assessment and supporting economic guidelines.

When viewed from the "top-down" climate change perspective, the motivation and scope of intervention can be different compared with the "bottom-up" view associated with individual assessments of platooning and/or other ITS technologies and/or alternative carbon-free fuels. A key lesson from HelmUK is that taking an overarching systems view gives a different perspective that focusing on the trial results in isolation.

Unless we significantly reduce greenhouse gas (GHG) emissions related to human activities, climate change will have a devastating impact on the planet, our societies, and the global economy. Transport accounts for a third of all carbon dioxide emissions in the UK [5] and became the largest emitting sector of GHG emissions in 2016. Calculating the monetary costs of mitigating climate change was beyond the scope of HelmUK, but it has demonstrated the potential of platooning to help us to act in the near to medium-term and accelerate the trajectory with regard to achieving net zero transport emissions by 2050.

Platooning could mitigate a relatively small proportion of GHGs today and it has the potential to contribute to a larger cumulative reduction, alongside other measures, over time. This would slow the global rate of temperature rise. This delay affords scientists and engineers valuable time for fundamental technological changes and breakthroughs to be made, alongside new supporting infrastructure, and affordable public transport services and favourable policies to be developed. All will be necessary to transform today's road transport system to one where people and goods experience net zero journeys. To address climate change we must simultaneously reduce GHG now and work towards the longer-term solutions.

The introduction of vehicle and infrastructure technologies to reduce transport related GHG emissions can be significantly accelerated by establishing an international "North Star" for transport innovation adoption. This would recognise an evolutionary process with improvements made over time. The HelmUK findings have significant value, and the challenge is to ensure that they are used as widely as practicable to help inform the next choices regarding connected vehicle development, ITS, and interventions to improve safety and sustainability.

Finally, setting a targeted "North Star" for innovation adoption with a focus on safe and sustainable transport, will help to facilitate the transition from fundamental ITS and fuel-based research to undertaking real-world commercially applicable product and service development. To close this gap will require greater coordination and leadership from governments, research funding bodies, industry, and academia.

7.8   Conclusion

HelmUK was the UK's first public-road trials of platooning technology and has provided valuable data and insights into the real-world operation of platooning. These insights enabled an assessment of the potential future of platooning and the effect it may have on various stakeholders, including Road Authorities, UK Government, and Freight Industry. The final conclusions and recommendations from the HelmUK project are based on the key findings as shown in Figure 7.2.

Based on these findings, the short-term business case for platooning is limited. Figure 7.3 summarises the overall case for platooning based on HelmUK's research

Platooning is as safe as ACC if risks at junctions are controlled

Little impact on driver workload

No increase in risk at 'platoonable junctions'

0.5% fuel savings over ACC for base scenario

Fuel savings increase to 2.5% if all junctions could be made "platoonable"

Maximum saving 4.1% in steady-state platooning

Figure 7.2   Key findings from HelmUK

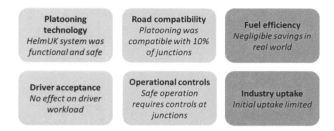

Figure 7.3   The overall case for platooning from HelmUK

and includes a "Red Amber Green" assessment of each element of platooning operation.

The main conclusions from this project are:

1. Across a three-vehicle platoon, in real-world conditions, fuel savings were 0.5% over current technologies such as ACC; there was no evidence this difference was statistically significant.
2. A key cause of low fuel savings is that only 10% of junctions along the trial route were assessed as safe for platooning as part of HelmUK's safety case.
3. Platooning poses a potential risk at many motorway junctions on because of the low headway between the vehicles.
4. Changes to infrastructure, or even junction design, could increase the percentage of "platoonable" junctions on motorways, but the business case for platooning doesn't support significant investment.
5. Even if all junctions were safe for platooning, HelmUK has shown that 2.5% fuel savings could be realised.
6. In perfect conditions this figure could rise to 4.1% across three vehicles.
7. The core challenge of platooning can be summarised as the need to balance road safety at junctions with realising fuel savings. The rest of this section is devoted to recommendations which could help achieve this balance.

7.9   Recommendation

Based on the findings and project experience of HelmUK, the driver support systems which enable the overall platooning concept would provide safety benefits if they were deployed at headways above current ACC minimums (approximately 1.4 seconds). It is also likely that platooning at larger headways would still save fuel over solo truck operation. This approach will enable a limited form of platooning on the motorways without risks at junctions being uncontrolled. Furthermore, any such deployments would not be affected by regulation of the full platooning concept.

HelmUK has shown the main risks from platooning are related to low headways. For closer following distances, as used in HelmUK, consideration should be given to regulating platooning at junctions over an unregulated approach to platooning which aligns with the "status quo" for driver assist systems. Regulation of platooning through junctions at close following distances will have a negative impact on the economic case for platooning but will mitigate the largest risk from platooning, i.e. conflict with merging vehicles.

Currently the technological maturity of platooning and the low fuel savings from real-world operation mean that there is not a short-term case for making changes to the road network to support platooning.

The longer-term future of platooning would benefit from National Highways, DfT or the freight industry (or a collaboration of these stakeholders) creating a risk management framework which would allow platooning through some junctions on English motorways. Initially, this would be based on the HelmUK junction assessment criteria but could be expanded to cover use cases which are likely to present a low-cost way of reducing the risks of platooning while increasing the fuel savings, such as night-time operation.

Beyond this, platooning should be considered as a use case for wider adaptions of English motorways (and the wider SRN) to CAV, especially with regard to "CAV friendly" junction designs.

References

[1]   TRL (2022). *HelmUK – HGV Platooning Trials: Final Report*. TRL (Transport Research Laboratory), Wokingham, Berkshire, UK. https://doi.org/10.58446/ozvr6006

[2]   ENSEMBLE Consortium. https://platooningensemble.eu/project. Accessed on 15 May 2024.

[3]   SARTRE Consortium (2013). SARTRE (Safe Road Trains for the Environment) Deliverable D5.1 – Commercial Viability. Brussels.

[4]   GG104 standard: www.standardsforhighways.co.uk/dmrb/search/0338b395-7959-4e5b-9537-5d2bdd75f3b9. Accessed on 15 May 2024.

[5]   2018 UK Greenhouse gas emissions, provisional figures, National Statistics, UK Department for Business, Energy & Industrial Strategy, UK's National Statistics service.

Chapter 8

C-ITS deployment in Australia – achievements and key learnings

Ada I. Lin[1], Luke A. Capelli[1] and Nicholas S. Brook[2]

8.1 Introduction

Lexus Australia conducted real-world Cooperative Intelligent Transport Systems (C-ITS) deployments within the Ipswich Connected Vehicle Pilot (ICVP) and the Australian Integrated Multimodal EcoSystem (AIMES). These deployments included the development and demonstration of safety applications for vehicle-to-vehicle (V2V), vehicle-to-infrastructure (V2I) and vehicle-to-network (V2N) communications, also referred to as use cases. V2V enables communication between C-ITS-equipped vehicles, and is used to alert drivers about potential collisions, such as where vehicles are emergency braking. V2V communications also facilitate communication with emergency service vehicles and public transport, extending existing applications to different transport modes and scenarios. V2I facilitates direct communication between C-ITS vehicles and roadside infrastructure, such as traffic signals broadcasting live signal status and road geometry information, and variable message signs broadcasting live roadworks information. V2N supports long-range communications with a cloud-hosted central facility to alert drivers to road hazards, traffic jams and roadworks, and provides information about static or variable speed limits.

This chapter discusses the importance, design and evaluation of C-ITS use cases, highlighting key findings and suggesting next steps for C-ITS deployment in Australia.

8.1.1 Road trauma

According to the World Health Organization, approximately 1.2 million people worldwide died in 2023 due to road traffic injuries [1]. Each year, about 1200 road users lose their lives across Australia, and around 40,000 people are admitted to hospitals due to serious injuries from traffic accidents [2]. Compared to other nations in The Organisation for Economic Co-operation and Development (OECD), Australia's fatality rate of 4.68 per 100,000 population in 2019 was

[1]Toyota Motor Corporation Australia (TMCA), Melbourne, Victoria, Australia
[2]Queensland Department of Transport and Main Roads (TMR), Brisbane, Queensland, Australia

ranked 16th out of 36 countries [3]. Over the last decade, the overall cost of road crashes to the Australian economy is estimated to be $300 billion [4]. In Australia, major cities account for 34% of all road crash deaths, with regional areas accounting for 55%, and remote areas accounting for 10% (rounded figures). Based on the fatality rate per 100,000 population, regional and remote areas have been rated 5 and 11 times higher, respectively, than major cities [2].

Since 1970, innovative approaches in road rule enforcement, driver behaviour, driver assistance, vehicle design and road design have significantly reduced the harm caused by road crashes, particularly those of high severity. Many counter-measures have focused on the protection of vehicle occupants and, more recently, on the avoidance of crashes altogether. The rollouts of technologies that detect and mitigate crash risks are now accelerating due to advances in sensing, connectivity and automation. Road incidents, such as rear-end collisions, can often be avoided by directly detecting other vehicles using onboard sensors, such as cameras and radars, to provide awareness to the driver to react with a safer behaviour. However, preventing accidents at intersections involves the identification of vehicles or pedestrians that onboard sensors may not capture. Enabling communication between vehicles, and between vehicles and infrastructure, facilitates safer driving by alerting drivers to dangerous situations that are not immediately visible.

8.1.2   C-ITS in road safety

Cooperative Intelligent Transport Systems (C-ITS) enable vehicle-to-everything (V2X) communications, allowing vehicles to communicate with each other through V2V communications, with roadside infrastructure through V2I communications, and with a cloud-based central facility through V2N communications. Based on the C-ITS messages exchanged, drivers receive safety alerts about immediate and upcoming hazards (see Figure 8.1). C-ITS increases drivers' situational awareness, thereby putting them in the best position to react to safety risks.

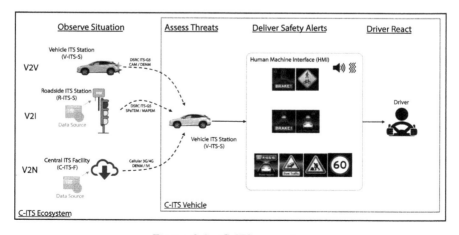

Figure 8.1   C-ITS ecosystem

Production vehicles equipped with C-ITS have been available in markets such as Japan since October 2015. As of 2023, Toyota/Lexus has sold more than 370,000 vehicles equipped with "ITS Connect" across multiple models. The evaluation results indicate a 38% reduction in near-miss events when turning right at cooperative intersections, and a 7.7% reduction in travel time between intersections for emergency service vehicles on major roads [5]. Furthermore, a 2019 survey of ITS Connect customers revealed that over 70% found the system helpful, especially in situations where the line of sight was obscured [6]. Other regions, including Europe, have also introduced C-ITS- equipped production vehicles and infrastructure.

Austroads has found that deploying C-ITS at scale in Australia could potentially lead to a reduction of up to 23% of fatalities and 28% of injuries [7].

8.2 C-ITS programs in Australia

In Australia, the journey towards the adoption and implementation of C-ITS has been progressive and research-driven. It involves several trials, research projects and collaborative efforts among government agencies, academic institutions and industry stakeholders. This section details the most recent programs and their contributions to the deployment of C-ITS, as well as Australia's direction towards nationwide deployment.

8.2.1 Ipswich Connected Vehicle Pilot (ICVP)

Several pilot projects were initiated across different states, with the most extensive being the Queensland Government Department of Transport and Main Roads (TMR) Ipswich Connected Vehicle Pilot (ICVP) [8,9].

8.2.1.1 ICVP Infrastructure

The ICVP area comprises 300 km^2 in the Ipswich local government area. The infrastructure for the pilot included: (1) *29 C-ITS-enabled intersections*: Each with its traffic signal fitted with a Roadside ITS Station (R-ITS-S), also known as a roadside unit (RSU), broadcasting C-ITS messages including intersection map and signal phase timing data; (2) *A cloud-based central ITS facility (C-ITS-F)*: Supported the integration of multiple traffic data sources and distribution of C-ITS messages, along with the road event information.

8.2.1.2 ICVP vehicle stations

Two Lexus cooperative vehicles, retrofitted by Lexus Australia's local engineering team, with a total of eight C-ITS use cases (including V2V), and 355 public participants' cooperative vehicles, fitted by TMR with a total of six V2I/V2N use cases participated in ICVP to evaluate the effectiveness of C-ITS, focusing on system performance, safety impact, and user acceptance of the technology, in real-world settings.

8.2.1.3 ICVP data collection

During the rotation of Lexus cooperative vehicles between TMR and Ipswich City Council, 8300 kilometres were driven, over 10 million messages were transmitted by Lexus vehicle stations, and 3.4 million messages were received from vehicle and roadside stations. From September 2020 to September 2021, the public ICVP participants covered 2,220,000 kilometres, over 48,000 hours and received 96,000 event warnings across the twelve months of the pilot [10].

8.2.2   Bruce Highway

Since the conclusion of the ICVP, the scope of the project was extended, with an additional 37 roadside stations installed, four new use cases added, and central facility extended to cover between Brisbane and Cairns (approximately 1600km of the Bruce Highway). These units were strategically positioned at traffic lights within major townships to broadcast crucial signal status and timing to connected road users. New V2N use cases were added to address key rural and regional crash causes of fatigue, dangerous overtaking, high speed intersections, and speed around curves.

8.2.3   Australian Integrated Multimodal EcoSystem (AIMES)

Australian Integrated Multimodal EcoSystem (AIMES) [11], a real-world platform for testing and demonstrating emerging connected transport technologies in complex urban environments, is located in Carlton, Victoria. AIMES incorporates over 100 kilometres of Melbourne's road-network, where partner organisations have installed diverse technologies for sensing, connecting, visualising and analysing mobility systems.

Lexus Australia extended its C-ITS deployment within AIMES to incorporate emergency and public transport vehicles, specifically, vehicle communication involving ambulances and trams.

8.2.3.1   Importance of vehicle communication with ambulances

In 2022, Ambulance Victoria provided emergency medical response to close to 6.5 million people in an area of more than 227,000 km^2 across the state.

As shown in Table 8.1, demand for ambulance services continued to rise to record levels. In 2022–2023, Ambulance Victoria responded to more than one million incidents in Victoria by road. October to December 2022 was the busiest quarter in Ambulance Victoria's history, with more than 100,000 Code 1 cases for the first time, which represented a 35.8% increase from 73,797 Code 1 cases for the same period in 2018–2019 [12].

Table 8.1   Number of emergency road responses

Road incidents[1]	2022–2023	2021–2022	2020–2021	2019–2020	2018–2019
Emergency Code 1	400,883	377,386	323,566	310,090	301,336
Emergency Code 2	213,821	217,114	240,836	236,933	230,890
Emergency Code 3	74,656	88,008	96,060	86,937	86,488
Non-Emergency	363,921	360,394	355,546	339,730	321,458
Total Road Incidents	**1,053,281**	**1,042,902**	**1,016,038**	**973,690**	**940,174**

[1]Code 1 incidents require urgent paramedic and hospital care, Code 2 incidents are acute and time sensitive, but do not require a lights and sirens response, and Code 3 incidents are not urgent but still require an ambulance response. Source: Ambulance Victoria Annual Report 2022–2023.

The risks for emergency vehicles responding to time-critical emergencies are higher in urban environments due to complex traffic conditions, noise pollution and limited visibility. In these environments, in-vehicle notification solutions can enhance drivers' situational awareness, such as by advising the direction of an approaching emergency vehicle. This can help reduce emergency response times and create safer working conditions for first responders—factors that have a tangible impact on survival rates in time-critical emergencies.

In Victoria, Road Safety Road Rule, Part-7, Division-4, Rule 79A [13] states that vehicles should not exceed 40 km/h when passing emergency vehicles that are stationary or moving slowly (less than 10 km/h). The road rule aims to ensure the safety of emergency service workers responding on the road or roadside and others at the scene [14]. Providing in-vehicle warnings when approaching a slow or stationary emergency service vehicle and instructing speed reduction to regulated speed is intended to encourage the drivers to comply with the road rule.

8.2.3.2 Importance of vehicle communication with tram

Melbourne's tram network, one of the key public transport systems in the city, is the largest in the world. It boasts 250 kilometres of double track and over 400 trams operating across various types on a typical weekday. Before the COVID-19 pandemic, the network recorded approximately 200 million boardings annually. Over 70% of the network shares its space with other road users; 80% of injuries on public transport in Victoria in 2018 were reported on the tram network.

Yarra Trams documented more than 1100 vehicle-to-tram collisions in 2018, with 97% being the fault of motorists [15]. Approximately, 70% of vehicle collisions result from vehicles merging midblock or making U-turns in front of a tram, often causing severe damage to the vehicles and severe injuries to road users. Moreover, vehicle-to-passenger collisions – where vehicles pass stationary trams while passengers are boarding or alighting – pose a significant risk. From 1 July 2014 to 30 June 2019, there were 128 passengers knocked down, some resulting in severe injuries that required emergency response [16].

Implementing a system that enables trams to communicate with surrounding vehicles can alert drivers to approaching trams and indicate passenger disembarkation/embarkation status, potentially reducing vehicle collisions with both trams and passengers, improving the efficiency and safety of public transportation.

8.2.4 National approach to C-ITS in Australia

Australian governments have set shared targets of halving road deaths by 2030, achieving zero road deaths by 2050, and reaching net-zero emissions by 2050. C-ITS technology has real potential to assist in these endeavours [17]. Since 2022, Australian governments have been collaborating to develop a framework to support C-ITS in Australia. In February 2024, the "Principles for a National Approach to C-ITS in Australia" were released by the Department of Infrastructure, Transport, Regional Development, Communications and Arts (DITRDCA) [18]. The document highlights the importance of government leadership, national consistency and harmonisation with international standards.

In October 2023, DITRDCA continued to seek public feedback on the draft "National Road Transport Technology Strategy" and the "2024–27 National Connected and Automated Vehicle Action Plan", which are intended to guide nationally coordinated and consistent rollouts of new road transport technologies across Australia [19].

Most recently, in April 2024, a new project of national importance, the C-ITS national harmonisation and pre-deployment research [17], was announced. With participation of various state, federal and industry partners, the objective is to provide a robust evidence base to support development of a harmonised national approach to C-ITS deployment in Australia. Nationally consistent and standardised C-ITS communications, digital infrastructure, and security systems are critical enablers for manufacturers to make vehicles available to the Australian market.

8.3   C-ITS safety applications (use cases) in ICVP and AIMES

Lexus Australia in close cooperation with government, industry and academic stakeholders, developed and evaluated several C-ITS applications in ICVP and AIMES projects, as presented in Table 8.2.

Table 8.2   Australian C-ITS applications

Application	V2X Type	Description
ICVP		
Emergency Electronic Brake Light (EEBL)	V2V via DSRC ITS-G5[1]	Alerts drivers that there is a risk of a rear-end collision with a cooperative vehicle braking hard ahead. (Lexus Australia Only)
Stopped/Slow Vehicle (SSV)	V2V via DSRC ITS-G5	Alerts drivers that there is a risk of a rear-end collision with a slow/stopped cooperative vehicle ahead. (Lexus Australia Only)
Advanced Red-Light Warning (ARLW)	V2I via DSRC ITS-G5	Alerts the driver that there is a risk of violating the red light at a signalised intersection unless the driver brakes.
Turning Warning Vulnerable Road-user (TWVR)	V2I via DSRC ITS-G5	Alerts the driver that there is a pedestrian or cyclist crossing at a signalised intersection.
Road Works Warning (RWW)	V2N via cellular 3G/4G (and V2I via DSRC ITS-G5)	Alerts the driver that driving speed is not appropriate for the roadworks speed condition.
Back of Queue (BoQ)	V2N via cellular 3G/4G	Alerts the driver that driving speed is not appropriate for a downstream back of queue.
Road Hazard Warning (RHW)	V2N via cellular 3G/4G	Alerts the driver that driving speed is not appropriate for a downstream hazard such as a crash or debris.
In-Vehicle Speed Limit (IVSL)	V2N via cellular 3G/4G	Provides the current regulatory speed limit, i.e., the default static, variable, school zone or roadworks speed limit.
Bruce Highway Extension		
Rest Area	V2N via cellular 3G/4G	Provides an alert of approaching rest areas if driver has exceeded safe fatigue limits of driving time.

(Continues)

Table 8.2   (Continued)

Application	V2X Type	Description
Curve Warning	V2N via cellular 3G/4G	Provides an alert of lower recommended speed for upcoming curve if the vehicle is over the recommended speed.
High speed intersection	V2N via cellular 3G/4G	Alerts drivers approaching from rural and regional side roads that a high-speed intersection is a head
Overtaking Lane	V2N via cellular 3G/4G	Alerts drivers if overtaking lane is approaching - to prompt safer choice of waiting rather than risky overtaking
AIMES Extension		
Emergency Service Vehicle Notification (ESVN)	V2V via DSRC ITS-G5	Notifies drivers of the direction of the approaching emergency vehicle and instructs drivers to comply with regulatory speed when approaching a stationary emergency vehicle in action.
Tram Awareness Alert (TAA)	V2V via DSRC ITS-G5	Alerts drivers about approaching trams.
Tram Passenger Warning (TPW)	V2V via DSRC ITS-G5	Alerts drivers to the status of tram passenger disembarkation/embarkation.

[1]ITS-G5 is a European standard for vehicular communications based on the IEEE-1609. x and IEEE-802.11p standards.

8.4   System implementations

This section details the system architecture and implementation of the Australian C-ITS deployment in Queensland and Victoria.

8.4.1   Radio communication interfaces

Both ICVP and AIMES adopt a hybrid communications model with two kinds of wireless radio communication technology to deliver C-ITS messages: (1) **ITS-G5 – Dedicated Short-Range Communication (DSRC):** An IEEE 802.11p-based technology that enables high-speed direct communication within the 5GHz (5.895GHz ∼ 5.905GHz) frequency band; (2) **Long-Range Communication:** Relies on a cellular 3G/4G network and a lightweight, publish/subscribe messaging transport protocol, Message Queuing Telemetry Transport (MQTT).

DSRC is used to exchange C-ITS messages for V2I and V2V communications in time-critical C-ITS use cases; long-range communication is used to deliver less time-critical traffic information through V2N communication.

Table 8.3   C-ITS message types

C-ITS message type	Description	Communication
Signal phase and timing (SPaTEM)	For a signalised intersection	DSRC ITS-G5
Map data message (MAPEM)	Digital representation of the road geometry at intersections	DSRC ITS-G5

(Continues)

Table 8.3   (Continued)

C-ITS message type	Description	Communication
Cooperative awareness message (CAM)	Speed, heading, positioning and status information of a vehicle	DSRC ITS-G5
Decentralised environmental notification message (DENM)	Generated by a vehicle or central facility to warn other users of a hazard such as hard-braking or roadworks	Cellular 3G/4G (and DSRC ITS-G5)
In vehicle information (IVIM)	Typically includes regulatory information such as a posted speed sign	Cellular 3G/4G

These C-ITS messages adhere to European Telecommunications Standards Institute (ETSI) standard, including message types summarised in Table 8.3.

8.4.2   Security credential management system

ETSI-compliant security credential management system (SCMS) (2019) established during ICVP, allows all vehicle and roadside ITS stations, as well as central cloud facility in the C-ITS ecosystem to communicate with each other in a safe, secure and privacy-protective manner while operating in a public space.

8.4.2.1   SCMS core elements

The SCMS comprises five core elements [20]:

1. **Authorisation authority (AA):** Issues tickets to ITS stations enrolled and authenticated by enrolment authority (EA), granting permissions within EA's domain based on AA's context as per **ETSI TS 102 940 v1.4.1**.
2. **Conformance manager:** Ensures ITS stations meet enrolment trust requirements stipulated in **ETSI 102 940**.
3. **Enrolment authority (EA):** Authenticates ITS stations and issues an enrolment credential as proof of identity, while concealing the canonical identifier from third parties, enabling ITS stations to request AA's service authorisation.
4. **Root certificate authority (CA):** Provides EA and AA with proof that they may issue enrolment credentials (EA) and authorisation tickets (AA).
5. **Trust list manager (TLM):** Issues a signed list of CA and TLM certificates.

8.4.2.2   Measures for secure communications

To ensure secure communications, the following measures were implemented:

- **Unique identification:** Each ITS station is assigned a unique Client ID to connect with the Cooperative-ITS Facilities (C-ITS-F). This ensures that only authorised stations with the correct security certificates and keys can communicate with other ITS stations.
- **Hardware security module (HSM):** This tamper-proof device at each station secures the transmission of messages. Every message sent is digitally signed by the sender's HSM and then checked for authenticity upon receipt using details from a hardware bootstrap certificate.

8.4.3   Positioning system

Accurate positioning is a fundamental enabler C-ITS. A cooperative vehicle uses the positioning information to compute its vehicle location in relation to the traffic events shared by other vehicles or infrastructures as part of threat assessment to provide driver safety alerts. The positioning system:

- Receives signals from multiple global navigation satellite systems (GNSS).
- Receives real-time kinematic (RTK) correction stream delivered by an RTK service provider via a cellular network. In Australia, RTK service is delivered by both public and privately owned organisations.
- Computes the positioning information of the vehicle for C-ITS applications.

In the future, positioning system can be extended to include other technologies, such as multi-sensor fusion, precision inertial measurement units and simultaneous localisation and mapping (SLAM).

8.4.3.1   AUSCORS

Geoscience Australia provides free and open access to real-time GNSS differential correction data streams from Continuously Operating Reference Stations (CORS) through Networked Transport of RTCM via Internet Protocol (NTRIP) Broadcaster. RTCM refers to Radio Technical Commission for Maritime Services that describes the protocol for transmission of differential corrections data allowing GNSS receivers to calculate their position with higher precision. AUSCORS provides RTCM3* (version 3) data from across its network of CORS stations in Australia. This data is shared by the central facility with the vehicle station, which can process it using RTK positioning augmentation methods to support improved positioning accuracy.

8.4.3.2   Commercial service providers

The service delivered by Geoscience Australia is free, and the correction data streams are delivered via a network of RTK mount points using the AUSCORS NTRIP Broadcaster. However, this service is limited to a single base station connection at a time – which means it is not possible to support automatic connection change-over from one base station to the next base station for a moving vehicle and is not suitable for the real-life deployment of C-ITS applications in Australia.

Therefore, to allow for seamless dynamic connection to the nearest base station for those vehicles moving from one to another reference station coverage area, a commercial service was used during the pilots. The cost of commercial RTK service remains high in Australia at this stage, and thus commercial concept for deployment requires further investigation.

8.4.4   Central ITS facility (C-ITS-F)

The central facility (Figure 8.2) is a cloud-based platform that served as a secure interface between C-ITS field devices, existing infrastructure, monitoring systems

---

*RTCM3 is a standard developed by the Radio Technical Commission for Maritime Services that describes the protocol for transmission of differential corrections data allowing GNSS receivers to calculate their position with higher precision.

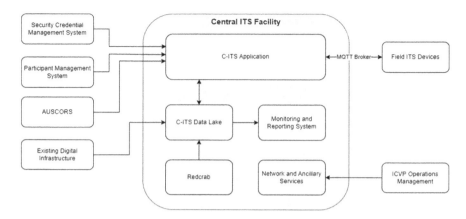

Figure 8.2   C-ITS-F system architecture

and configuration tools for ICVP [20] and other C-ITS pilot projects. This section summarises the major components of the facility.

8.4.4.1   Application layer

The application layer served multiple roles as the receiver, manager, transformer and publisher of C-ITS communications and data. The application was the primary interface between field devices, the SCMS and other elements of the C-ITS-F, including: interface with external devices and data sources; generate, sign and publish C-ITS messages; perform data collection and logging; perform device configuration management and software updates; and handle ITS field device information.

Generally, the application layer provided a single interface for ITS devices to retrieve information from or transmit information to the central cloud service.

8.4.4.2   Data lake

The function of the data lake was to ingest, store, curate and stage all data used by the central facility. Broadly speaking, the data lake was used as a repository for storing structured or unstructured data for use by any ITS stations or service connected to, or internal to, the central facility. All ICVP participant vehicle (and roadside) data generated in the field were collected and stored in the C-ITS-F data lake for processing. Any data of interest was post-processed through an Amazon Web Services (AWS) analytics pipeline. In an actual C-ITS deployment, it is not expected that drivers' individual data would be directly collected by road authorities. This was done just to enable research of C-ITS by academic partners.

8.4.4.3   Monitoring and reporting system

The monitoring and reporting system (MRS) was used to manage and perform the Extract, Transform, Load (ETL) process on data stored in the data lake, serving the purpose of system performance analytics and diagnostic data analysis.

8.4.4.4   MQTT Broker

Broadly, the MQTT broker was used to perform several roles related to the handling of MQTT-protocol messages flowing in/out of the central facility. These include:

1. **Data access management**: Define the access mechanisms, structures and restrictions for inbound and outbound data.
2. **Data collection**: Collect, validate and aggregate inbound data for storage and/ or use by the C-ITS-S.
3. **Publish/subscribe**: Provide a one-to-many message distribution service intrinsic to the MQTT protocol.

The MQTT broker was the key to transmission of the C-ITS messages to vehicles (V2N). By providing multiple broker topics the vehicle was able to connect and subscribe to those relevant to its location, type, messages of interest and receive them on request and when the information on the broker updated. In future deployments, this has also been tested with a different broker protocol that operates similarly, Advanced Message Queuing Protocol (AMQP).

8.4.4.5   Redcrab

Redcrab is a purpose-built program that was used for managing the location and timing of roadworks. Data from Redcrab was ingested via the data lake to generate C-ITS messages, DENM and IVIM, for the RWW use case.

8.4.4.6   Network and ancillary services

Other services/functions that supported the operation of the central facility, include: Continuous integration and continuous deployment (CI/CD) pipelines; Networking, security and application analytics systems; Project and resource management systems; and Virtual Private Network (VPN) and secure virtual desktops.

8.4.5   Roadside ITS stations (R-ITS-S)

Intersections in Ipswich were integrated with roadside stations connected to the traffic light controller via a STREAMS[†] Field Processor (FP), to allow the transmission of real-time traffic light signal status (SPaTEM) and intersection geometry information (MAPEM), matching with signal group assignment via ITS-G5.

Similarly, in AIMES, C-ITS intersections were also fitted with roadside stations. However, these traffic light controllers are integrated using the Sydney Coordinated Adaptive Traffic System (SCATS), which adapts traffic signal timing in real-time to match the traffic conditions [21].

In ICVP and AIMES, the detection of the vulnerable road users was based on a pedestrian button press request for a green light to cross. During COVID-19, the usual press buttons to request a green crosswalk were always set to "on" to reduce surface contact. This meant that a TWVR alert would still be shown even without pedestrians being present, subsequently experienced as "false alarms" leading to a high dislike of the use case by participants. For future implementation, it would be

---

[†]STREAMS is an ITS (intelligent transport systems) platform in Australia.

advantageous to detect pedestrians using image recognition, video analytics or sensors at intersections to decrease the number of unnecessary alerts.

For widespread deployment, C-ITS intersection fitment in Australia and beyond, a system compatibility and rollout plan should be identified.

8.4.6   Vehicle ITS stations (V-ITS-S)

Vehicle stations in ICVP and AIMES were retrofitted by TMR and Lexus Australia. ICVP participant vehicles were fitted with an antenna mounted on a roof rack, as well as an in-vehicle communications box placed under the driver's seat and a display on the dashboard to display warnings to the participants.

The Lexus RX was the first production vehicle released with C-ITS technology in Japan in 2015. During the ICVP and AIMES projects, Australian RX450h F-Sport SUVs were retrofitted to:

- Interact with R-ITS-S, V-ITS-S and C-ITS-F within the C-ITS ecosystem, incorporating SCMS.
- Assess potential threats based on the information received and configured parameters.
- Deliver the C-ITS warnings to the driver via the Human-Machine Interface (HMI) using a combination of visual, audio and haptic alerts.
- Perform positioning, in-vehicle and cellular network communications, automated system startup and shutdown, and data logging functionalities.
- Capture the driver's behaviour as well as the situation inside and outside of the vehicle.

The C-ITS in-vehicle components were installed on the roof, inside the vehicle cabin and in the rear luggage compartment (Figure 8.3).

Lexus C-ITS Vehicles

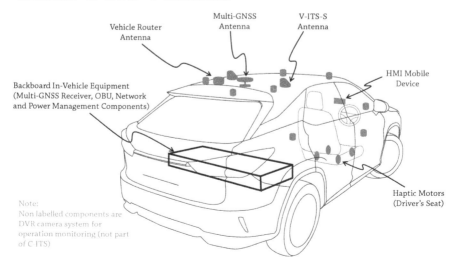

Figure 8.3   In-vehicle components fitted on Lexus Cooperative Vehicles

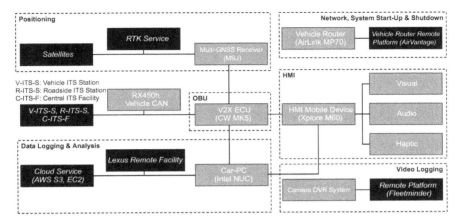

Figure 8.4 Lexus Australia C-ITS system architecture

Remote facilities to support Lexus cooperative vehicles' system updates, operation monitoring and data analysis were also established (Figure 8.4).

8.4.6.1 Positioning

Lexus positioning system consists of a multi-GNSS receiver evaluation unit and its GNSS antenna. The receiver computes positioning information using satellite signals from GNSS and RTK correction streams to calculate the vehicle's positioning for C-ITS use case threat assessment. Lexus cooperative vehicles used the correction service delivered by commercial service providers for dynamic connection (Figure 8.5).

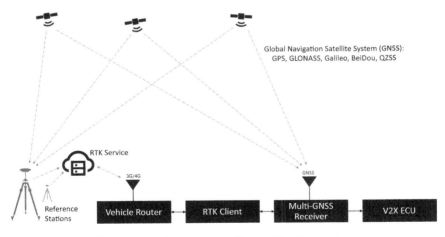

Figure 8.5 Lexus Australia positioning system

8.4.6.2 Network architecture, system startup and shutdown

There were two separate networks in each of the Lexus cooperative vehicles. The main C-ITS subsystem network integrated all the essential system components to perform the C-ITS operations, and there was an independent network for the video logging functionality.

System startup was triggered when the drivers of the Lexus cooperative vehicles turned on the vehicle ignition. During startup, in-vehicle components were started in sequence to ensure smooth operations. System shutdown was triggered when drivers turned off the vehicle ignition. During shutdown, logs collected during the drive were uploaded to a remote facility for post-processing.

8.4.6.3 Onboard unit

The onboard unit (OBU) is the core of the cooperative vehicle's C-ITS operation. Within the electronic control unit (ECU) of the OBU, the use case application software processes the C-ITS messages, vehicle controller area network (CAN) signals and positioning inputs, performs threat assessment according to the use cases' algorithms and parameters, and sends the alert requests to the HMI.

Different vendors design and supply OBUs in different shapes, some allow installation within the car while others allow mounting on the top of the roof.

Integrating vehicle CAN

The Lexus cooperative vehicles additionally used signals from the vehicle CAN bus to reflect vehicle status and driver intention more closely.

For ARLW and TWVR, the turn signal indicator status was used to indicate driver intention in a shared turn/straight lane to assist in displaying the suitable use case based on the vehicle's intended movement. For future implementation, it would be advantageous to include navigational inputs as an indicator of the driver's intended route.

EEBL used the internal emergency stop signal (ESS) to determine an emergency braking event. The vehicle ESS was generated when the vehicle has a hard-braking event from a speed above 50 km/h.

For vehicle-to-vehicle use cases, one vehicle is defined as the Target Vehicle, and all others are defined as Subject Vehicles. When describing use cases, the Target Vehicle is a vehicle with some attribute that affects the Subject Vehicle, from Subject Vehicle's frame of reference.

For SSV, the subject vehicle incorporated the hazard light status of the target vehicle during its threat assessment:

- **Target vehicle with hazard lights "on"**: The alert was always generated by the Subject Vehicle.
- **Target vehicle with hazard lights "off"**: The priority of the alert was determined by considering the subject vehicle current speed, driver response time, braking performance and distance to target vehicle.

IVSL read vehicle speed from the CAN bus to compare speed limit data and displayed speed alerts differently, with white background speed icons when the driver complied with the speed requirements and with pink background speed icon to indicate speeding. Different background colours were used to further con-textualising driver alerts.

Parameterisation
Parameterisations for individual use cases were done to improve the relevance of the driver alerts. These alerts were triggered by comparing the status of the vehicle and traffic event information delivered from V-ITS-S, R-ITS-S and C-ITS-F in ICVP.

SSV speed condition parameters were tuned to reduce unnecessary warning triggers. An alert was triggered if the Target Vehicle is very slow or stopped, with *speedMax* set to 35 km/h, and with at least 10 km/h speed difference between the two vehicles, with Subject Vehicle *speedMin* set to 45 km/h, to avoid alerts being triggered continuously in normal traffic flow.

V2N use cases, RHW, BoQ, IVSL and RWW, computed alert triggers using event location information broadcasted from the central facility. While the functions of the use cases were successfully verified during the testing, some of the lateral positioning errors were mitigated by increasing the lateral offset parameters as none of the events were at lane level. However, increasing the lateral offset could lead to increased sensitivity of the vehicle to events that were not relevant for its drive path. For example, in a highway situation, an event 3 lanes across from the subject vehicle may receive a warning irrelevant to the driver's current behaviour.

8.4.6.4   Human–machine interface
Human–machine interface (HMI) presents C-ITS safety alerts to indicate potential traffic hazards. With increasing technology applications in vehicles, distracted driv-ing presents a challenge. Drivers' attention towards the road competes with a range of distractions within the vehicle. Driver fatigue or behaviours that impair the driving task can benefit from C-ITS alerts to refocus driver attention on the most urgent task at hand.

It is critical that a driver awareness alert be immediately understandable, useful and not a distraction to the driving task; hence, Lexus Australia deems industry trials, testing and user experience feedback on warnings as valuable. In each Lexus cooperative vehicle, a dedicated mobile screen was placed in the eye-line of the driver in the centre of the instrument panel to provide visual and audio alerts. Haptic alerts were delivered to the driver by placing four vibrating motors in the cushions of the driver seat. When there were multiple traffic threats, the HMI prioritised the alerts. The most safety-critical warning occupied the centre of the screen as the primary alert; simultaneously, audio and haptic alerts were triggered, and less critical alerts were displayed using smaller icons to the side. The alert screen could simultaneously display up to five use case warnings (Figure 8.6).

The colour of icons designated the urgency of the safety alerts – those requiring immediate action were coloured red (high priority), those requiring increased awareness were coloured amber (medium priority) and those are for general information were coloured blue (low priority). These safety alert icons

Figure 8.6   Visual alert screen layout (source: Lexus Australia)

were designed with two visual representations and in a way to be familiar to drivers of Lexus cooperative vehicles to facilitate appropriate reactions.

- **In-vehicle view for EEBL, ARLW and RHW primary alerts:** often used in multifunction displays (MFD) of production vehicles for driver assistance functionalities for safety-critical warnings, such as lane departure or forward-collision warnings.
- **Traffic sign view for SSV, TWVR, RWW, IVSL and BoQ or all secondary alerts**: often used as part of the navigation system display to highlight road rules or current conditions.

When any of the EEBL, SSV, ARLW, TWVR, RHW or BoQ primary visual alerts were triggered, the audio and haptic alerts were activated simultaneously with the low, medium or high-profile configuration. Following initial testing, the combination of sound and vibration was replaced with a spoken audible warning that described the hazard. For example, "Caution! Red light ahead.", "Caution! Emergency vehicle approaching from the right!" or "Warning! Approaching emergency vehicle. Reduce speed to 40 km/h". A visual warning with distance information and an audio warning with directional information in a human speech made the warning contextual and relevant, keeping the drivers' focus on the road.

8.4.6.5   Data logging and analysis

Similar to ICVP participant vehicle and roadside data generated in the field, Lexus cooperative vehicles collected multiple data sets from system components. Collected data was used for system validation and system performance analysis. There are two avenues of access to the vehicle data: (1) **Direct Vehicle Access**: accessed in-vehicle components through remote login and via the in-vehicle network while the vehicles were in operation; (2) **Cloud Database Access**: after each vehicle operation cycle, the Car-PC collected all log data from the various system components, such as OBUs, HMI and positioning logs, and uploaded them to an Amazon Simple Storage Service (S3) bucket for further analysis.

8.4.6.6   Vehicle security and video logging

On top of the implementation in SCMS, an additional secure mechanism, periodically changing media access control (MAC) addresses, was configured for the V2X electronic control units (ECUs) of ICVP participants' vehicles, while for Lexus cooperative vehicles, MAC addresses remained static during the pilot. Additionally, general vehicle security measures implemented for system access protection include:

- **Mechanical security**: System components were mechanically secured against removal or tampering by installing a cover or a tether; these included a back cover, an HMI mobile device tether, a haptics cover and a camera hard drive lock, which only the assigned personnel can access.
- **Password protection**: Secure passwords were implemented to restrict access to the in-vehicle network, ITS components and the Car-PC as well as all web access to any data stored remotely.
- **Restricted user access to HMI mobile device**: The mobile device was running in a locked screen mode during the operation of C-ITS application, any access to other applications or the internet connection was prevented by Kiosk mode.

The video logging system consisted of a mobile digital video recorder (MDVR) and eight cameras. The data captured by the cameras was stored locally on a hard drive which was only accessible to Lexus Australia personnel. The system allowed remote access to the live feed via a web interface. In the production or commercial C-ITS deployment, this level and collection of data will not be done given C-ITS's inherent security, privacy and anonymity posture.

8.4.7   *Public transport and emergency vehicle V-ITS-S*

Enabling vehicle-to-vehicle communications with specialised vehicles, such as public transport vehicles, trams or buses, or emergency vehicles, ambulances, police cars or fire trucks involved installing the OBU onto these specialised vehicles.

The vehicles exchanged their status using cooperative awareness messages (CAM), which includes important parameters such as location, speed, direction, heading and type/role of the vehicles. In the case of special vehicles such as emergency vehicles, lightbar and siren information is also a part of CAM. For example, the emergency service vehicle notification (ESVN) application in the vehicle can detect an emergency service vehicle based on the vehicle type/role, classify its location and warn the driver if the lightbar and siren are active. For an ambulance, the lightbar is integrated with the OBU using an I/O module to convert 12V DC signal from the lightbar to a digital signal. The ambulance transmitted the vehicle type as emergency service vehicle, and details of which were populated in a container called "emergency container" within the CAM.

8.4.8   Road events and road network model

For V2N use cases relying on long-range communications, cooperative vehicles with connectivity to C-ITS-F determined the active tile according to their location and subscribed to messages only for the currently active tile. The road network model included speed limit information based on the actual road speed limits, including school zones and variable speed limits. The road network model is a critical component of the central facility. It is used as the base map in the spatial service, allowing the messaging engine to generate C-ITS messages based on triggers from any other data sources ingested by the central facility. All road events (road hazards, roadworks and back-of-queue events) were derived from real-time traffic data. These events were generated from the synthesis of live, field-detected data and digital representations of the road network and its characteristics.

8.4.8.1   Field-detector data

Various existing roadside hardware were integrated into the C-ITS ecosystem using an ITS management platform (in this case, the Australian STREAMS Platform) to collect real-time information about the status of the transport network. For existing trials, these were limited to managed motorways, and include: (1) **vehicle detector sites**: In-ground loop detectors which provided speed / volume / occupancy of traffic in real-time at their installed location. These are used by STREAMS algorithms to determine motorway queuing that was used for the BOQ application; (2) **lane use management system (LUMS) and variable speed limit signs (VSLS)**: Variable speed limit is applied on Australian managed motorways. The central facility handled real-time speed reductions from the LUMS and provided this data to vehicles to ensure the accuracy of the IVSL application.

8.4.8.2   Geospatial data

Events must refer to locations or features of the road network through a standardised spatial data library. This enables any spatial data provided to the system to be uniformly handled. This data library includes:

1. **Tiles**: Map tiles geographically delineate the topics to which ITS-S can subscribe to. These were created to balance coverage and data load constraints. Originally, these were produced manually but are now using a quadtree algorithm.
2. **Variable speed limit zones**: Locations of LUMS and VSLS used to assign speed reductions to relevant areas.
3. **Detector sites**: Locations of motorway detector sites, used to associate queue detection and queue protection alerts with their relevant location to generate BoQ DENM.
4. **Road network model (RNM)**: A carriageway-level data source containing details of road alignment and attributes.
5. **MAPEM data**: A High Definition (HD) surveyed digital representation of road geometry is used to generate the MAPEM for each intersection.
6. **School zones**: Calendars for school zones are stored, and references to them were supplied as an attribute of relevant links on the RNM.

7. **Road hazard information**: Collected through a hazard aggregation platform (for Queensland, this service is called Queensland Traffic).
8. **Roadworks information**: Manually created and activated in RedCrab application by road workers on site when the physical roadworks were activated).

8.5   System evaluation

Reliable radio communications, accurate positioning and granular road network modelling on top of quality system implementation, were found to contribute to timely, accurate and relevant use case deployment. This section discusses the position and communications performance.

8.5.1   DSRC ITS-G5 radio communication performance

The allocation of the 5.9 GHz band to ITS-G5 communications for safety purposes is an important step to allow the future adoption of the 802.11bd standard for DSRC. DSRC ITS-G5 performance is vital for vehicles to effectively communicate with other vehicles and roadside stations, delivering V2V/V2I time-critical and safety-critical messages. The measurements include: (1) **latency**: The time taken for a packet to travel from source to destination; (2) **packet error rate (PER)**: The reliability of the packet exchange, which represents the proportion of sent messages that were received without error; (3) **range**: The distance that transceivers can exchange packets while the vehicle is in motion.

V2V was measured in Victoria through a designed experiment based on the CAM messages, which were periodically broadcasted at 10Hz, with each packet size of 122–133 bytes with security disabled (compared to secure packets of 215–364 bytes). Note ETSI (2014) specifies that the rate of CAM may be dynamic, depending on the heading, speed and position of the communicating devices to conserve bandwidth.

8.5.1.1   Latency

ETSI standard (2013) for longitudinal collision risk warning (LCRW) specifies a maximum end-to-end latency of 300 milliseconds, labelled as T0 to T6 in Figure 8.7. This experiment measured radio communication latency, T2 to T3 in the figure. Most

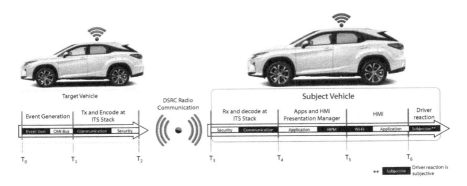

Figure 8.7   V2V end-to-end communication stack

transmissions took between 0.35 to 1.35 milliseconds, less than 2.2 milliseconds for more than 95% of the 152,000 CAMs exchanged over two days. This concluded that the technology is capable of supporting safety-critical applications.

8.5.1.2   Packet error rate and range

Packet error rate (PER) and range measurement was done by having one cooperative vehicle approach another stationary cooperative vehicle, pass it and then move away at 60 km/h. As the antenna was fixed on a sloping roof near the rear spoiler due to the panoramic moon roof, the radio communication pattern was identified to be non-omnidirectional.

Although the radio transmission range was shortest in front, creating an unfavourable antenna orientation, it still achieved a communication range of 600 metres (more than the desired range of 500 metres for DSRC).

Correcting the ground plane of the antenna can make the radio transmissions omnidirectional, which allows vehicles to communicate reliably with low PER with almost double the range, around 1,200 metres in all directions (Figure 8.8).

For V2I, the roadside stations in Ipswich were placed on the traffic-light poles transmitting SPaTEMs at 10Hz and MAPEMs at 2Hz with packet size between 273–532 bytes and 904–1789 bytes (secured packets), respectively, depending on the intersection.

The performance of V2I communication range varied at different intersections. The range of ITS-G5 short-range communication message reception is affected by obstacles that impede line-of-sight communication between the RSU and OBU. These measurements were based on the drives done on public roads. For example, at two different intersections, V2I communication at Brisbane Road/Hopener Road was reliable for up to 300 metres in front of the vehicle and 700 metres for the rear of the vehicle; at Warwick Road/Chelmsford Avenue, the signal from the roadside station is limited by the curved road and a tall building to the northern side of the intersection,

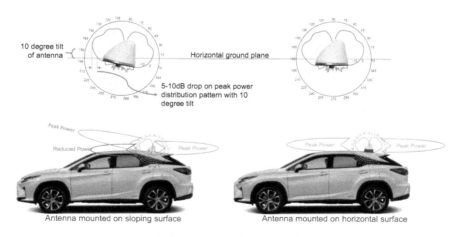

Figure 8.8   Impact of antenna placement

where the range was only 120 metres on the approach. Although in this case, the communication range is sufficient, for future new use cases, improved range might be needed. It must be noted that successful packet transmission between two ITS-S was observed in many cases at over 1.5km, where line-of-sight was maintained.

Since radio signal obstruction can be alleviated by relocating equipment to maximise line-of-sight communication, the location of the RSU installation should be carefully selected to optimise communication range in all directions.

8.5.2   *Cellular performance*

Availability of cellular communication is critical for the reliable delivery of the cellular use case messages to the vehicle stations. Poor connectivity can lead to missed or false alarms due to V2N message update failures, and degradation of RTK position augmentation services. The city of Ipswich represents a very typical regional/suburban Australian environment and is a good candidate to assess availability of high-quality, low-latency cellular connections.

Of present concern, however, is inconsistent cellular coverage in regional and remote Australia, areas which experience disproportionately high road trauma (see: 8.1.1). Advancements in, and use of, Low Earth Orbit (LEO) cellular connectivity systems could improve network accessibility, and should be considered for future wide-area trials in Australia and other sparsely population regions.

Cellular reception was assessed for the ICVP using 12 months of the vehicle stations' Receive Signal Strength Indicator (RSSI) data. Measuring by absolute road length, it was found that 38% of the road network had good cellular strength (an RSSI > -75 dBm), 57% of the road network had fair cellular strength (an RSSI of -75 to -85 dBm) and 5% of the road network had poor cellular strength (RSSI < -85 dBm).

8.5.3   *Positioning performance*

Cooperative vehicles assess threats to alert drivers using the data received from multiple external sources and based on time-to-action (TTA) to the hazardous events computed by the distance to the event and its vehicle speed. Use case performance is dependent on the positioning accuracy. Inaccurate positioning information can result in false or early/delayed warnings.

Longitudinal accuracy is important for a cooperative vehicle to determine its location relative to events in its traveling direction, such as the distance from the stop bar at an intersection. Lateral accuracy is crucial for determining which lane the vehicle is in (Figure 8.9).

GNSS-based positioning solutions are performant in regional and suburban environments with little overhead obstruction that would affect the reception of satellite signals. Generally, GNSS solutions fail where satellite signals are interrupted, either through obstruction or reflection. These effects are commonly observed in urban environments where tall and tightly spaced buildings (urban canyons), dense urban greenery and bridges/covered roadways are common.

Positioning performance can also degrade due to other factors such as the loss of the RTK solution when the cellular coverage is poor or when a cooperative

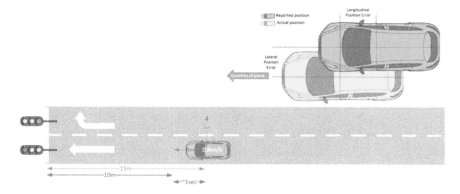

Figure 8.9   Impact of longitudinal and lateral positioning error

vehicle is far away from RTK reference stations. Different makes, models and grades of GNSS receivers also have different levels of positioning performance.

Using a GNSS receiver equipped with an Inertial Navigation System (INS), which utilises rotation and acceleration information from an Inertial Measurement Unit (IMU) for dead reckoning, in conjunction with connection to an RTK correction service that can dynamically connect to the nearest reference stations, would offer the best positioning capability to deliver appropriate driver alerts.

Positioning and other enabling technologies will evolve between now and any prospective implementation in vehicles for Australia. Deeper integration with existing vehicle systems and advanced driver assistance systems will lead to greater accuracy and information presented to the driver and minimise false warnings.

8.5.4   Road network model (RNM) granularity

The granularity of the RNM also affects how an event can be interpreted. For example, for IVSL, both RNM and positioning need to be accurate to display accurate speed information, especially where diverging road segments have different speed limits. As mentioned, the existing RNM used in the Queensland C-ITS deployment is developed at the carriageway level, and as such had some limitations where more complex geometry exists on the road. Furthermore, the RNM was assembled from various data sources of varying quality – the interface of these data sources had posed a challenge to the efficient update and maintenance of the RNM. The process for expanding and maintaining the RNM is becoming more efficient with further development from Queensland TMR. As more vehicles become available with C-ITS and higher-precision positioning, vehicle-generated data may become a viable source of road survey data and enable rapid creation of high-definition maps for network wide RNM. An accurate "digital twin" of the road network is a critical component of the C-ITS ecosystem, as the model is the backbone for any spatially implicated data. Governments, as road stewards, must be increasingly aware of, and involved in the collection, processing, and dissemination of quality (correct, accurate, timely) asset data that can be trusted as a true representation of the road network.

8.6 C-ITS user experience

Post-drive collection of user feedback [10] on the C-ITS system is a way to assess user acceptance of the technology. Drivers vary in their expectations and use of in-vehicle technology, and a key objective is to ensure that users do not switch off safety features they feel have no utility, as this would negate their safety purpose.

Lexus's vehicles were rotated between Queensland Government, Ipswich City Council and Lexus staff during the operational phase, and 81 drivers responded to a request to complete an online questionnaire regarding their experience with the C-ITS system. It should be noted in what follows that the 30–50 age group was over-represented in the sample, as was a more highly educated demographic (Figure 8.10).

Drivers generally found C-ITS warnings easy to understand, clear and with just the right amount of information. However, some warnings were a little distracting. Haptic alerts were particularly polarising, with 61% of respondents finding them distracting, with 38% finding them useful (Figure 8.11).

The Following represents the summary of the most commonly observed warnings, ARLW, IVSL and TWVR, and the usefulness of these warnings based on the ranking list in the driver feedback. Other V2N use cases, RWW, BoQ and RHW and the two V2V use cases, EEBL and SSV, were experienced less frequently in the Ipswich environment, which is reflected in the data below (Figure 8.12).

Figure 8.10   User feedback demographics

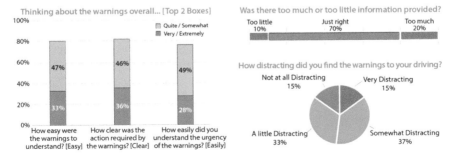

Figure 8.11   Alert quality rating

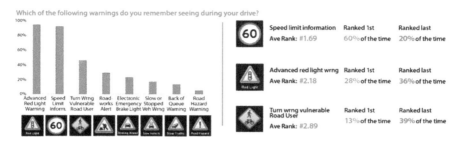

Figure 8.12   Most commonly noticed warning and ranking on the usefulness

IVSL (in-vehicle dynamic speed limit information) was commonly rated as most useful (ranked 1st) because *"It could help me avoid a fine"* and *"It was relevant to my driving"*. TWVR was often rated as least useful (ranked last) for being too late or irrelevant.

As a COVID-19 precaution, the pedestrian phase was called each cycle so that pedestrians could avoid using the buttons to call a green crosswalk signal. As a result, drivers often saw a TWVR alert without a pedestrian. Per the standards, infrastructure detection of the pedestrian is needed, which would have improved the trustworthiness of the alert.

ARLW was polarising, with some users praising the potential of it, but others put off by it being too distracting:

- It was rated as most useful because *"It could help me avoid a crash"* and *"It was easy to understand"*.
- It was rated as least useful because *"I'd already reacted to the situation"* and *"The brake warning sounded too frequently when I was already starting to brake. It was quite aggressive and distracting, which made me worried I had missed a more urgent reason to brake."*

Calibration of the timing of the alerts, potential customisation and less aggressive tone could have improved driver acceptance. Most respondents believed the biggest potential benefit of C-ITS technology was *"Saving lives by preventing accidents"* (60%), followed by *"Protecting vulnerable road users"* (~16%). Overall, 66% of drivers believed it to be moderately to extremely important to have this system in their next vehicle, 19% believed it to be slightly important. In comparison, 15% considered it unimportant, and about 2-in-5 drivers would be willing to pay extra for these systems, with the most common price in the $501 to $1,000 range. In addition to the above study, the ICVP research partners sought feedback from the 355 participants and found they were generally positive towards C-ITS and could see the potential for safety benefits. The most highly rated use case was also IVSL, whilst recognising the conceptual value for the other use cases [21]. Generally, the participants felt they had increased awareness of their surroundings. Some of their requests for future implementation include greater customisability of warnings and display; integration within existing vehicle systems; and personalised driving feedback.

8.7 Safety evaluation

Following the completion of the ICVP trial, a safety evaluation was undertaken to assess the effectiveness of the C-ITS applications to improve road safety [22]. This section summarises the study methodology, how data gathered was used to understand safety impacts of each use case, and to estimate a crash reduction factor for each use case.

8.7.1 Study methodology

The safety evaluation focussed on measuring the effectiveness of C-ITS warnings to improve road safety. The 355 public participants in the trial were split into three groups based on the timing and order of the system components (in particular, HMI) being enabled in their vehicles: (1) **treatment first**: Participants started with the HMI enabled, then later disabled; (2) **baseline first**: Participants started with the HMI disabled, then later enabled; (3) **control**: Participants did not have the HMI enabled.

The safety impact of the messages was assessed by the following measures:

Average speed: the speed while approaching/within an event.
Celeration: The average absolute acceleration/deceleration, used as an indicator of driving smoothness:

$$C = \frac{\sum_{i-1}^{n-1} |s_{i+1} - s_i|}{n - 1}$$

where C is average celeration, and s is speed.
Near-crashes: The count of near-crash events.

When applied comparatively between the treatment-first, baseline-first and control groups, these measures are used to assess the safety effectiveness of C-ITS.

8.7.2 Safety impact findings

8.7.2.1 Average normalised speed

Reduction in average normalised speed of $1-2\%$ was observed as participants approached the safety event for RWW, TWVR and ARLW, indicating improved driver behaviour. Specifically, within roadworks zones, the overall speed was reduced by 3.1% when RWW was active, suggesting that the participants in Treatment (HMI-on) group were consistently driving 2.1 km/h slower than those in the control group (HMI-off).

Conversely, there was no significant changes for BoQ and RHW use cases, likely due to a combination of environmental and system factors related to the relevance, accuracy and event-clearance criteria for these applications (Table 8.4).

8.7.2.2 Average celeration

Significant reductions in celeration were observed for RWW (in-event), BoQ, TWVR and ARLW, indicating less-aggressive, smoother driving. An increase in celeration by 4.7% in RWW approach was also observed, which was attributed to sharper braking before roadwork when drivers were made aware of upcoming speed reductions (Table 8.5).

Table 8.4   Speed comparison between HMI-on and HMI-off conditions

Use case	Impact of C-ITS safety warning on average speed	
	Approach zone	**Event zone**
RWW	1.1% Reduction	3.1% Reduction
BoQ	Nonsignificant change	Not applicable
RHW	Nonsignificant change	Not applicable
TWVR	1.2% Reduction	Not applicable
ARLW	1.9% Reduction	Not applicable

Source: Ipswich Connected Vehicle Pilot safety evaluation – summary report.

Table 8.5   Celeration comparison between HMI-on and HMI-off conditions

Use case	Impact of C-ITS safety warning on average speed	
	Approach zone	**Event zone**
RWW	4.7% Increase	3.7% Reduction
BoQ	2.8% Reduction	Not applicable
RHW	Nonsignificant change	Not applicable
TWVR	7.9% Reduction	Not applicable
ARLW	0.9% Reduction	Not applicable

Source: Ipswich Connected Vehicle Pilot safety evaluation – summary report.

8.7.2.3   Near crashes

The BoQ use case showed a potential 30–50% reduction in near crashes when HMI warnings are enabled, indicating improved driver responsiveness to changing traffic conditions following alerts. The same impact on near crashes was not observed for other use cases, however. The researchers suggest that this is likely due to the rarity of such events within the specific time windows assessed. Motorway queuing occurred multiple times per day in the pilot area, thus many events existed. Additionally, these events commonly involve significant speed changes for approaching vehicles.

8.7.2.4   Additional ARLW observations

The ARLW use case found a 22% reduction in red-light running instances when comparing HMI-on and HMI-off conditions. Furthermore, HMI-on-group participants were 27% less likely to see an escalating warning, suggesting that early notification is effective at reducing risk of red-light violation.

8.7.3   Potential crash reductions

8.7.3.1   Crash reduction factor calculation methods

In the ICVP safety evaluation, there were two methods by which Crash Reduction Factor was estimated for each use case: (1) **Nilssen power model**: This model computes a relationship between vehicle changes in vehicle speed and crash frequency/severity. It was used to estimate crash reduction rate for each FOT use case. (2) **Red-light running likelihood**: This method was used exclusively for the ARLW use case to estimate the likelihood of a driver running a red light.

These methods apply the observed change in average normalised speed identified in section 1.7.2.1 to compute the crash reduction factor for each use case.

8.7.3.2   Crash reduction factor sources

There were three data sources used by the methods in section 8.7.3.1 to estimate Crash Reduction Factor, those being: (1) **Field Operational Testing (FOT)**: Data collected from participant vehicles as part of the ICVP public trial. (2) **Simulator**: A vehicle simulator was used to evaluate driver response to a variety of road conditions and use case combinations. (3) **Literature**: Learnings from previous trials, simulations and other research that is applicable and transferrable to the context of this safety evaluation.

8.7.3.3   Crash reduction factor results

The methods and data sources in sections 8.7.3.1 and 8.7.3.2 above were applied to the ICVP applications with significant speed changes to estimate the reduction in crashes that may be realised by their respective implementation. Table 8.6 captures groups crash reductions into two categories: (1) **relevant crash types**: Estimated reduction in crashes of specific types targeted by a relevant ITS use case. For example, RWW is expected to be effective in reducing crashes in the proximity or within road works zones, while TWVR is expected to reduce crashes involving pedestrians at signalised

Table 8.6   Crash reduction factor summary

Use case	Data source	Crash reduction (relevant)		Crash reduction (network)	
		Fatalities	S. Injuries[a]	Fatalities	S. Injuries[a]
RWW (approach)	FOT	5.3%	3.0%	0.0%–0.2%	0.0%–0.2%
RWW (event)	FOT	11.4%	6.8%	0.0%–0.6%	0.05–0.2%
TWVR	FOT	6.1%	4.3%	0.2%	0.1%
ARLW (approach)	FOT	9.7%	5.9%	0.2%–0.4%	0.4%–1.0%
ARLW (event)	FOT	22.0%	22.0%	1.0%–5.6%	1.2%–5.1%
SSV	Simulator	7.5%	4.5%	0.6%–1.2%	0.9%–1.4%
EEBL	Simulator	25.3%	15.1%	1.4%–3.6%	2.6%–4.4%
IVSL	Literature	7.7%	7.7%	7.7%	7.7%
Cumulative total				**11.1%–19.3%**	**12.9%–20.1%**

[a]S. Injuries: serious injuries.
Source: Ipswich Connected Vehicle Pilot safety evaluation – summary report.

intersections. (2) **South-East-Queensland network**: Estimated overall impact of each C-ITS use case on total crashes across the entire South-East-Queensland network.

In a broader context, with a total of 526 fatalities and 20,826 serious injuries being reported from crashes on the South-East-Queensland road network over the five-year period of 1 July 2016 and 30 June 2021, the eight C-ITS use cases implemented in the ICVP could have prevented up to 101 fatalities and 4,198 serious injuries from crashes. This equates to an average of 20 fatalities and 840 serious injuries prevented each year in South-East-Queensland.

8.8   Conclusion

ICVP and AIMES offered exciting opportunities to investigate the requirements for the large-scale deployment of C-ITS technologies. In particular, the ICVP represents the largest on-road public trial of C-ITS in Australia and successfully created a real-world C-ITS ecosystem at scale to validate the impacts and benefits of C-ITS use cases and user perception.

All use cases were successfully tuned and evaluated in real-life traffic. According to user response to their driving experience, C-ITS is beneficial for road safety as well as assisting users to comply with the road rules through improved situational awareness. These deployments highlighted strong collaboration between governments, industry stakeholders, public service and academic sectors to form the base for a significant step towards safer roads where multimodal transport and road users are mixed.

ITS-G5 is an effective and reliable medium for transferring safety information between field devices; thus, preservation of the 5.9 GHz frequency band and cross-border interoperability are required for C-ITS to be commercially viable. Positioning and mapping are essential factors for C-ITS technology, and there should be further studies in these areas. By seeking to integrate real-life traffic events such as live signal status and traffic status, all stakeholders have recognised the need for robust, meaningful and current data streams to support use cases.

In ICVP, positive impacts on road safety were observed, including reductions in average speed and smoother driving. Specific warnings, like those for road works and advanced red-light warnings, proved particularly effective. The safety evaluation findings suggest the potential for significant reductions in crashes with broader C-ITS deployment across Southeast Queensland, estimating a reduction of 13–20% in serious crashes if widely adopted.

ESVN is an important safety-related C-ITS service and can be extended to other emergency services such as police and fire response. Extending C-ITS technology to all traffic participants, not only passenger vehicles but also specialised vehicles like ambulances/fire trucks, trams/buses, scooters/bikes, among others, would greatly increase the penetration rate of C-ITS and maximise its benefits. This can be supported by evaluating retrofit solutions for existing long/medium-life assets.

Harmonisation with the European Union is required to ensure that C-ITS systems can be developed and deployed for Australia with minimal additional cost. Australia- wide consistency including the standardisation of technology and systems is required before automakers can bring the product to Australia. A national harmonisation project involving multiple jurisdictions has recently been announced to drive the nationally consistent C-ITS deployments. Ongoing collaboration across government, academia, and industrial sectors is critical to maintain momentum of works and investment into C-ITS. Targeted, standardised deployment of C-ITS technology will support road user expectations and maximise progress towards associated safety, efficiency, and sustainability objectives.

At this stage, there is enough evidence from this activity to recommend continuing with the development of C-ITS for the Australian market. These technologies will enhance in-vehicle safety systems and complement both existing and planned road infrastructure safety measures. Ongoing investment in projects and infrastructure is required to keep moving towards commercialisation, bringing this lifesaving technology to all Australians and creating a safer, cleaner and more mobile future.

References

[1]   World Health Organization (2018). *Global Status Report On Road Safety 2018*. https://www.who.int/publications/i/item/9789241565684. Accessed on 1st September 2024.

[2]   Commonwealth of Australia (2023). *Bureau of Infrastructure and Transport Research Economics (BITRE) Road Trauma Australia—Annual Summaries* https://www.bitre.gov.au/publications/ongoing/road_deaths_australia_annual_summaries. Accessed on 1st September 2024.

[3]   Commonwealth of Australia (2021). *International Road Safety Comparisons 2019* https://www.bitre.gov.au/sites/default/files/documents/international_comparions_2019.pdf. Accessed on 1st September 2024.

[4]   Commonwealth of Australia (2021). *National Road Safety Strategy 2021-30*. https://www.roadsafety.gov.au/sites/default/files/documents/National-Road-Safety-Strategy-2021-30.pdf. Accessed on 1st September 2024.

[5]   ITS Connect Promotion Council (2023). *To Enhance Vehicle Safety through the Development of Aftermarket Onboard Devices and Further Promote ITS Connect*. https://www.itsconnect-pc.org/wp/wp-content/uploads/2023/11/20231115_PR.pdf. Accessed on 1st September 2024.

[6]   Toyota Motor Corporation (2021). *TOYOTA ITS Web Exhibition 2021-V2X-*. https://www.toyota.co.jp/its/en/2021/. Accessed on 1st March 2024.

[7]   Logan, D. B., Young, K., Allen, T. and Horberry, T. (2017). *Safety Benefits of Cooperative ITS and Automated Driving in Australia and New Zealand, Austroads Research Report*. AP-R551-17.

[8]   Queensland Government (2021). *Ipswich Connected Vehicle Pilot*. https://www.qld.gov.au/transport/projects/cavi/ipswich-connected-vehicle-pilot. Accessed on 1st September 2024.

[9]   Queensland Government Department of Transport and Main Roads (2022). *Ipswich Connected Vehicle Pilot*. https://www.tmr.qld.gov.au/business-industry/Technical-standards-publications/Ipswich-Connected-Vehicle-Pilot. Accessed on 1st September 2024.

[10]   Lexus Australia (2020). *Cooperative and Automated Vehicle Initiative Ipswich Connected Vehicle Pilot – Final Report*. (Unpublished Report Submission to TMR)

[11]   The University of Melbourne Faculty of Engineering and Information Technology (2023). *Australian Integrated Multimodal EcoSystem (AIMES)*. https://eng.unimelb.edu.au/industry/aimes. Accessed on 1st September 2024.

[12]   Ambulance Victoria. *Annual Report 2021–2022*. https://www.ambulance.vic.gov.au/wp-content/uploads/2022/12/Ambulance-Victoria-Annual-Report-2021-22.pdf. Accessed on 1st September 2024.

[13]   Road Safety Road Rules 2017 S.R. No. 41/2017 Authorised Version Incorporating Amendments as of 4th November 2020 (Authorised Version No. 009)

[14]   Victoria State Government (2021). *Law Enforcement & Emergency Vehicles*. https://www.vicroads.vic.gov.au/safety-and-road-rules/road-rules/a-to-z-of-road-rules/law-enforcement-and-emergency-vehicles. Accessed on 1st September 2024.

[15]   Victoria State Government (2022). *Tram Collisions on the Rise*. https://transport.vic.gov.au/about/transport-news/news-archive/tram-collisions-on-the-rise. Accessed on 1st September 2024.

[16]   Yarra Trams (2020-05). *Tram Stop Road Safety Data Insights Overview* (unpublished Yarra Trams analysis).

[17]   iMove (2024). *C-ITS National Harmonisation and Pre-Deployment Research*. https://imoveaustralia.com/project/c-its-national-harmonisation-and-pre-deployment-research/. © iMOVE 2024. Accessed on 1st September 2024.

[18]   Australian Government The Department of Infrastructure, Transport, Regional Development, Communications and the Arts (2023). *Principles for a National Approach to Co-operative Intelligent Transport Systems (C-ITS) in Australia*. https://www.infrastructure.gov.au/infrastructure-transport-vehicles/transport-strategy-policy/office-future-transport-technology/intelligent-transport-systems/principles-national-approach-co-operative-intelligent-transport-systems-c-its-australia. Accessed on 1st September 2024.

[19]   Australian Government The Department of Infrastructure, Transport, Regional Development, Communications and the Arts (2024). *Draft National Road Transport Technology Strategy and 2024–27 National Connected and Automated Vehicle (CAV) Action Plan*. https://www.infrastructure.gov.au/have-your-say/draft-national-road-transport-technology-strategy-and-2024-27-national-connected-and-automated. Accessed on 1st September 2024.

[20]   The State of Queensland (Department of Transport and Main Roads) (2022). *TMRD26 Master System Architecture and Design* https://www.tmr.qld.gov.au/business-industry/technical-standards-publications/ipswich-connected-vehicle-pilot. Accessed on 1st September 2024.

[21]   Transport for NSW (2022). *SCATS and Intelligent Transport Systems*. https://www.scats.nsw.gov.au/. Accessed on 1st September 2024.

[22]   The State of Queensland (Department of Transport and Main Roads) (2022). *Ipswich Connected Vehicle Pilot Safety Evaluation – Summary Report*. https://research.qut.edu.au/carrsq/wp-content/uploads/sites/296/2022/11/IVCP-Safety-Evaluation-1.pdf. Accessed on 1st September 2024.

Chapter 9

C-ITS evaluation on C-Roads – findings from C-Roads Germany

Orestis Giamarelos[1] and Jan Schappacher[1]

9.1 Introduction

The C-Roads Platform is a collaborative initiative involving European Member States and road operators aimed at advancing C-ITS across Europe. The initiative focuses on harmonising C-ITS deployments to facilitate seamless vehicle-to-vehicle (V2V) and vehicle-to-infrastructure (V2I) communications across borders. By doing so, C-Roads aims to enhance road safety, improve traffic efficiency and ensure interoperability of transport systems through standardised technological solutions and collaborative efforts across member states. The platform's strategy is structured around harmonising functional and technical specifications for C-ITS services, addressing both short-range communications (ETSI ITS G5) and long-range cellular solutions.

The C-Roads Platform comprises several working groups (WGs), each dedicated to specific aspects of C-ITS implementation (see Table 9.1).

This chapter focuses on work accomplished within WG3, which is responsible for the evaluation and assessment of C-ITS deployments. WG3 is integral to the C-Roads Platform, ensuring that C-ITS deployments effectively meet operational and policy objectives. This group's work is structured around three main pillars:

1. Transmission of achievements: Facilitates the application of results from various working groups into practical, operational environments to achieve service harmonisation and interoperability.
2. Methodology definition and cross-site testing: Develops and implements methodologies for Europe-wide testing, ensuring consistency and comprehensiveness in evaluations.
3. Impact Assessment: Critically assesses the broad impacts of C-ITS deployments on safety, efficiency, interoperability and environmental sustainability. This includes evaluating the interoperability between different countries, partners and technologies.

---

[1]Federal Highway Research Institute (BASt), Section Traffic Management and Road Maintenance Services, Bergisch Gladbach, Germany

Table 9.1   An overview of the C-Roads Platform working groups

No.	WG title	Description
WG1	C-ITS organisation	Focuses on the coordination between public authorities and private entities to foster C-ITS deployment, addressing organisational and legal challenges, including privacy issues, to ensure seamless service delivery.
WG2	Technical aspects	Deals with the standardisation of technical interfaces and specifications necessary for interoperable C-ITS services across Europe. This includes harmonising communication profiles for road infrastructures and addressing security aspects related to C-ITS service provision.
WG3	Evaluation and assessment	Dedicated to assessing the effectiveness of C-ITS implementations, ensuring they meet safety, efficiency and environmental goals.
WG4	Urban C-ITS harmonisation	Focuses on integrating C-ITS applications within urban settings, ensuring that city traffic management systems can effectively utilise these technologies to improve urban mobility.
WG5	Digital transport infrastructure	Works on developing digital infrastructure that supports the deployment and operation of C-ITS services, focusing on the technological backbone necessary for these systems to function efficiently across different regions.

The activities of this group provide evidence-based recommendations to enhance the functionality and integration of C-ITS services across Europe.

9.2   The C-Roads Evaluation and Assessment Plan

The C-Roads Evaluation and Assessment Plan [1], established under WG3, serves a pivotal role in the structured assessment of C-ITS within the framework of the C-Roads projects. Drafted early in the project's lifetime, this comprehensive document outlines the evaluation protocols and methodologies designed to assess the deployment and operational efficiency of C-ITS across diverse European pilot sites. The plan's primary objective is to facilitate a standardised approach to evaluation, aligning with the overarching goals of enhancing road safety, improving traffic efficiency and increasing the sustainability of transport infrastructure across Europe.

The scope of the Evaluation and Assessment Plan extends to a detailed examination of Day 1 C-ITS services and use cases, focusing on their impacts across multiple dimensions (see Table 9.2).

These areas are the unanimously guideline for assessment in pilot tests, although exceptions are made if certain areas are deemed inapplicable or beyond the scope of contractual obligations. This comprehensive scope ensures that each pilot not only adheres to a uniform evaluation framework but also contributes meaningfully to the collective understanding and advancement of C-ITS technologies.

Table 9.2   C-Roads impacts evaluation and assessment dimensions

Impact assessment dimension	Description	
User acceptance	Assessing how end-users perceive and adopt the C-ITS technologies.	
Functional evaluation	Measuring the functionality and performance of C-ITS services.	
Impact assessment	Safety	Evaluating the contribution of C-ITS to road safety improvements.
	Traffic efficiency	Determining the effects of C-ITS on traffic flow and congestion management.
	Environment	Analysing the environmental benefits resulting from C-ITS deployment, such as reductions in emissions.
	Socio-economic impacts	Gauging the broader economic implications and benefits of C-ITS technologies.

Table 9.3   Indicators and methods for evaluating user acceptance

Indicators and methods	Description
Observed rate of use	This indicator measures the frequency and extent of use of the system or specific parts of the system, providing insights into the system's acceptance and perceived usefulness.
Perceived system consequences	It captures user impressions and attitudes regarding the potential positive or negative consequences of using the system. Such data are typically gathered through interviews and exploited further in focus groups.
Motivation and behavioural intention	These are assessed through partially harmonised questionnaires, gauging the level of motivation to use the system and the intention to continue using it.
Response to perceived social control	This sociological indicator assesses the social benefits or disapproval perceived by users related to system usage, reflecting societal norms and values associated with ITS.

The method for evaluating **user acceptance** within the C-Roads Evaluation and Assessment Plan encompasses various indicators and methods to measure how well users perceive and interact with C-ITS technologies. These are summarised in Table 9.3.

Data collection methods include surveys and interviews, which are used to directly capture user feedback on their experiences and perceptions, as well as focus groups, which offer a dynamic setting to explore collective user attitudes and deeper insights into acceptance factors. For all these data collection methods compliance with the European General Data Protection Regulation (GDPR) is mandatory, ensuring that all user data is collected and handled with proper regard to privacy and data protection regulations.

The C-Roads Evaluation and Assessment Plan also provides a structured approach to the functional evaluation of C-ITS services. The process is focused on assessing the real-world performance of services as experienced by end-users, particularly emphasising how the human–machine interface (HMI) performs in delivering C-ITS services. The evaluation framework involves several key components. The scope of the evaluation includes the following aspects:

1. Content detection: Assessing the accuracy and reliability of the information detected by C-ITS sensors and systems.
2. Content processing: Evaluating the efficiency and correctness of the processing algorithms that interpret the raw data from the sensors.
3. Service provision: Examining the adequacy and timeliness of the services provided to the end-users based on the processed data.
4. Service presentation: Focusing on how information is presented to the user through the HMI, evaluating the clarity, timeliness and appropriateness of alerts and messages.

In addition, the functional evaluation assesses the presentation quality, which includes the clarity of information displayed, the user interface design and the responsiveness of the system to user inputs. Although the C-Roads platform does not standardise HMI designs, the evaluation looks at how these designs impact the effectiveness of service delivery.

Finally, an important component for capturing insights from the implementation of C-ITS services involves documenting the challenges encountered and the successes achieved during the deployment phase. These lessons learned are vital for refining future implementations and enhancing the overall design and functionality of C-ITS services.

The impact assessment within the C-Roads Evaluation and Assessment Plan aims to quantify the effects of C-ITS services on safety, traffic efficiency and the environment. The methodology is comprehensive, involving several analytical approaches of baseline and impact comparisons:

1. Before and after: Comparing the state of traffic, safety and environmental conditions before and after the implementation of C-ITS services. This approach requires the collection of comprehensive baseline data before the services are deployed.
2. Difference in differences: This method involves comparing the changes in traffic conditions, safety and environmental impacts between areas or times with C-ITS services and those without, to isolate the effects of the interventions.
3. Randomised control trials: Where feasible, this approach involves randomly assigning participants to either receive C-ITS services or be part of a control group without such services. This method helps in determining the causality of observed changes.

To complete the evaluation, data is collected from various sources, also including vehicle telemetry (e.g. ITS Station, CAN Bus data, GPS), which provides detailed information on vehicle behaviour and environmental conditions. This data is valuable for assessing the direct and indirect impacts of C-ITS services on driving behaviour and traffic conditions.

The impact assessment covers various C-ITS services such as road works warning, in-vehicle signage and others. For each service, specific use cases are evaluated to determine their individual impacts on safety, traffic flow and environmental conditions.

This structured approach ensures that the evaluations not only reflect the direct impacts of the services but also consider broader social and environmental implications, providing a holistic view of the benefits and challenges associated with C-ITS deployments. These methods, as outlined in the C-Roads Evaluation and Assessment Plan, provide a robust framework for assessing the various dimensions of C-ITS implementations, ensuring that the deployments are effective, user-friendly and beneficial from a socio-economic perspective.

9.3 The German pilot sites

Within the framework of C-Roads Germany and C-Roads Germany Urban Nodes, several C-ITS use cases are being tested in various German pilot locations.

The primary goals of these pilot sites include fostering a standardised deployment pattern for C-ITS services across Germany that conforms to EU regulations and the recommendations of the C-ITS platform. This effort aims to not only improve the functionality of current transport systems but also ensure their compatibility and interoperability with systems across other EU member states. By encouraging the German automotive industry to integrate C-ITS compatible systems, the initiative seeks to increase the adoption rate of such technologies among end-users, thereby enhancing overall traffic safety and efficiency.

Additionally, these pilot sites serve as a model for the potential harmonisation of C-ITS deployments at a European level, providing crucial insights into the organisational patterns required for successful implementation. They also contribute to developing the necessary infrastructure, technical specifications and evaluation methods to assess the effectiveness of deployed C-ITS services.

A short description of the German pilot sites is provided below. More detailed information is provided on the website of C-Roads Germany urban nodes [2]. In addition, the C-Roads platform publishes an annual pilot report [3], which includes the latest developments from the pilot sites from all participating countries.

9.3.1 C-Roads Germany pilot sites

In the already completed first phase of C-Roads Germany, two German pilot sites participated: Niedersachsen and Hessen. Both pilot sites were integral components of the broader C-Roads Germany initiative, which seeks to implement C-ITS technologies that can communicate across different systems and borders, ultimately aiming to enhance road safety, manage traffic more efficiently and reduce environmental impacts.

The pilot site in Niedersachsen had as primary focus areas sections of the A2 motorway, as well as federal roads connecting Hannover, Brunswick and Wolfsburg. The pilot focused on enhancing traffic management and safety through the deployment of three key C-ITS services: Slow or stationary vehicle warning (SVW), in-vehicle signage (IVS) and probe vehicle data (PVD). These services were integrated into existing traffic management centres to test and demonstrate how C-ITS can enhance the capabilities of traffic management operations and provide additional benefits to road operators by deploying and utilising these advanced services.

The Hessen pilot site focused on the deployment of C-ITS around Frankfurt am Main. It was dedicated to advancing cooperative, connected and automated traffic technologies through several innovative services that enhance traffic efficiency and safety across the region. Key services tested and implemented include PVD, road works warning (RWW) and traffic jam ahead warning (TJW).

9.3.2   C-Roads Germany urban nodes pilot sites

In the second phase of C-Roads Germany (urban nodes), three German pilot sites participate: Hamburg, Hessen/Kassel and Dresden.

The Hamburg pilot site focuses on deploying innovative cooperative intelligent transport system services to enhance urban mobility and safety. Key services implemented include the green light optimal speed advisory (GLOSA), traffic signal priority (TSP) and probe vehicle data (PVD). These technologies aim to improve traffic flow efficiency, reduce emissions and enhance overall traffic safety by enabling more intelligent vehicle-to-infrastructure communication. The pilot operates along the TAVF test track for automated and connected driving, which provides a complex environment for testing and developing these C-ITS services within the city's main road network, including heavily loaded main arteries. This pilot site exemplifies the integration of C-ITS into a complex urban environment, showcasing the benefits of real-time, data-driven traffic management solutions.

The pilot site in Hessen/Kassel integrates C-ITS in both urban and interurban settings, with a focus on services that enhance traffic safety and efficiency. Key services include emergency vehicle approaching (EVA), which optimises routes for emergency responses and vulnerable road user (VRU) protection, which enhances safety measures for pedestrians and cyclists. In addition, the route advice service is deployed to manage traffic flows on highways near the city, using variable message signs for network control and providing route recommendations to drivers, with the goal to enhance traffic capacity and reduce congestion through better information dissemination and cooperative traffic management strategies. The infrastructure includes numerous roadside ITS stations, as well as several vehicle ITS stations installed in public transport, emergency and road operator vehicles, that facilitate effective communication across the traffic network over ETSI ITS G5.

The Dresden pilot site focuses on urban C-ITS applications, aiming to enhance traffic safety and efficiency across the city's main roads. The test track includes heavily loaded main streets and access roads, providing a challenging and representative framework for testing C-ITS services. Services such as TSP and GLOSA are utilised to improve the efficiency of traffic flows and reduce waiting times at intersections, while further implemented services include EVA, PVD and VRU protection.

Each of these pilot sites illustrates the unique applications of C-ITS technologies tailored to the specific needs of urban and interurban environments. They serve as testbeds for innovative traffic management solutions that can be scaled and adapted for broader use across Germany and Europe, aligning with the goals of the C-Roads platform to enhance the interoperability and efficiency of transport systems in Europe.

9.4   C-Roads Germany Urban Nodes Evaluation and Assessment Plan

The C-Roads Germany Urban Nodes (CRG-UN) Evaluation and Assessment Plan [4] is a comprehensive document designed to outline the methodologies and frameworks for evaluating C-ITS across several pilot sites in Germany. This plan aims to provide a structured approach to assessing the efficiency and impact of C-ITS implementations within urban nodes, leveraging lessons from previous projects and aligning with current technological and policy developments in intelligent transportation.

The primary objective of the CRG-UN Evaluation and Assessment Plan is to ensure that all C-ITS pilot implementations in CRG-UN (Hamburg, Hessen/Kassel and Dresden) are systematically evaluated against a set of predefined metrics and criteria. As previously mentioned, the evaluation can consist of confirming the functionality and user acceptance of these systems as well as measuring the impact of the implementations on traffic efficiency, safety and environmental sustainability. By doing so, the plan supports the broader goals of the C-Roads platform to enhance mobility and safety across European road networks through harmonised C-ITS services.

The CRG-UN plan is linked with the broader C-Roads platform by providing a harmonised framework that aligns with the evaluation and assessment guidelines established at the platform level. The plan incorporates inputs from national harmonisation efforts and integrates insights from the German pilot sites, thus ensuring a cohesive approach to evaluating C-ITS across different jurisdictions and contexts. The document scope and input dimensions are visualised in Figure 9.1.

Structured around the insights gained from the predecessor project C-Roads Germany, the CRG-UN plan adopts a refined methodological approach to evaluation and assessment. It acknowledges the distinct requirements and focuses of the German pilot sites, allowing for tailored evaluation strategies that address the specificities of each site. While each pilot site explores a subset of the C-Roads catalogue of C-ITS services, the evaluation plan ensures comprehensive coverage by focusing on aspects particularly relevant to the operational goals of each pilot.

Overall, this evaluation framework facilitates the synthesis of data and insights across different pilots and enhances the ability to draw broader conclusions about the effectiveness and scalability of C-ITS services. Therefore, by aligning pilot-specific evaluations with national and European C-ITS objectives, the plan not only

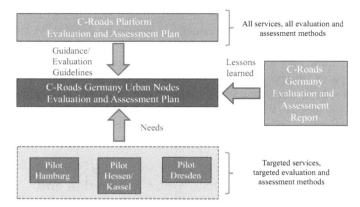

Figure 9.1   Document scope and input dimensions of the CRG-UN Evaluation and Assessment Plan (adapted from [4])

supports the operational success of current implementations but also contributes to the ongoing development and harmonisation of intelligent transportation systems across Europe.

9.5   Results from the German C-Roads Pilots

The evaluation results from the pilots of all C-Roads partners were regularly exchanged with each other, discussed in WG3 meetings and they are published at the end of each project phase in the WG3 Final Report. The final report of the second phase of the C-Roads platform will be published at the end of 2024. Therefore, in this section results from the C-Roads Germany pilots are presented, based on the WG3 Final Report [5] of the first phase of the C-Roads platform: A combined user survey and viewpoint analysis study conducted in Germany, as well as a summary of the evaluation of the Roadworks Warning – Lane Closure (RWW-LC) use case.

9.5.1   *User survey with combined viewpoint analysis of the C-ITS Service GLOSA*

This study focused on evaluating the GLOSA system implemented in specific urban settings. GLOSA systems are designed to provide speed recommendations to drivers to minimise stops at red lights, thereby enhancing traffic flow and reducing emission levels. This study explored the effectiveness of the GLOSA service in real-world conditions on a route containing six equipped signalised intersections, its impact on driving behaviour and the implications for traffic management and environmental sustainability.

The primary objectives of the study were to assess how effectively the GLOSA system influences driver behaviour, particularly in terms of compliance with speed recommendations, and to evaluate its impact on overall traffic efficiency and environmental factors like fuel consumption and emissions. Additionally, the study aimed to understand user acceptance of the GLOSA system and its potential integration into broader traffic management strategies.

The study compared the effects of a widely available Google Navigation App as baseline and a specialised GLOSA app as the target of evaluation. The test persons drove the same route twice, once using the Google Navigation App and on the second round using the GLOSA app. This comparison aimed to assess which application caused more distraction to the drivers. By testing both apps under similar conditions, the researchers could also evaluate the relative driving performance during the use of each app to analyse the changes based on the speed advisories and overall influence on driving patterns. An additional feature of the test person collective was the setup of two small groups, one using the GLOSA app for the first time on the testing day, and one group already familiar with the service beforehand.

The test persons were equipped with eye tracking glasses, in order to analyse gaze behaviour and understand how often and for how long drivers looked at the GLOSA system interface compared to the Google Navigation App. This evaluation helped assess whether the GLOSA system distracted drivers or if it was integrated into the driving process in a way that maintained safe driving practices.

At the same time, it was monitored if the drivers adhere to the speed recommendations provided by the GLOSA system. The key evaluation metric was whether the speed advice allowed drivers to synchronise with green lights, thus minimising stops at red lights. The effectiveness of these recommendations was gauged by observing changes in driving speed as vehicles approached intersections.

The evaluation of user behaviour analysed changes in driving habits resulting from using the GLOSA system. This involved monitoring how drivers adjusted their speed, acceleration and deceleration in response to the system's recommendations. By examining these behaviours, the study aimed to determine if the GLOSA system led to smoother driving patterns and whether it influenced drivers to follow the speed advisories consistently over time.

These methods allowed the researchers to gather both quantitative data with the use of onboard video recording and GPS tracking as well as the eye tracking device (speed compliance and gaze patterns) and qualitative feedback by filling in questionnaires (user satisfaction and system usability). This combination provided a holistic view of the GLOSA system's impact on traffic flow, driver behaviour and overall system efficacy in real-world driving conditions as well as a comprehensive understanding of both the impact of GLOSA on driving performance and driver interaction with the system.

The results showed that the GLOSA system effectively provides speed recommendations that help drivers synchronise with green lights, thus reducing the frequency of stops at red lights. The evaluation highlighted a significant level of driver compliance with the speed recommendations, indicating the system's reliability and the accuracy of its predictions. The gaze behaviour analysis revealed that the GLOSA system did not particularly distract drivers, see Tables 9.4 and 9.5.

The system interface was deemed adequately intuitive, allowing drivers to quickly glance for information without significant diversion from the road. The

Table 9.4 Gaze allocations of the test persons for Google App baseline (adapted from [6])

GOOGLE APP		GAZE ALLOCATIONS TOTAL			
		Driving		Stopping	
		Number	Avg. Duration [s]	Number	Avg. Duration [s]
VLSA 1	Experienced	20	0.77	19	0.66
	Inexperienced	36	0.41	19	0.57
VLSA 2	Experienced	25	0.73	26	0.50
	Inexperienced	34	0.62	19	0.65
VLSA 3	Experienced	21	0.57		
	Inexperienced	17	0.56		
VLSA 4	Experienced	9	0.70		
	Inexperienced	14	0.49		
VLSA 5	Experienced	35	0.60	1	0.70
	Inexperienced	35	0.59		
VLSA 6	Experienced	30	0.56	24	0.40
	Inexperienced	15	0.66	7	0.44
SUM EXPERIENCED		140	3.93	70	2.26
SUM INEXPERIENCED		151	3.33	45	1.66

Table 9.5 Gaze allocations of the test persons for GLOSA app (adapted from [6])

GLOSA APP		Driving		Stopping	
		Number	Avg. Duration [s]	Number	Avg. Duration [s]
VLSA 1	Experienced	42	0.46	22	0.45
	Inexperienced	93	0.77	24	0.87
VLSA 2	Experienced	19	0.39	36	0.66
	Inexperienced	58	0.67	34	0.67
VLSA 3	Experienced	21	0.40		
	Inexperienced	45	0.54		
VLSA 4	Experienced	13	0.23		
	Inexperienced	37	0.62		
VLSA 5	Experienced	14	0.34		
	Inexperienced	58	0.61		
VLSA 6	Experienced	33	0.53	9	0.71
	Inexperienced	51	0.61	1	0.40
SUM EXPERIENCED		142	2.35	67	1.82
SUM INEXPERIENCED		342	3.82	59	1.94

GAZE ALLOCATIONS TOTAL

summary results highlighted in red refer to the total number of gazes on the respective system. While users familiar with the GLOSA interface (experienced group) take a similar total number of gazes towards the HMI, the group of first time GLOSA users (inexperienced group) take a higher number of gazes on the HMI. In relation to the total duration of gazes resting on the respective system, indicated by the blue highlight, low totals were calculated for both systems, indicating negligible distraction from the main driving task.

Further analysis focused on how the GLOSA recommendations influenced driving speed as vehicles approached intersections, see Table 9.6.

The results indicated that the system effectively aided drivers in adjusting their speeds in anticipation of changing signals, thus improving their ability to pass through intersections without stopping. With a free choice of speed, assumed for intersections (VLSA) one to five, the use of GLOSA results in a speed reduction. Due to the traffic conditions on the study day, i.e. at intersection number six, the choice of speed was influenced by other factors and therefore deviates from this tendency.

Overall, the impact of GLOSA on traffic efficiency was evident in the reduced number of stops at traffic signals, which typically contribute to delays and increased fuel consumption. By advising drivers on optimal speeds to match green light phases, GLOSA contributed to a more fluid movement through intersections, thereby enhancing the overall efficiency of the network. Also from an environmental perspective, GLOSA demonstrated potential benefits by reducing the occurrences of vehicle idling and frequent stops, which are significant sources of emissions in urban areas.

User feedback was predominantly positive, with many drivers appreciating the system's assistance in making more informed speed decisions. This feedback underscores the user-friendliness and practical utility of the GLOSA system, aspects that are crucial for its acceptance by the users.

Finally, based on the findings of the study, several recommendations were made to enhance the GLOSA system's effectiveness and user experience. These include improvements to the system's predictive accuracy, particularly in dynamically changing traffic conditions, and further refinements to the user interface to ensure even greater ease of use and minimal driver distraction.

9.5.2   Roadworks Warning – Lane Closure (RWW-LC)

In the C-ITS Use Case Road Works Warning – Lane Closure (RWW-LC), the road user receives information about the closure of one or several lanes, either partially or completely, due to a road works site. The use case was evaluated focusing primarily on assessing its safety impacts within the context of its economic viability. This assessment was part of a broader evaluation framework used to determine the effectiveness of the RWW service in enhancing road safety at short-term road works sites.

The safety impact assessment began with a scoping step that involved analysing the most frequent accident patterns observed near short-term road works. These accidents typically involve collisions with road works safety trailers, merging and weaving before lane closures, and rear-end collisions further upstream. This analysis highlighted the inherent risks associated with these environments, with an estimated 100 accidents per year occurring at short-term road works on motorways in the

Table 9.6 *Driving speed analysis (adapted from [6])*

DRIVING SPEED ANALYSIS				
		Free choice traffic flow		
SPEED TOTAL		Avg. Speed [km/h]		Avg. Speed Difference [km/h]
		Google App	GLOSA App	
VLSA 1	Experienced	41.95	34.85	-7.10
	Inexperienced	41.60	34.15	-7.45
VLSA 2	Experienced	43.08	40.42	-2.67
	Inexperienced	47.50	42.17	-5.33
VLSA 3	Experienced	47.25	44.83	-2.42
	Inexperienced	49.50	48.33	-1.17
VLSA 4	Experienced	51.33	48.25	-3.08
	Inexperienced	54.50	52.00	-2.50
VLSA 5	Experienced	46.50	45.92	-0.58
	Inexperienced	50.08	47.08	-3.00
VLSA 6	Experienced	35.42	41.17	5.75
	Inexperienced	37.00	46.25	9.25
AVG. EXPERIENCED		44.26	42.57	**-1.68**
AVG. INEXPERIENCED		46.70	45.00	**-1.70**

Federal State of Hessen, Germany. Extrapolating these figures nationally, approximately 800 related accidents occur annually across all German motorways.

The evaluation methodology did not involve new data logging or measurements; instead, it leveraged data from previous large-scale field operational tests (FOTs) in Germany, some of which contributed to European C-ITS FOTs. This existing data provided a basis for assessing the impact of the RWW-LC service, which offers earlier and more explicit warnings to drivers.

The earlier warnings are designed to prompt changes in driver behaviour, specifically in terms of speed and acceleration adjustments. Complementing these objective measures, subjective data from user surveys indicated that two-thirds of users perceive the RWW service as enhancing road safety. This subjective feedback supports the objective data suggesting significant safety improvements.

The evaluation extended to consider the economic implications of deploying the RWW service. The safety impacts were viewed within the larger scope of economic viability, which considers likely market penetration and hybrid communication strategies. The resulting benefit-cost ratios (BCRs) for the years 2018–2032 of the three penetration scenarios researched are presented in the diagrams in Figure 9.2. A distinction was made between the number of traffic control centres required as base stations for the management of the C-ITS service (0.5 or 1 traffic control centre per

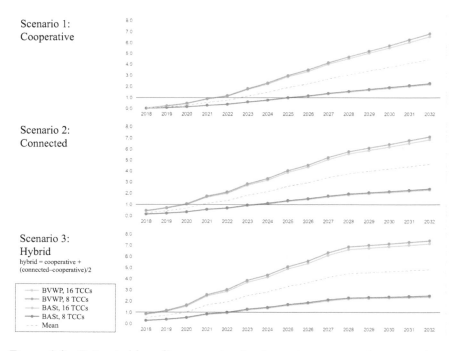

Figure 9.2   Estimated benefit–cost ratios for the various penetration scenarios, accident cost estimation methodologies and number of traffic control centres (adapted from [7])

German federal state, thus a total of 8 or 16 traffic control centres, as well as between two widely used methodologies for the accident cost estimation. The two accident cost methodologies considered differ in the scale of socio-economic costs included into the computation. The BVWP method used for the national socio-economic assessment for infrastructure projects results in higher accident costs than the more conservative methodology to compute accident costs established by BASt. In Figure 9.2, a mean value of these four combinations is displayed in the graph, while the red line shows the year that a positive BCR (BCR>1.0) has been achieved.

It can be observed that the number of traffic control centres required has only a marginal influence on the results, while the effect of the different accident cost estimation methodologies is reflected in the results to a much greater extent.

However, it is apparent that significantly positive BCRs are achieved in all scenarios after a certain time. Using the BVWP accident cost estimation methodology, a BCR>1.0 is already achieved within one year in the "connected" scenario. On average across all scenarios, a positive BCR is achieved in a maximum of four years, while by 6–10 years all scenarios exceed a BCR of 4.0. Using the BASt accident cost estimation methodology, a positive BCR is achieved in 4–7 years. At the end of the forecast horizon (14 years), a BCR of approximately 7.5 is estimated for the BVWP methodology and a BCR of 2.5 for the BASt methodology, leading to a mean BCR of approximately 5.0.

As an example, for the year 2025 the safety benefits in the hybrid scenario were estimated at approximately €87.5 million annually. These benefits primarily stem from reductions in personal damage (casualties, injuries), accounting for over 90% of the total benefits, with an additional 5% related to property damage and approximately 3% due to reduced congestion caused by accidents. The annual costs in the same year were estimated at approximately €25.5 million, divided as approximately 40% operational costs, 30% equipment costs, 20% communication costs and 10% costs related to the traffic control centres.

The comprehensive assessment underscores the potential of the RWW-LC service to significantly reduce casualties and enhance road safety, contributing to positive benefit–cost outcomes from the onset of its deployment. This evaluation highlights the critical role of advanced warning systems in mitigating risks at short-term road works, thereby supporting broader initiatives to improve traffic safety and efficiency on motorways.

9.6 Conclusion

The C-Roads platform's structured approach to evaluating Cooperative Intelligent Transport Systems (C-ITS) across various European pilot sites has provided crucial insights into the effectiveness and scalability of these technologies. The evaluation and assessment conducted by C-Roads WG3 have played a pivotal role in determining the operational efficiency and policy alignment of C-ITS deployments. By focusing on a standardised evaluation framework, the C-Roads initiative ensures that the technological deployments enhance road safety, improve traffic efficiency and contribute positively to the environmental sustainability of transport infra-structure across Europe.

References

[1]   C-Roads Platform WG3. *C-Roads Evaluation and Assessment Plan* [Online]. 2019. Available from https://www.c-roads.eu/fileadmin/user_upload/media/ Dokumente/C-Roads_WG3_Evaluation_and_Assessment_Plan_version_June19_ adopted_by_Countries_Final.pdf [Accessed 19 April 2024].

[2]   C-Roads Germany. *Pilots in Germany* [Online]. 2024. Available from https:// www.c-roads-germany.de/english/c-its-pilots/ [Accessed 19 April 2024].

[3]   C-Roads Platform. *Annual Pilot Report 2022* [Online]. 2023. Available from https://www.c-roads.eu/fileadmin/user_upload/media/Dokumente/Annual_ pilot_overview_report_2022.pdf [Accessed 19 April 2024].

[4]   C-Roads Germany Urban Nodes. *CRG-UN Evaluation and Assessment Plan*. 2021. Internal working document.

[5]   C-Roads Platform WG3. *Final Report*. 2021. Available from https://www. c-roads.eu/fileadmin/user_upload/media/Dokumente/C-Roads_WG3_Final_ report_March_2022.pdf [Accessed 19 April 2024].

[6]   Stiegler R., and Zinser J. *User Survey with Combined Viewpoint Analysis of the C-ITS Service GLOSA (Green Light Optimal Speed Advisory)*. 2021. [Unpublished].

[7]   Spangler M., Bachmann F., Busch F., Grawenhoff S., and Vierkötter M. *Wirtschaftlichkeitsanalyse von kooperativen fahrbaren Absperrtafeln*. 2018.

Chapter 10

Impact assessment of large-scale C-ITS services in Greece

Areti Kotsi[1] and Evangelos Mitsakis[1]

10.1 C-Roads Greece project overview

Greece, under the frame of the C-Roads Platform initiative, participated as a member state with its national pilot in June 2019. The Greek pilot took place at two test sites, one in Northern Greece, i.e., Egnatia Odos Tollway, and one in Attica Region, i.e., Attica Tollway, with the objective to conduct a specific set of Day-1 and Day-1.5 C-ITS services by using a balanced mixture of ETSI ITS G5 and cellular communication technologies. A breakdown of the selected Day-1 Cooperative Intelligent Transport Systems (C-ITS) services of the Greek pilot contain: road works warning (RWW), with special focus on lane closure and other restrictions; hazardous locations notification (HLN), which includes the cases of stationary vehicle, weather condition warning, and obstacle on the road; in-vehicle signage (IVS), focusing on variables message signs (VMS) of "free text" as well as speed advice messages to avoid shockwave damping; and probe vehicle data (PVD), regarding cooperative awareness message (CAM) aggregation. In addition, Day-1.5 C-ITS service smart routing is selected. The above-mentioned services were tested in two test sites.

In Northern Greece, the test site included a 30 km section of the motorway named "Egnatia Odos" covering an area with specific attributes of geometry, traffic volumes, rural environment, mountains, successive tunnels, and bridges. More specifically, the average annual daily traffic (AADT) of the road section of Egnatia Odos motorway is 11,230 (HGV: 16%). The area is managed by a traffic control center and the existing intelligent transport system (ITS)-related equipment includes VMS, CCTV traffic cameras, traffic detection inductive loops, and meteorological and smoke sensors. Twenty-five roadside access points (road side units – RSUs) plus one mobile were installed. The RSUs enabled communication and the messages' transfer between vehicles and road infrastructure.

In the Attica region, the test site included a 21 km road section of the central sector of Attica Tollway (motorway consisting of the ring road of the greater metropolitan area of Athens) along the main axis of the road, between the

[1]Centre for Research and Technology Hellas (CERTH), Hellenic Institute of Transport (HIT), Greece

Dimokratias Interchange and Paiania Interchange (from 21 km to 42 km of the Attica Tollway), corresponding to the section of the motorway with the heaviest traffic (AADT: 53,000 vehicles as per data registered in 2018). This section is managed by a traffic management center. For the purposes of the project, ten roadside access points plus one mobile were installed.

The project started in June 2019, with the pilot activities beginning in January 2020. All the C-ITS services were launched by the end of 2021 and the project was completed in December 2023.

10.1.1   *Greek pilot*

10.1.1.1   **Attica Tollway pilot**

The Attiki Odos Motorway (Attica Tollway) in Athens, Greece, is a 70 km long urban Motorway, fully access-controlled through 39 toll barriers (6 mainline barriers at the extremities plus 33 entry ramps). Attica Tollway is the Athens Ring Road, providing free-flow traffic conditions in the city center-periphery, linking the downtown areas by radial connections and main arterials. It provides a link between the national motorway network to the south and the north of the Greek Capital and connects the city and its suburbs with the new Athens International Airport. It also provides connections with mass transit mode facilities (metro, suburban rail, and buses).

Attica Tollway is an urban motorway with two directionally separated carria-geways, each consisting of three lanes and an emergency lane (hard shoulder), and with tolls paid at all entry points to the Tollway. Being a closed motorway, it is fully controlled at its access points and consists of two sections, which are per-pendicular to one another: (1) Elefsina-Stavros-Spata Airport motorway (ESSM) and Aigaleo Western Peripheral Motorway (AWPM), extending along approxi-mately 57 km; and (2) Imittos Western Peripheral Motorway (IWPM), extending along approximately 13 km.

The pilot was deployed on the central sector of Attica Tollway, a road segment of 20 km with the heaviest traffic. The already installed ITS-related equipment (VMSs, CCTV traffic cameras, traffic detection inductive loops, meteo & smoke sensors) was utilized alongside the C-ITS field equipment that was installed during the pilot. The field equipment included:

- Ten ITS-G5 RSUs along the pilot road segment. The RSUs were mainly installed in existing VMS gantries as well as CCTV poles, so power cabling and network connectivity was readily available.
- One mobile ITS-G5 RSU mounted on a patrol vehicle.
- Ten On-Board Units (OBUs) operating on ITS-G5 communication protocols.
- Ten OBUs operating on cellular network.
- In both pilots, all OBUs contained a Human Machine Interface (HMI) subsystem.

10.1.1.2   **Egnatia Odos Tollway pilot**

Egnatia Odos Motorway extends to 660 km and is part of the TEN-T Core and Comprehensive Road Network Corridors. It crosses the north part of Greece from its westernmost edge (Igoumenitsa port, Ionian Sea) to its easternmost borders with Turkey (Kipoi, Evros). It's a dual carriageway motorway, with each driving direction consisting of two lanes (in few sections of three lanes) plus an emergency lane. Egnatia Odos motorway is equipped with many ITS devices and safety systems that aim at providing travel safety and comfort to the end users. Along with its five vertical axes of 330 km it connects Greece with all the north neighboring countries (i.e. Albania, North Macedonia, and Bulgaria).

The section of Egnatia Odos motorway called "Kastania bypass", which was included in the project pilot, is a rural road section with continuous bridges and tunnels, which spans 26 km between two approaching intersections, plus additionally 2+2 km on either side of them, with AADT of 11,230 (HGV: 16%). The only alternative route to the pilot road section is the old National Road "4", which travels across the mountainous area of Vermio mountain.

The pilot site was equipped with ITS related equipment such as VMS, blank out signs (BOS), lane control signs (LCS), traffic signals, traffic counters with inductive loops, CCTV traffic cameras, over-height vehicle detectors (OHVD), road weather information systems (RWIS) and air quality, and visibility sensors in tunnels (CO/NO/VIS). This equipment was utilized as possible data gathering sources alongside the new C-ITS field equipment (RSUs) that was installed during the pilot. Also, four additional virtual VMS (2 per carriageway, 2 km before each interchange) were utilized, to provide additional messages to road users. All VMS messages (physical and virtual) were provided in real time through DATEX II publication, from Egnatia Single Access Point (Egnatia SAP) to the National Access Point (NAP). The new field equipment that was deployed on Egnatia Odos Tollway pilot included:

• Twenty-five ITS-G5 RSUs along the pilot road section. The RSUs were mainly installed in existing equipment suspension infrastructure (VMS gantries, CCTV poles, lighting poles, tunnel walls), so power supply, communication, and control through active network equipment and fiber optic cables were readily available.
• One mobile ITS-G5 RSU mounted on a patrol vehicle.
• Ten OBUs operating on ITS-G5 communication protocols.
• Ten OBUs operating on cellular 4G network.

10.1.2   C-ITS services

The C-ITS services and use cases that were deployed in the Greek pilot are summarized in Table 10.1.

10.1.2.1   RWW

The service describes warnings on road works that may be either mobile or static, short- or long-term. They may affect the road layout and the driving regulations and frequently come unexpectedly to road users. This may lead to unsafe situations and sometimes even accidents, which involve both road users and workers. Moreover, the attention of the driver can fade with regular or longer road works.

Table 10.1   Services and use cases deployed in the Greek pilot

Service	Use case	Egnatia Odos Tollway	Attica Tollway
RWW	Lane closure (LC)	x	x
HLN	Weather conditions warning (WCW)	x	x
HLN	Obstacle on the road (OR)	x	
HLN	Stationary vehicle (SV)		x
IVS	Embedded VMS "free text" (EVFT)	x	x
IVS	Shockwave damping (SWD)		x
PVD	CAM aggregation	x	
Traffic management (TM)	Smart routing	x	x

Use cases involving road operator vehicles (salting, ploughing, bypassing towards the incident, protecting accident zone, vehicle recovery by road operator) support the safety of the involved road operators and road users. The objective is to cause more attentive and adjusted driving while approaching and passing a work zone or road operator vehicles in operation by providing in-car information and warnings about road works, changes to the road layout and applicable driving regulation.

RWW aims at reducing the number of collisions with road vehicle safety objects and road operator vehicles near road works. It will inform the road user when approaching a work zone and will simultaneously provide information on the changes in the road layout. Moreover, it may offer better traffic flow and fewer accidents. The use case developed in C-Roads Greece concerning was the Lane closure and other restrictions (RWW-LC). This use case informs the road user about the closure of part of a lane, whole lane, or several lanes (including hard shoulder), but without the road closure. The closure is due to a static road works site. In many cases road users either enter the road works sites or strike the pro-tection equipment causing casualties. The objective is to allow road users to anticipate the closure of lanes due to a road works site on the road ahead and to adapt their speed and lane on the road, well in advance. With this use case the road users are expected to adapt their speed, change lanes (if needed), and keep an increased level of vigilance. They will be informed about risks of discomfort on the road slowing down or maneuvering. Consequently, fewer accidents and risky situations will exist for road users and workers, safeguarding the traffic flow.

10.1.2.2   HLN

This C-ITS service describes an infrastructure-to-vehicle (I2V) warning message related to one or a series of potentially hazardous events on the road, where the approaching road users receive information and therefore warning about the loca-tion and type of hazard they are approaching and – if available – also the duration of the event.

Hazardous locations/situations create a risk for road users, potentially causing (more) accidents resulting in injuries/fatalities. This C-ITS service has the potential to directly inform involved and relevant road users so they can adapt their driving behavior accordingly. The objective is to inform road users of hazardous locations on their route to enhance overall road safety by providing in-vehicle information about hazards, including the location and type of hazard, whenever possible also the remaining distance to the location, the duration of the event(s) creating the hazard, as well as lane and speed advice.

The use cases developed in C-Roads Greece were: Stationary Vehicle (HLN-SV), Weather Condition Warning (HLN-WCW), Obstacle on the Road (I2V) (HLN-OR), and Stationary Vehicle (HLN-SV). The HLN-SV use case includes two distinct use cases. Slow vehicle warning as a use case of cooperative awareness application and stationary vehicle as a use case of road hazard warning application. Stationary vehicle (s) service warns approaching drivers about stationary/ broken down vehicles ahead, which may represent obstacles on the road. It is a preventive safety service, as drivers will have advanced notice and more time to prepare for the hazard. The road operator could have an event management system to initialize a conventional (non-C-ITS) broken-down vehicle event and trigger an I2V message to warn other vehicle drivers. An interesting variant of this use case, which adds to the quality of the information, is when the stationary vehicle information is also processed by a nearby roadside unit and then, to further distribute the same warning via the roadside infrastructure, other RSUs connected via the road operator distribute the warning via resending it. The objective is to avoid collisions (mostly rear-end) with stationary vehicles on the road and enhance road safety. The vehicle drivers adapt their driving behavior, slowing down or changing lanes. Because the I2V warning is targeted and accurate from the management system of the road operator, the reliability is high and improves the driver's attention near these areas. In addition, driver awareness is raised to the possible presence of vulnerable road users (VRUs) on the road. The HLN-WCW use case shows both static and dynamic information of weather conditions and road status (related to weather conditions) in-vehicle, providing accurate and up-to-date local weather information. Drivers are informed about dangerous weather conditions downstream, especially when it is difficult to perceive the danger visually, such as black ice or strong gusts of wind. The objective is to improve traffic safety via additional means of C-ITS messages to inform drivers about weather conditions and road status information in a more accurate way. The vehicle drivers adapt their driving behavior compliant to the applicable driving regulations and any advice or guidance provided. The primary expected impact is more attentive driving by providing actual and continuous (expected) information on road conditions (poor road traction condition, visibility, wind, rainfall), which improves traffic safety as it reduces the numbers and the severity of accidents. The HLN-OR is a use case where the road operator identifies that there are one or several obstacles on one or several lanes on the network and broadcasts this information to road users. However, traffic can still bypass the lane(s) occupied by obstacles (not a blockage). The objective of this use case is to alert road users of a potential danger, hence increasing driver attention. Since there is no automatic detection system, the accuracy of the localization may be negatively affected. Drivers are expected to adapt their speed, change lanes, and keep increased levels of attention, reducing the risk of accidents thus helping road operators to improve traffic management.

10.1.2.3 IVS

The service provides a means to inform road users via in-car information systems on "static" road signs (mirroring physical road signs along the road) or dynamic road signs. Additionally, further information (virtual road signs or additional free text) can be provided. The IVS service can target either specific vehicle types or individual vehicles. The IVS information is sent out by means of I2V communication. VMS systems are also used by road operators to provide operational, tactical, or strategic information to road users. Different types of traffic sign systems are used, utilizing static and variable pictograms, as well as predefined or dynamic text on VMS, variable text panels (VTP), and variable direction signs (VDS). Such information can also be delivered by IVS to inform road users. The objective is to increase attentive driving and awareness on the content of road signs, by providing sign information directly in the vehicle where it can potentially be displayed in a more driver friendly manner as well as throughout its validity period. This will severely reduce observation problems attributed to physical road signs, such as limited line of sight, obstructions obscuring sight of a sign, limited attention by drivers passing signs. Another attractive characteristic is that messages may be displayed in the driver's preferred language.

The use cases developed are: Embedded VMS "Free Text" (IVS-EVFT) and Shockwave Damping (IVS-SWD). The goal of the IVS-EVFT is to display to the road user in-vehicle information of type "free text". The information will either reproduce what is displayed on a physical VMS or display a completely new message that does not mirror a physical VMS (a virtual VMS). The objective of this use case is to transmit to road users' information in "free text" format that is not provided by other (in-vehicle signage) use cases, like adding details (in preferred language) to existing messages to provide more precise and comprehensible information to the road users to achieve their desired behavior. The information may already be displayed on a physical VMS or other means of signaling on the road. The road users adapt their driving behavior compliant to the applicable driving regulations and any advice or guidance provided. It occurs through greater traffic management control and the comfort of constant display of in-vehicle information. The IVS-SWD is related to the shockwave phenomenon. Shockwaves can occur in dense traffic when, while on the highway, vehicles drive relatively close to each other and/ or change their speed or driving lane abruptly. In addition, it can be a result of a temporary overload on a highway ramp. In such situations, a braking action by a single vehicle may lead to a disturbance that propagates or is even amplified in the upstream direction, ultimately bringing the following vehicles to a full stop, or creating an incident. In shockwaves traffic flow can constantly alternate between "free flow" and "congested" traffic. The objective of this Use Case is to provide I2V in-car information to avoid emerging or ideally even accomplish the elimination of shockwave situation in highway traffic through C-ITS. The main goal is to mitigate or prevent shockwaves and to avoid occurrence of a traffic jam.

10.1.2.4 PVD

PVD is a service in which vehicle or road user data is collected by the road operator or service provider. Modern vehicles acknowledge at any time their position, speed and direction, vehicle information, such as type and length, and other traffic information, along with events affecting driving experience. This data could be used by the road operator to get a more comprehensive knowledge of its network and to enhance the road operator's knowledge of events, complementing patrol and other existing sources. The objective is to collect data from vehicles and/or from road users to improve knowledge of traffic conditions, incidents and for statistical and modelling purposes.

CAMs are sent from the vehicles either to the Roadside ITS units via G5, which are transmitted in turn to the Central ITS unit ending up to the Traffic Control Center or via cellular to the Traffic Control Center. The CAM Aggregation involves the above transmission path either for a single vehicle or for a group of vehicles. The objective of this Use Case is to ultimately collect traffic data from vehicles providing information on speed, type of vehicle, weather conditions and other, that may be processed in the Traffic Control Center. Passing by vehicles send out traffic data automatically in specific detection zones of a Roadside ITS unit on the highway. Either in short and/or long-term aggregation intervals the information is forwarded to the Central ITS unit, and then to the Traffic Control Center. Alternatively, CAM messages are sent via cellular. The Traffic Control Center may acknowledge specific traffic data per single vehicle, types of vehicles, weather conditions or other aggregated traffic information.

10.1.2.5 Traffic management – smart routing

Smart routing service is an information service informing road operators and road users in real-time about the traffic conditions and facilitating the selection of the most appropriate traffic management decision or route. This service has the potential to directly inform the relevant road users so they can decide their departure time and choose their preferred route. In situations where (recurrent or not) congestion is observed, smart routing could play a critical role because it would instantly provide the Traffic Control Center with precise real-time information about the quickest route for road users. This knowledge will help promote the most suitable traffic management decision for the road operator.

The expected benefits are saving energy, money, and time, and reaching optimum use of road capacity. At the same time, traffic efficiency will be improved, and the driving conditions will be smoother. Environmental benefits will be achieved, and the personal benefits will be more comfortable driving and better driving experience. Also, homogeneous traffic flow with no or less jams/ congestion is expected.

10.2 Impact assessment and evaluation

The impact assessment and evaluation of the C-ITS services is associated with user acceptance and simulation experiments conducted for the services.

User acceptance was evaluated through the development and distribution of questionnaires to the users of the C-ITS services (drivers having access to the services via a mobile application and road operators having access to the services via a traffic management software). The questions included in the questionnaires were related to personal information, to driver profile, to the general expectations of the C-ITS services (traffic efficiency, road safety, environment), to expectations of each C-ITS service, and to the usefulness and user friendliness of the software for the road operators. The objective of the questionnaires was to assess the expectations of the users regarding the use of the C-ITS services as the answers were received prior to the launch of the C-ITS services.

The simulation experiments were conducted to assess the C-ITS services and generate results indicating their impacts. Simulation experiments were conducted for both pilots, Attica Tollway and Egnatia Tollway, and for various demand scenarios (number of vehicles in the network). The resulting indicators included traffic measures (such as average vehicle speed, travel time), safety measures (such as collisions), and environmental factors (such as CO_2 emissions).

10.2.1   Subjective evaluation

The impact assessment for the project involved subjective evaluation through questionnaires which were collected in February 2022. Two types of questionnaires were created.

The first type of questionnaires aimed to assess the willingness of the traffic management personnel at the traffic management centers of Attica Tollway and Egnatia Odos Tollway to use the C-ITS services and the traffic management software that was developed specifically for the project purposes. The traffic management software had the objective to enable the C-ITS services handling from the personnel in the traffic management centers. Four representatives of the traffic management personnel provided their feedback to the questionnaires. The participants expressed their strong agreement towards the potential positive impacts of the software on road safety and the environment. Moreover, the participants stated that the software shows a potential in improving the traffic conditions in the network and in providing more accurate information compared to the existing systems (i.e., VMS). Concerning the visualization elements provided by the software, they were characterized as useful and more specifically for displaying roadworks and extreme weather conditions. Concerning the ease of use of certain features of the software, the participants had mixed opinions about the functionalities of creating and sending C-ITS messages and displaying the field equipment locations. Opinions also varied on how complex the monitoring capabilities of the software were perceived.

The second type of questionnaire aimed to assess the willingness of the drivers to use the C-ITS services. In total, 59 participants provided their answers, stating their preferences and expectations for the use of the C-ITS services. Most of the participants were professional drivers. The participants showed a general interest in using the C-ITS services, and more specifically the ones targeting road safety improvement and travel times reduction. Most of the constraints were expressed towards potential distractions that could be caused by the services while driving and towards issues related to data privacy/security when using the services. Concerning the willingness to pay for use of the services, it was relatively low among all participants.

10.3 Objective evaluation

10.3.1 Simulation experiments

The simulation experiments were executed in SUMO [1] which is an open source, highly portable, microscopic, and continuous multi-modal traffic simulation package designed to handle large networks.

10.3.2 Simulation methodology

The methodology for the services simulation relies on the implementation of a traffic management strategy, a Long-Term Evolution data communication network to route vehicle-to-everything (V2X) data communications, and CAMs that ensure communications among C-ITS enabled vehicles by exchanging continuously data packets with information such as location, speed. Moreover, the acquired information is integrated to infrastructure through a traffic control server (TCS) that generates proposed messages which are taken over by the infrastructure and transmitted to vehicles via Vehicle-to-Infrastructure (V2I) communications. As CAMs indicate abrupt deceleration or sudden stop of the transmission of messages could suggest to the TCS that a crash or a hazardous incident has taken place. Finally, when the TCS detects such a conflict, it distributes the corresponding messages to the vehicles.

A traffic management logic was developed for modelling the TCS as well as the drivers' responses to warning messages using the microscopic traffic simulation SUMO. The traffic management strategy is presented in the flowchart in Figure 10.1.

First, when the TCS detects that a crash has occurred, it regularly sends RHW messages to all vehicles in predefined influence zones through broadcast communication. At this point it is important to highlight that the driver's response to the traffic conflict depends on the type of information that is provided by the TCS, as well as on the location of the C-ITS enabled vehicles regarding the road hazard and the other implicated vehicles of the considered network. Taking the above into consideration, it is assumed that the vehicles are grouped to drive in specific zones, hence the behavior of a vehicle will be defined with respect to the zone or zones in which it belongs. Vehicles which travel on the hazard lane and enter the dangerous zone receive a RHW message to adjust their desired speed as well as to try to change lane at every time step, if feasible. Next, vehicles that drive in the near crash zone receive an emergency electronic brake light (EEBL) warning to inform drivers that a vehicle in front may crash or brake abruptly and therefore the drivers are advised to increase their headways appropriately via a newly developed open-gap function, to avoid a collision. Briefly, this gap control mechanism, facilitates the creation gap between two specific subsequent vehicles and has been modelled to increase the desired time headway of the following vehicle, and determines the minimum space headway that must be maintained between the two vehicles for a predefined duration. Regarding the EEBL, the

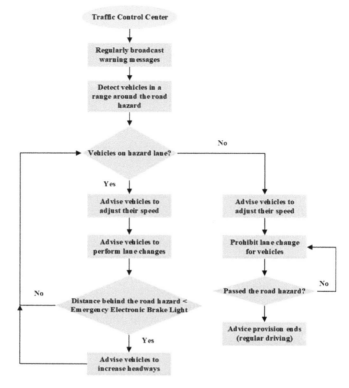

Figure 10.1   Flowchart of the traffic management strategy

TCS decides to which vehicles an EEBL warning will be sent. On the other hand, vehicles that travel in a lane different than the hazard one pertains to the safe zone. These vehicles are advised to reduce their desired speed with respect to the speed limit of the freeway, whereas they are concurrently restricted from entering the hazard lane. Finally, after passing the conflict area, vehicles' speed and lane change operation are no longer under the control of the TCS. Thus, vehicles that have received no information from the TCS are considered to belong in the standard zone; the behavior of drivers in this state is entirely determined by the default car-following (Krauss) and lane change (LC2013) model implemented in SUMO.

The KPIs based on the experiments results are: average vehicle speeds, CO_2 emissions, collisions, lane changes, throughput and travel time. Regarding the number of vehicles considered in the network for the simulation, two scenarios were tested for both pilot locations:

- Baseline: Actual traffic demand (based on collected traffic volumes).
- Egnatia Odos Tollway pilot: 500 vehicles.
- Attica Tollway pilot: 4500 vehicles (maximum capacity).
- High C-ITS penetration rate - Egnatia Tollway pilot: 1500 vehicles.

10.3.2.1   Attica Tollway network

The experiments executed in the Attica Tollway network use cases HLN-WCW, Traffic Jam Ahead Warning, RWW-LC, and HLN-OR. In the C-ITS Scenario the vehicles' speeds drop down to 0.8 500 m upstream and to 0.6 when entering the edge where the event is located (speed decreases smoothly). In the manual scenario, the vehicles' speeds drop down to 0.6 150 m upstream.

For the service HLN-WCW lane changes were observed to decrease in the C-ITS scenario, which is an expected result as in this case vehicles are informed in advance about the slippery road; hence drivers tend to avoid lane changes as it is risky. The number of collisions showed a significant decrease in the C-ITS scenario. This is an anticipated result due to the timely provision of the C-ITS messages that make drivers aware of the slippery conditions on the road and they drive more carefully. HLN-WCW could be considered that contributes to road safety at an important level. CO_2 emissions were decreased also in the C-ITS scenario. Average vehicle speed showed only a very slight increase in the C-ITS scenario; hence it could be considered that no important changes happen with the provision of C-ITS messages. Travel time remained similar in the C-ITS scenario, which is an anticipated result when considering that there is no change in vehicles speeds.

For the service Traffic information and smart routing (Traffic Jam Ahead Warning) the lane changes decreased in the case of C-ITS message provision. This could be justified since vehicles in the C-ITS scenario are advised to decrease their speed in a timely manner as there is a traffic jam ahead and hence drivers do not need to make many lane changes, to avoid the event. CO_2 emissions decreased as well in the C-ITS scenario. This could be considered as an anticipated result due to smoother fluctuations (smoother deceleration) in vehicles speeds in the case of the C-ITS information provision. Average vehicle speed showed a very slight increase in the C-ITS scenario; hence it could be considered that no changes are observed in this indicator. The same applies for travel time. No significant changes were observed, which is something logical when considering that vehicles speeds didn't change as well.

For the services RWW-LC the number of lane changes decreased in the C-ITS scenario. This could be justified by the fact that drivers are aware in advance about the closed lane ahead, hence they can adjust their driving behavior timely, and they don't need to perform any sudden changes, such as many lane changes. The number of collisions was reduced significantly at the C-ITS scenario, indicating that drivers have a more attentive driving behavior due to the C-ITS message provision, con-tributing this way to road safety increase. There was also a significant reduction to CO_2 emissions in the C-ITS scenario, leading to the conclusion that RWW-LC could have a positive impact to the environment as pollutant emissions show a reduction due to smoother driving (smoother speed fluctuations). The indicator of average vehicle speed was lower in the C-ITS scenario. This is an anticipated result and drivers are timely informed about the lane closure and they tend to decelerate and drive at lower speeds. Regarding travel time, this indicator showed an increase, which is something logical considering the speed decrease.

For the services HLN-OR the indicator of lane changes did not show a significant difference in the C-ITS scenario as it has increased only slightly. This could be justified by the fact that since drivers are aware in advance of the existence of the obstacle on the road, they are able to perform lane changes timely and earlier than in the case of having no information in advance. On the other hand, the number of collisions showed a very significant decrease in the case of the C-ITS scenario indicating that the service could have a high contribution to road safety since drivers would show more attentive driving reducing this way the possibility for a collision. CO_2 emissions were observed to have a slight increase in the C-ITS scenario compared to the manual one. The average vehicle speed showed a decrease in the C-ITS scenario. This is since drivers are informed timely about the event and they can adjust appropriately their driving behavior by reducing their speeds. The travel time increased significantly in the C-ITS scenario. This is due to the vehicle speed increase which is observed as well in this case.

The comprehensive results of the simulation experiments are presented in Table 10.2.

Table 10.2   Comprehensive results of the simulation experiments conducted for the C-ITS services in the Attica Tollway network

Use case	Key performance indicator (KPI)	Manual scenario	C-ITS scenario
	Attica Tollway pilot		
HLN-WCW	Throughput (veh/h)	4537.60	4544.80
	Lane changes (#/km)	0.67	0.58
	Collisions (#)	14.40	7.65
	CO_2 emissions (gr/km)	315.69	297.61
	Average vehicle speed (km/h)	69.55	69.97
	Travel time (min/km)	0.88	0.87
Traffic jam ahead warning	Throughput (veh/h)	4584.30	4587.90
	Lane changes (#/km)	0.67	0.60
	Collisions (#)	9.70	5.53
	CO_2 emissions (gr/km)	314.52	308.30
	Average vehicle speed (km/h)	75.43	75.88
	Travel time (min/km)	0.82	0.81
RWW-LC	Throughput (veh/h)	4404.40	4363.40
	Lane changes (#/km)	0.72	0.67
	Collisions (#)	16.20	8.70
	CO_2 emissions (gr/km)	344.12	317.83
	Average vehicle speed (km/h)	64.00	58.00
	Travel time (min/km)	1.05	1.15
HLN-OR	Throughput (veh/h)	4583.70	4322.30
	Lane changes (#/km)	0.84	0.86
	Collisions (#)	1257.60	5.90
	CO_2 emissions (gr/km)	321.42	330.69
	Average vehicle speed (km/h)	80.01	55.68
	Travel time (min/km)	0.76	1.210

10.3.2.2 Egnatia Tollway network

The experiments executed in the Egnatia Tollway network refer to use cases HLN-WCW, traffic jam ahead warning, RWW-LC and HLN-OR. In the C-ITS Scenario the vehicles' speeds drop down to 0.8 (500 m upstream) and to 0.6 when entering the edge where the event is located (speed decreases smoothly), while in the Manual scenario the vehicles' speeds drop down to 0.6 150 m upstream.

HLN-WCW

The number of vehicles used in the simulation is around 500 for the baseline scenario, both for C-ITS and manual scenario, and around 1200 for the high C-ITS penetration scenario respectively, both for C-ITS and manual scenario. The throughput in each scenario is presented in Figure 10.2.

Concerning lane changes performed by the vehicles, a higher number of lane changes has occurred in the manual scenario for both baseline and high C-ITS penetration rate scenario, while the number of lane changes was observed to be lower in the C-ITS scenario (see Figure 10.3). This could be explained because in the C-ITS scenario the vehicles are advised in the simulation to remain in the same lane and not to perform changes as all lanes are considered slippery, hence any change could be considered risky.

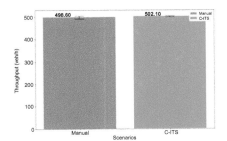 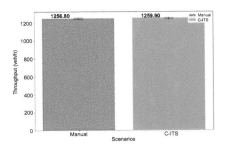

Figure 10.2   HLN-WCW throughput for manual and C-ITS scenario in baseline (left) and high C-ITS penetration rate scenarios (right)

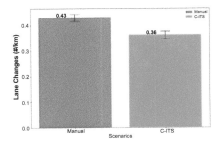 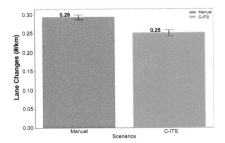

Figure 10.3   HLN-WCW lane changes for manual and C-ITS scenario in baseline (left) and high C-ITS penetration rate scenarios (right)

Concerning the indicator of CO_2 emissions, not a significant difference has resulted in the two scenarios, but still there is a slight decrease in the C-ITS scenario for both penetration rates (500 and 1200 vehicles), leading to the conclusion that HLN-WCW does not contribute at an important level to CO_2 emissions reduction (see Figure 10.4).

Similarly for average vehicle speed, a slight difference was observed in the two scenarios. More specifically, there is a slight increase in the C-ITS scenario for both penetration rates. This could be justified because the provision of information about the slippery road could contribute to smoother fluctuations in speed, hence speed remains higher (see Figure 10.5).

Traffic jam ahead warning

The number of vehicles used in the simulation is around 500 for the baseline scenario, both for C-ITS and manual scenario, and around 1200 for the high C-ITS penetration scenario respectively, both for C-ITS and manual scenario. The throughput is for each scenario is presented in Figure 10.6.

Concerning the indicator of lane changes, there is an increase in the high C-ITS penetration rate scenario compared to the baseline in the case of 500 vehicles, while the opposite happens in the case of 1200 vehicles (see Figure 10.7).

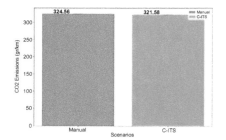 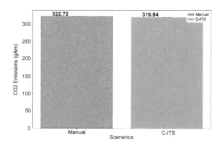

Figure 10.4   HLN-WCW CO_2 emissions for manual and C-ITS scenario in baseline (left) and high C-ITS penetration rate scenarios (right)

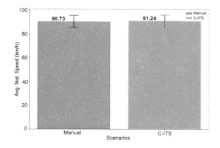 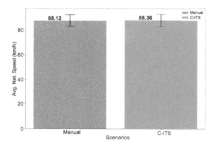

Figure 10.5   HLN-WCW average vehicle speed for manual and C-ITS scenario in baseline (left) and high C-ITS penetration rate scenarios (right)

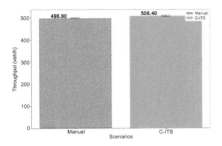 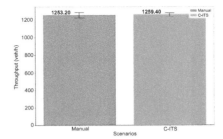

Figure 10.6 *Traffic information and smart routing throughput for manual and C-ITS scenario in baseline (left) and high C-ITS penetration rate scenarios (right)*

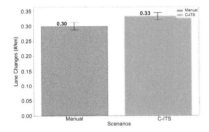 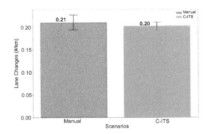

Figure 10.7 *Traffic information and smart routing lane changes for manual and C-ITS scenario in baseline (left) and high C-ITS penetration rate scenarios (right)*

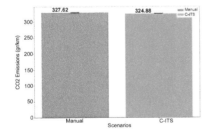 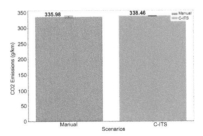

Figure 10.8 *Traffic information and smart routing CO_2 emissions for manual and C-ITS scenario in baseline (left) and high C-ITS penetration rate scenarios (right)*

CO_2 emissions are slightly increased in both scenarios, 500 and 1200 vehicles, in the case of high C-ITS penetration rate (see Figure 10.8).

Regarding vehicle average speed, there is quite a significant decrease in the baseline scenario (case of 500 vehicles), while a slight increase is observed in the same scenario for the case of 1200 vehicles (high C-ITS penetration rate) (see Figure 10.9).

The same happens for travel time. Travel time is slightly increased in the baseline scenario but there is a very slight decrease in the high C-ITS penetration rate scenario (see Figure 10.10).

RWW-LC

The number of vehicles used in the simulation is around 500 for the baseline scenario, both for C-ITS and manual scenario, and around 1200 for the high C-ITS penetration scenario respectively, both for C-ITS and manual scenario. The throughput for each scenario is presented in Figure 10.11.

The number of lane changes has increased in both scenarios (500 and 1200 vehicles) in the case of the C-ITS scenario. However, the increase is very low for the high C-ITS penetration rate scenario (1200 vehicles). The significant increase in lane changes in the case of the baseline scenario could be justified by the fact that in the C-ITS scenario vehicles are advised timelier about the lane closure compared to vehicles in the manual scenario, hence drivers are aware in advance

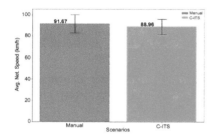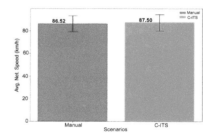

Figure 10.9   Traffic information and smart routing average vehicle speed for manual and C-ITS scenario in baseline (left) and high C-ITS penetration rate scenarios (right)

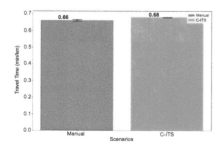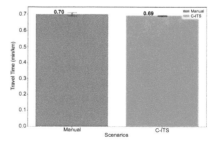

Figure 10.10   Traffic information and smart routing travel time for manual and C-ITS scenario in baseline (left) and high C-ITS penetration rate scenarios (right)

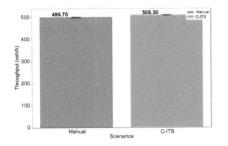
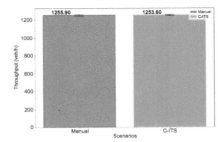

*Figure 10.11   RWW-LC throughput for manual and C-ITS scenario in baseline
(left) and high C-ITS penetration rate scenarios (right)*

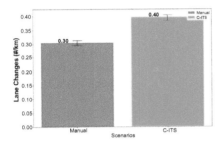
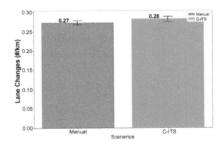

*Figure 10.12   RWW-LC lane changes for manual and C-ITS scenario in baseline
(left) and high C-ITS penetration rate scenarios (right)*

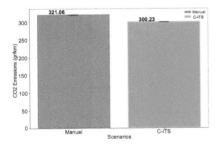
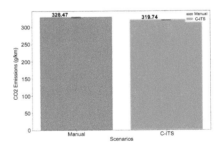

*Figure 10.13   RWW-LC CO_2 emissions for manual and C-ITS scenario in baseline
(left) and high C-ITS penetration rate scenarios (right)*

about the event and try to avoid it earlier by performing more lane changes (see
Figure 10.12).

CO_2 emissions are decreased in the C-ITS scenario in both cases, 500 and 1200
vehicles, but not at a significant level (see Figure 10.13).

Average vehicle speed shows a decrease in the C-ITS scenario in both cases
(500 and 1200 vehicles). This could be due to the earlier provision of information

in the case of the C-ITS messages provision where drivers start slowing down earlier and more smoothly as they are aware of the lane closure in advance (see Figure 10.14).

Travel time is increased in the C-ITS scenario in both cases, baseline and high C-ITS penetration rate. This is logical as the decrease in vehicle speed could lead to higher travel time (see Figure 10.15).

HLN-OR

The number of vehicles used in the simulation is around 500 for the baseline scenario, both for C-ITS and manual scenario, and around 1200 for the high C-ITS penetration scenario respectively, both for C-ITS and manual scenario. The throughput for each scenario is presented in Figure 10.16.

The number of lane changes shows a significant increase in the C-ITS scenario for 500 vehicles, but the opposite happens in the same scenario for 1200 vehicles. Lane changes are expected to increase in the C-ITS scenario as drivers are aware in advance of the event through the provision of the C-ITS messages (see Figure 10.17).

A very significant decrease is observed in the number of collisions in the C-ITS scenarios for both cases, 500 and 1200 vehicles, hence it could be considered that HLN-OR contributes at an important level to road safety (see Figure 10.18).

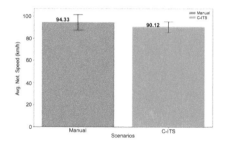 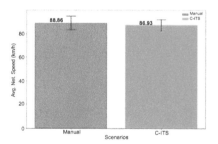

Figure 10.14   RWW-LC average vehicle speed for manual and C-ITS scenario in baseline (left) and high C-ITS penetration rate scenarios (right)

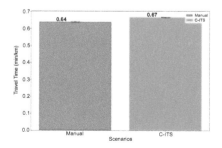 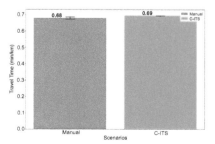

Figure 10.15   RWW-LC travel time for manual and C-ITS scenario in baseline (left) and high C-ITS penetration rate scenarios (right)

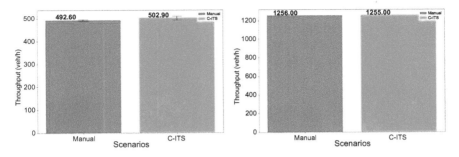

Figure 10.16   HLN-OR throughput for manual and C-ITS scenario in baseline (left) and high C-ITS penetration rate scenarios (right)

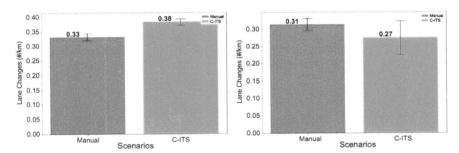

Figure 10.17   HLN-OR lane changes for manual and C-ITS scenario in baseline (left) and high C-ITS penetration rate scenarios (right)

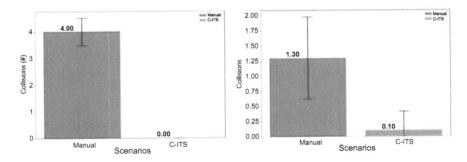

Figure 10.18   HLN-OR collisions for manual and C-ITS scenario in baseline (left) and high C-ITS penetration rate scenarios (right)

Regarding the indicator of CO_2 emissions, a decrease is observed in the C-ITS scenario in both cases. It should be mentioned that CO_2 emissions decrease for HLN-OR is higher than for the abovementioned services (see Figure 10.19).

Average vehicle speed shows a decrease in the C-ITS scenarios, and this could be explained by the fact that since drivers are informed timely of the existence of an obstacle on the road, they are able to start decreasing their speeds earlier and more smoothly (see Figure 10.20).

Travel time is increased in the C-ITS scenarios which is anticipated as in the same scenarios vehicles speeds show a decrease (see Figure 10.21).

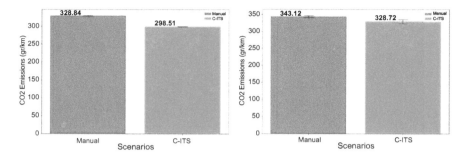

Figure 10.19   HLN-OR CO_2 emissions for manual and C-ITS scenario in baseline (left) and high C-ITS penetration rate scenarios (right)

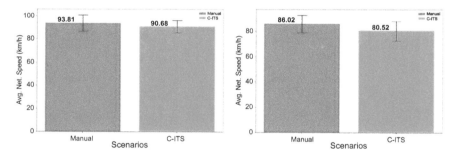

Figure 10.20   HLN-OR average vehicle speed for manual and C-ITS scenario in baseline (left) and high C-ITS penetration rate scenarios (right)

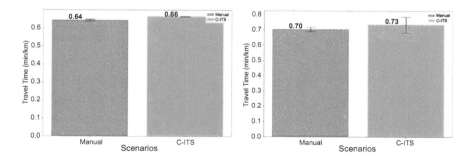

Figure 10.21   HLN-OR travel time for manual and C-ITS scenario in baseline (left) and high C-ITS penetration rate scenarios (right)

10.4 Driving simulator

Tests in a driving simulator were executed as well. The tests were carried out by four individuals in a driving simulator developed by CERTH-HIT. The driving simulator included a screen which displayed the layout of the road network environment, a special driving seat with a driving wheel, and a smartphone located next to the screen which displayed the C-ITS messages.

The SUMO-Unity integration driving simulator is a tool that combines multiple components to create a realistic driving experience. The simulator leverages the OpenStreetMap (OSM) file format, which serves as the common basis for both the SUMO simulation environment and the 3D environment developed in Unity. By utilizing Esri CityEngine, the OSM file is imported to construct a detailed 3D representation of the simulated world, and the resulting model is exported as an FBX file for seamless integration into Unity.

To establish bidirectional communication and control between Unity and the SUMO simulation, the simulator employs the Traffic Control Interface (TraCI) module provided by SUMO. Within the Unity environment, a dedicated class utilizes the TraCI module to issue commands and enable interaction with the SUMO simulation. These commands contain a range of functionalities, including retrieving real-time information on other vehicles present in the simulation, displaying this data within the Unity environment, accessing traffic light data, and manipulating the position and behavior of the ego-vehicle (user-controlled vehicle) within the SUMO simulation.

To achieve these capabilities, the tool incorporates a C# port of the TraCI module, ensuring efficient and effective integration. Users can use traditional keyboard controls (WASD* for moving and spacebar for braking), or for a more realistic driving experience, users can use a Logitech driving wheel, specifically the Logitech Driving Force GT wheel, which has been integrated into the simulator using a C# script. This driving wheel enhances the authenticity and realism of the driving simulation, providing users with a more engaging control mechanism. Also, a C# script to track the ego-vehicle position every second of the simulation was added to the tool.

Concerning the road network displayed on the screen, it included the same road segments of Attica Tollway and Egnatia Odos Tollway pilots with the ones used in the SUMO simulation experiments described in the previous section.

The service which was tested was RWW and more specifically the use case of lane closure. In the baseline scenario, the participants were provided with no C-ITS messages (no information display in the user interface of the smartphone) and they were heading to the road segment where the lane closure existed. In the C-ITS scenario, the participants were provided with a C-ITS message (information display in the UI of the smartphone) 500 meters before the exact location of the lane closure event. During these tests, data was collected and processed for evaluation. The results are presented in Figures 10.22 and 10.23, respectively.

*WASD refers to a set of four keys on a keyboard: W, A, S and D represent up, left, down and right, respectively.

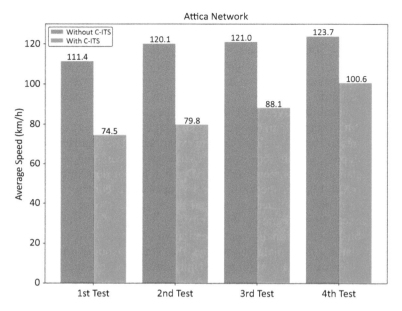

Figure 10.22   Driving simulator results in the Attica Tollway network

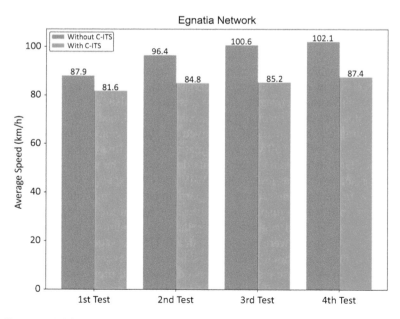

Figure 10.23   Driving simulator results in the Egnatia Tollway network

The KPI calculated based on the test logs was average speed (km/h). As it can be noticed in the diagrams, all participants decreased their average speeds in the C-ITS scenario. This result indicates that the display of the C-ITS message in the smartphone HMI led to more attentive driving and could contribute to safety increase in the case of a realistic environment.

10.5 Conclusions

The Greek pilot implemented and tested a variety of C-ITS services including both Day-1 and Day-1.5 applications. The RWW service aims to inform drivers about road works, changes in road layout, and driving regulations to enhance safety. It includes use cases like lane closure warning drivers about lane closures due to road works, aiming to reduce accidents and ensure smoother traffic flow. The HLN service alerts drivers about hazardous events on the road, such as accidents or weather conditions, to enhance safety. Use cases include warnings about stationary vehicles, weather conditions, and obstacles on the road. The IVS service delivers road sign information directly to in-car systems, improving driver awareness and reducing reliance on physical road signs. Use cases involve displaying free text messages and addressing shockwave phenomena to prevent traffic jams. The PVD service collects data from vehicles to enhance road operator knowledge of traffic conditions and incidents. CAMs are sent from vehicles to Roadside ITS units, which transmit the data to traffic control centers for analysis. The traffic management – smart routing service provides real-time traffic information to road operators and users, facilitating optimal route selection. Benefits include energy and time savings, improved traffic efficiency, and a more comfortable driving experience with reduced congestion.

The evaluation and impact assessment results were in general positive concerning both user acceptance and the impact of the services at transport level. Regarding the expectations of the users, the majority showed that they are in favor of using the C-ITS services and they expect that many aspects related to driving and transport will be improved by using the services. The only factor that remained low was the willingness to pay for the services, as most of the users had a neutral opinion about it or were negative to such an option. Concerning the impact of the services at transport level, the results generated by the simulation experiments indicated that most of the services could have a positive contribution to road safety, as the services led to collisions' reduction and more careful lane changes. Also, a positive impact could be expected with regards to traffic efficiency, since C-ITS messages provision led to smoother speed fluctuations. Finally, the use of the services contributed also in some use cases to the reduction of CO_2 emissions, indicating the potential of C-ITS services to lead to environmental impacts' reduction.

C-ITS deployment in Greece holds significant promise for enhancing road safety, improving traffic efficiency, and providing valuable real-time information to both traffic management personnel and drivers. The positive reception of C-ITS services, as indicated by the strong agreement among participants regarding its potential benefits, underscores the importance of furthering deployment efforts. However, the assessment also revealed areas where improvements and refinements are necessary, particularly in user interface design and ease of use for certain features. Therefore, for the next steps of C-ITS deployment in Greece, it is imperative to focus on enhancing user experience, addressing concerns related to data privacy and security, and fostering greater awareness and acceptance among drivers. Collaboration among stakeholders, continued investment in infrastructure, and ongoing evaluation and adaptation based on user feedback will be critical in realizing the full potential of C-ITS technology and ensuring its successful integration into the Greek transportation system.

Acknowledgments

The authors would like to acknowledge the "C-Roads Greece" project [2] and the consortium for all the efforts in implementing and executing the project. Project partners provided support, guidance, and resources throughout the duration of the project. The participants in the pilot demonstrations provided their valuable data and insights contributing to the extraction and formulation of detailed C-ITS services evaluation results and to an overall in-depth analysis of the impact of the project. Special thanks are due to the EU funding program Connecting Europe Facility (CEF) under which "C-Roads Greece" was funded.

References

[1]   Krajzewicz, D. (2010). Traffic simulation with SUMO – simulation of urban mobility. In: Barceló, J. (eds) *Fundamentals of Traffic Simulation. International Series in Operations Research & Management Science*, vol. 145, pp. 269–293. Springer, New York, NY. https://doi.org/10.1007/978-1-4419-6142-6_7

[2]   Kotsi, A., and Mitsakis, E. (2023). Large scale deployment of C-ITS: Impact assessment results of the C-Roads Greece pilots. *arXiv preprint arXiv:2311. 10734.*

Chapter 11

A system architecture for the deployment of autonomous mobility on-demand vehicles

Arunkumar Ramaswamy[1] and Javier Ibanez-Guzman[1]

11.1 Introduction

Mobility services using automated driving (AD) technology have emerged as a promising solution to both urban and rural transportation [1]. The rapid advances in AD technology, along with growing economic and societal interest in mobility-on-demand (MoD) systems, has sparked intense discussions regarding the potential of autonomous mobility-on-demand (AMoD) systems. These consists of a fleet of automated vehicles (AVs) that pick up passengers and transport them to their destination. A manager oversees the fleet by coordinating the assignment of passengers to AVs, planning their routes, and adjusting the fleet's balance by relocating unoccupied AVs to match transportation demand across different locations [2]. The self-balancing feature of AMoD systems helps to efficiently manage the fleet's operations without human intervention, including tasks such as monitoring its energy levels, scheduling recharging when necessary, and optimizing its availability for passenger service based on demand patterns and operational constraints. The automated system-wide coordination leads to increased revenue for stakeholders [3]. Additionally, they offer passengers freedom from driving responsibilities, offer a personal mobility solution for those unable or unwilling to drive, and could potentially enhance safety [4]. These advantages have spurred numerous companies and traditional automotive OEMs to actively pursue AD technology. OEMs that want to enter the robotaxi industry as vehicle providers need to develop purpose-built vehicles with best-in-class durability, low maintenance costs, and high up-time availability [5]. This pursuit includes designing vehicles optimised for AD operations and to diversify their business to various mobility services.

The AMoD vehicle, commonly known as robotaxi, relies on four key industry trends: automated driving, connectivity, electrification, and shared mobility [6]. Robotaxis are self-driving electric vehicles that provide on-demand, point-to-point transportation for passengers. Similar to contemporary ride-hailing services, users

---

[1]Ampere Software and Technologies, Renault Group, Guyancourt, France

would request robotaxis through a smartphone app, with fares determined by factors like vehicle category, trip duration, and distance traveled [7].

In general, automotive companies adapt and transform a conventional human-operated vehicle into a computer-controlled vehicle. Subsequently, they transfer it to an affiliate responsible for outfitting the vehicle with essential sensors and software, thereby converting it into an automated vehicle. Currently, several automotive OEMs globally are engaged in active research and development in the domain. Every year, robotaxi prototypes are frequently exhibited at popular technology events, drawing large crowds. Furthermore, certain OEMs operate subsidiary organisations that provide commercial robotaxi services. This chapter is an aggregated presentation of the engineering approach and cumulative results regarding system architecture of robotaxi service, from the various experimental projects and field operational tests (FOTs) [8]. Figure 11.1 showcases several AV prototypes deployed by Renault in different experimental mobility services.

This chapter offers practical perspectives on systems engineering and architectural considerations essential for constructing autonomous mobility services. The purpose is to present the engineering approach and a proven reference architecture for autonomous mobility systems such as automated shuttle and robotaxi use case. The focus is limited to a comprehensive examination of functional architectures and systems engineering considerations pertinent to automated driving. Given the experimental nature of the projects, special attention is placed on

*Figure 11.1   Vehicle prototypes utilised in various automated driving based
mobility experimentation at Renault research division*

ensuring flexibility and scalability, with a focus on reusing interfaces to test various partner organisations in different projects. Topics exclusive to the engineering of safety-critical systems, metrics for architecture and systems engineering, and AD technology as such are not addressed.

11.2   Systems engineering approach

A new set of complex functionalities, obscure and often conflicting stakeholder interests, pose a major problem in the construction of automated vehicle-based mobility systems [9]. This issue is further elevated by the fact that there is no proven business model in the industry [10]. As new players enter, the ecosystem is consolidating with new partnerships, sometimes bypassing tier 1 suppliers and directly interacting with OEM companies. To overcome these challenges, a systematic systems engineering approach in architecture design that provides clear functional boundaries and at the same time being agnostic to business models is necessary [11]. These functional boundaries relate to both technical aspects such as interfaces, integration, and testing, and management aspects such as governance and stakeholder involvement.

Systems engineering consists of a succession of phases that include numerous iterations of functional analysis, synthesis, optimisation, and design to ensure the compatibility of the different interfaces between the various subsystems and between the system and the environment in which it will operate. A model-based systems engineering (MBSE) approach has been applied for designing mobility systems based on automated vehicles. MBSE provides common platform for collaborating among multidisciplinary teams, including domain experts, software architects, developers, network engineers, and vehicle prototyping experts.

In this chapter, we employ a MBSE method called Arcadia for designing AMoDS. To facilitate the approach, a graphical modelling workbench called Capella is utilised [12]. The Arcadia method enforces an approach structured on successive engineering phases that establishes clear separation between needs (operational need analysis and system need analysis) and solutions (logical and physical architectures). Different engineering levels in the Arcadia method are illustrated in Figure 11.2. The highest level is operational analysis which focuses on the identification of the needs and objectives of different entities of the target system. The next stage of system analysis involves the identification of capabilities and functions of the system that will satisfy the operation needs. The boundary of the proposed system is defined at this stage. The level of logical architecture aims to identify the logical components of the system, their relations, and data exchanges independent of any technological implementation. Finally, how the systems will be developed and built will be modelled in physical architecture level. It must be noted that the method is an iterative process and does not always have to be a top-down in nature [13]. The architecture results presented in the following sections have been refined and applied over a five-year period to the construction of prototype mobility systems based on automated vehicles in the research department of Renault.

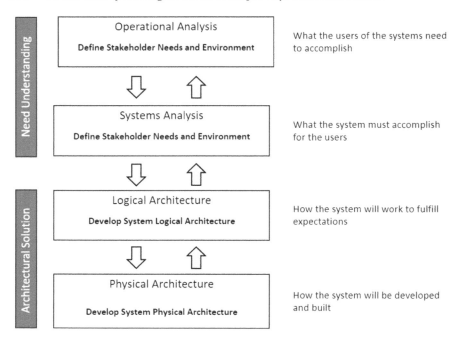

Figure 11.2   Different levels in the Arcadia method [13]

By leveraging MBSE for early design validation and optimisation, empirical results show that it can reduce development costs associated with rework, late-stage design changes, and system failures. This leads to a more efficient use of resources and shorter deployment time for such systems.

11.2.1   Operational analysis

The operational analysis level can be considered as the most abstract stage. The goal is to show what a user of the system wishes to achieve. The focus is on the stakeholders of the system. The models at the operational analysis layer allow to model the required high-level operational capabilities and perform an operational needs analysis without even defining the system-of-interest. The models are designed by considered the following general concepts of MoD systems. The automated vehicle receives the destination and navigates autonomously to its requested destination. The vehicle will perform the service by navigating autonomously to the destination transporting the users and/or items onboard. Destination details can be received from passengers on board the vehicle using a specialised interface or can be received from a remote server through the internet. In a simple use case, the passengers/client access a mobile application to request a trip to go to a destination. The application may propose different options depending on the number of requested seats, cost, comfort, etc. In addition, the vehicle can serve multiple clients by sharing the trip. There are multiple entities involved in an automated vehicle-based mobility service system. These entities can be vehicle OEMs, AD software providers, network operators, fleet operators, multi-modal transportation providers, mobile application providers, etc.

Different components of AMoDS can be broadly classified into three main categories: autonomous navigation, ride sharing, and mobility services as shown in Figure 11.3. The identified high operational activities are allocated to each operational entities. The first operational entity, autonomous mobility deals with point-to-point autonomous navigation using on-board perception, localisation, and decision-making capabilities. It ensures that platforms can drive unmanned while operated for ride sourcing. In order to achieve this, it may need the support from the cloud for receiving the mission and for supporting route planning. Depending on the technology used, the autonomous navigation function can interact with intelligent infrastructure such as connected traffic lights and off-board cameras. When the vehicle is stuck and could not proceed, it can also request support from the cloud, for which it may have to change to its mode of operation from autonomous to teleoperation mode. System safety is of high priority in this category for passengers in the vehicle and for other road users. It may include on-board fault detection and redundancy system to improve the system safety.

The second operational entity of AMoD system is ride sharing that connects users with automated vehicles. Since vehicles are shared in an AMoD system, it typically requires smaller fleet sizes and have lower static land consumption compared to systems that use privately owned and individually operated vehicles. It also implies higher vehicle utilisation, better fuel efficiency, low land utilisation, etc. Depending on the strategy of the stakeholder that provides ride sharing, it uses an algorithm to coordinate the usage of the fleet. The algorithm may consider the

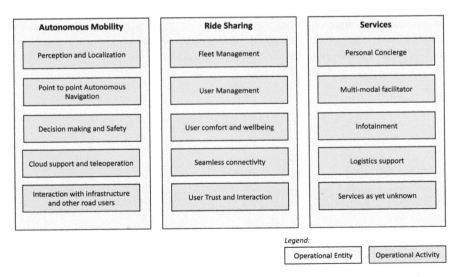

Figure 11.3   *Main operational entities and operation activities of a typical autonomous mobility on-demand system*

capabilities of the technology used in each vehicle. For example, if the AD vehicle can operate in a specific geographical area, the algorithm will not allocate the vehicle to the user where the user's destination is outside this geographical area. In addition to vehicle utilisation, this category may include services provided to the user when they are on-board. The main tasks of ride sharing provider include fleet management, user management, user on-board comfort, and seamless connectivity services as shown as operational activities in Figure 11.3.

The third operational entity is mobility services that provide end-to-end mobility services that are not necessarily using automated vehicles. It includes multi-modal transportation services, personal concierge, and other personalised services. The resulting categorisation into these three silos is not only based on functionality but also includes business rules, value proposition, etc. During the process of identifying the operational activities, there are several intermediate activities involved. These activities include modelling the interaction between the user and smart phone application, high-level vehicle to user allocation, etc. These are modelled using sequence diagrams, state machine diagrams, functional data flow diagrams, etc. In addition, these models are the result of several iterations including the modelling and design activity in the subsequent levels.

11.2.2   Systems need analysis

The grouping of functions into three operational entities in the previous section is to facilitate conceptual understanding and the objective is not to impose any functional boundary at this stage. The whole purpose is to clearly understand the different components and their functions at a high level within the whole system. Using all the operational activities identified in previous level, the next task is decide whether the activity will be realised by the system in the entirely or not. At this system need analysis level, the boundary of the system is defined and the expected behaviour is modelled a system functions. Figure 11.4 shows a functional breakdown of autonomous mobility service (AMS) function of the AMoD system. AMS is classified into autonomous navigation, cloud support, and mobility

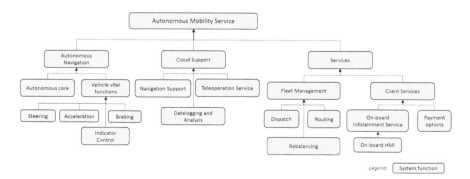

Figure 11.4   Functional breakdown diagram of an AMoD system

services. It is to be noted that there is no one-to-one mapping between these three functions and the three operational entities presented in the previous section.

Autonomous navigation is a system within a larger transportation system. As shown in Figure 11.4, it is divided into autonomous core and vehicle vital functions. The automated driving core refers to the part of the system embedded in individual vehicles responsible for the dynamic driving task. The vehicle vital function comprises functions such as steering, braking, acceleration, and indicator control. The automated vehicle can perceive its surroundings and act independently to attain an assigned navigational goal. It is understood that this goal is imposed from outside the vehicle, by a user, passenger, or a larger fleet management system. There may be further interaction, the navigation may be performed in the vehicle or off board, and in some cases the contextual information may be enhanced by external systems in the cloud. In general, the autonomous navigation function comprises of an autonomous core which provides intelligence to the vehicle and the vehicle's default vital functions. The autonomous core controls the basic vehicle functions that physically manoeuvre the vehicle base. The autonomous core senses other road users, vehicle and pedestrians, and drives the vehicle to have safe and predictable interactions with them. The system responds to traffic lights and other infrastructure signals. The route that the system follows is generated internally in the near future based on a destination that comes from an external source, typically fleet management, but possibly also direct intervention from an onboard passenger. The system may or may not depend on on-board a-priori maps for determining the route, or it may be directed from an external service, but sufficient information in proximity to the vehicle is loaded onboard to safely end a mission by exiting the road or at least pulling over to a stop without reference to external resources. The system supplies current status, route, and vehicular information to other external functions.

The control tower support provides assistance to the automated vehicle in its autonomous and teleoperated mode of the vehicle in case of emergencies. In addition, it includes datalogging and analysis support for further improvement of the AD system. It is important to understand that the control support consists of mainly AD technology dependent functions. For example, depending on the technology of AD used, the way in which the teleoperation works can be different. For example, in one kind of AD technology, you just need to provide high-level waypoints in order to teleoperate the vehicle. There exists greater amount of intelligence on-board the vehicle even during its teleoperation mode. While in another kind of AD technology, more low level actuator command needs to be send in order to teleoperate.

A mobility service includes fleet-related and client-related services. The fleet management functions comprise dispatching, routing, and re-balancing. These functions will be reconfigured according to the model chosen by the service provider. For example, based on a first-in-first-out (FIFO) model, the nearest available vehicle will be assigned to each customer, determined by the shortest path distance between the customer and the available vehicle. The autonomous vehicles routes will follow the most cost-effective route between two locations, with the cost being proportional to the distance travelled. To ensure the availability of cars when needed, the fleet of

AMoD vehicles must be properly sized and managed. Fleet sizing for mobility-on-demand systems is a highly researched area, with numerous studies evaluating the optimal fleet sizes for AMoD systems [14]. The optimal size of an autonomous fleet depends on several critical factors: (a) Network size and trips. Longer trips across wider areas require more vehicles to maintain service levels. (b) Demand. Higher demand necessitates a larger fleet to serve more passengers or deliveries efficiently. (c) Service level. A more frequent and faster service requires a larger fleet to be readily available. (d) Routing policy. Efficient routing algorithms can optimise fleet size by minimising travel distances and idle time. In addition, client-focused features to enhance the passenger experience in mobility services are needed. These include (a) On-board HMI and infotainment. Interact with the vehicle's controls, access entertainment options, and get real-time trip information. (b) Seamless payment. Pay for your trip conveniently, often integrated with the on-board controls. (c) Security. Means to ensure the security of passengers.

11.2.3   Logical reference architecture

The tasks expected from a mobility service are quite clearly defined in the system needs analysis stage. The next step is to devise a logical reference architecture considering the technological and business constraints. A reference architecture is, in essence, a predefined architectural pattern, or set of patterns, possibly partially or completely instantiated, designed, and proven for use in particular business and technical contexts, together with supporting artefacts to enable their use. Often, these artefacts are harvested from previous projects [15]. These architectures would encompass definitions of the diverse architectural components required by an AMoD system, its interaction, data flows, and provide interfacing guidelines tailored to specific business rules.

Having introduced the necessary functional components in the systems needs analysis level, we now combine them into a suggested reference architecture. A high-level reference architecture for AMoD system is shown in Figure 11.5. Different components of the reference architecture AD fleet providers receive support from intelligent infrastructure. AV fleet providers communicate to a common cloud platform called 'Common Mobility Platform for Automated Vehicles (CoMPAV)'. The goal of CoMPAV is to transmit the mission to the allocated vehicle and to support it during its operation. More details regarding the functions of CoMPAV are explained in the next section. CoMPAV communicates with ride sourcing provider (RSP) to receive the mission-related information. RSP is a platform that connects passengers with drivers through a digital app or platform.

The principal function of a RSP is fleet management, which in our case involves assigning, dispatching, and routing autonomous vehicles. The main difference between AMoD systems and conventional transportation systems is their demand-responsiveness. In the AMoD architecture, the RSP is primarily responsible for deployment. While scheduled systems like buses and trains operate periodically, AMoD vehicles operate solely on a demand basis, a crucial aspect of the mobility as a service (MaaS) concept. This feature highlights the strong potential of AMoD as it provides accessibility to individuals living in areas poorly served by

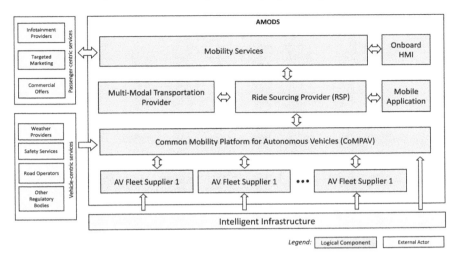

Figure 11.5   Reference architecture of autonomous mobility on-demand system

public transport such as peri-urban and rural regions. For users with limited mobility, like the elderly and young, such a service can significantly enhance their daily lives.

The AMoD service can operate alongside regular taxis and private and public transportation vehicles. Once passengers are dropped off, the vehicles can return to their base, the closest station, or simply wait at the drop-off location for the next service request. This optimisation process, known as fleet rebalancing, is critical. An unbalanced fleet can result in service availability problems for potential passengers, especially during periods of high demand. It is a multidimensional optimisation problem where fleet management depends on the context and vehicle capabilities. The context includes information on the workspace such as road network geometry, current and future operating conditions, traffic density, and the estimated arrival times of the nearest vehicles. Through automated rebalancing, AMoD systems can redistribute cars to better meet demand. It is possible to optimise the routes of autonomous vehicles based on the current and future state of the road network. The ability to build twin models and use connectivity as crowd-sourcing technologies allows AMoD systems to use existing road infrastructure more efficiently, for example, by reducing the distance headway and routing vehicles through less-congested roads. Proper rebalancing has been shown to positively affect system performance, resulting in smaller fleet sizes [16].

For non-automated mobility on demand (MoD) systems, the free-floating scheme is arguably preferable for consumers as it alleviates the costs associated with returning vehicles to their base or waiting point. For autonomous vehicles, self-returning to base is possible, but this return trajectory means the vehicle will be empty, potentially increasing road congestion and energy usage. In the free-floating model, vehicles can become severely unbalanced, leading to longer waiting times

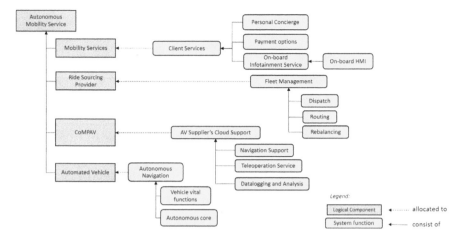

Figure 11.6   Reference architecture of an autonomous mobility on-demand system

for consumers. The RSPs can redistribute cars to better meet demand and thereby increase revenue. These providers typically do not own the vehicles or directly employ the drivers but act as intermediaries. They can be considered facilitators of transportation services between users and independent contractors or drivers. The RSP is responsible for allocating vehicles to satisfy passenger demand. A typical interface for on-demand services could be the use of a mobile application to establish the necessary communications. The RSP can communicate with mobility services available to passengers while they are using the vehicles. This allows for the provision of information on the trip status and access to additional services, which could include data for multi-modal transit or specific demands.

Figure 11.6 illustrates the allocation of functions to the logical components in the AMoD reference architecture. Autonomous navigation is allocated to automated vehicle and cloud support is allocated to CoMPAV. Fleet management is allocated to RSP and client services are allocated to mobility service component. It is to be noted that the sub-functions of service function are allocated to two different components, namely RSP and the mobility services. This is just an example to explain that some components in control tower are tightly coupled with the technology used in the AD vehicle. However, there are components that are more generic such as traffic information and route planning.

11.2.3.1   Common mobility platform

Common mobility platform for automated vehicles (CoMPAV) is a layer between the RSP and AV fleet. It mainly consists of an AD provider's cloud support component, common event management component, and layers required for interfacing with external services. This common layer is also responsible for interfacing with RSP. In general, CoMPAV acts as a vehicle-centric control centre and RSP can be viewed as a passenger-centric control centre (Figure 11.7).

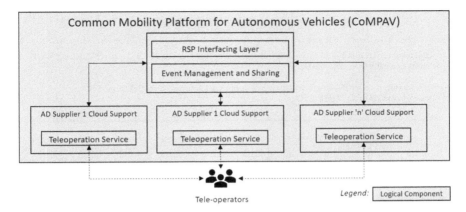

Figure 11.7   Logical architecture of a CoMPAV module

The goal of the AV supplier's control tower is to support its fleet of AD vehicles to navigate safely in its autonomous or teleoperation mode. An analogy can be made with control tower for a fleet of aircrafts. Even when equipped with autopilot, a fleet of aircrafts still depends on a control tower to be operated. Compared to automated driving, the aircraft autopilot systems are more mature and integrated into the operation routine in the aviation domain for many years. Despite the maturity in the aviation control, the control tower is still central in safety and efficiency operation, and is not replaceable [17]. To ensure smooth operation of the fleet, certain contractual interfaces between the cloud support of the AV supplier and the RSP must be enforced. This contractual interface includes both technical APIs and operational agreements. As an example of an operational agreement, the stop locations where the vehicle are allowed to stop for boarding and deboarding passengers are flagged as parking capable or not. RSP are allowed to stop the vehicle at a parking capable location for more than 2 minutes and in other stop locations, they are used just for serving passengers only. Two APIs and their data structures provided by AV supplier's cloud service are shown in Figure 11.8. The adMissionSet function is utilised by RSP to send mission to AV cloud service and gets the mission acceptance status. The data structure of mission and its acceptance are shown in Figure 11.8. Using the second API function, the RSP will be able to get the current status of the vehicle. The status includes its position, speed, availability, etc.

Independent of the supplier of the AV, the CoMPAV maintains the state of each vehicle for fleet management. Figure 11.9 shows the states and their transitions as state machine diagram. The availabilityStatus indicates if the vehicle can accept a mission or not. The vehicle can stop at predefined stop locations only. The state machine consists of two states: vehicle disconnected and vehicle connected state, at a high level. The vehicle will be in disconnected state when it does not receive any messages for more than 5 seconds. All the operational states of the vehicle are in the vehicle connected state. When the vehicle accepts a mission while in parking location, the vehicleState is changed from PARKED to ON_MISSION. If the destination

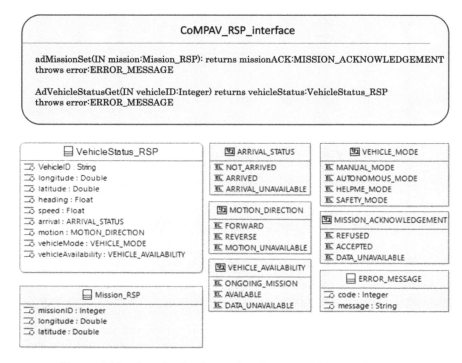

Figure 11.8   Standardised interface between RSP and CoMPAV

stop is parking capable, the state variables after reaching the destination will be arrivalStatus = ARRIVED, vehicleState = PARKED. It is to be noted that the vehicleState will be in ON_MISSION even when it has reached pickup/dropff location and as PARKED when the vehicle has reached parking capable location. It is the responsiblity of the RSP to make sure vehicle is always send to the parking capable location when it is not servicing passengers.

11.2.4   Physical architecture

The component interfaces are further refined in the physical architecture, considering the constraints imposed by the allocation of components to cloud servers or edge computing units and the bandwidth of inter-component communications. Decisions regarding the vehicle's connectivity options, such as cellular, Wi-Fi, or DSRC, are determined at this stage. Detailed electronic and electrical connection diagrams are then developed based on the design constraints established in the preceding stage. As previously mentioned, this process is iterative rather than strictly hierarchical. Constraints identified at the physical architecture level may prompt modifications to the logical architecture. For instance, the limited bandwidth of communication channels may restrict the volume of data they can transmit within the required latency parameters.

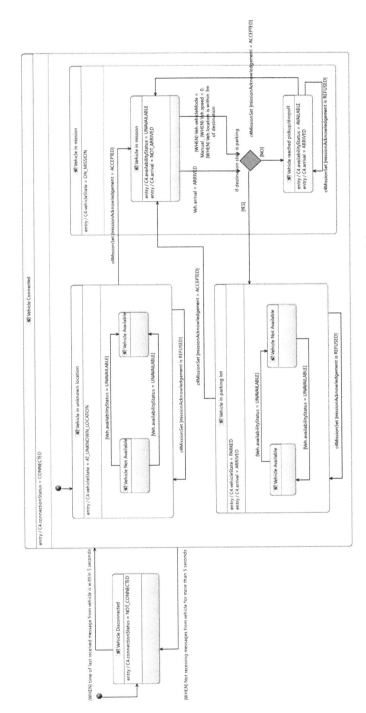

Figure 11.9 State machine of CoMPAV module

11.3   Value proposition and revenue streams

In pursuit of maximising revenue, several OEMs also aim to enter the robo-taxi market themselves. However, they must prepare for significant transformations by transitioning into mobility service providers and collaborating with partners, investors, and AV technology providers to offer a comprehensive solution. The influence of product architecture on the organisation of individual firms has been widely studied [18]. Correlation between industry structure and the functional interdependence in architectural components has been verified in several empirical studies. With respect to the functional range of stakeholders in AMoDS, a value chain structure will be of remarkable interest in order to align the design decisions and to experiment with different business models. This method will also help to position the technologies in general and to do a suitability analysis for deployment. The proposed reference architecture can be represented as a design structure matrix (DSM) representing provided value and incurred costs. A DSM is a tool used in engineering and project management to represent the relationships between components or activities within a complex system or project. It provides a visual representation of the interactions and dependencies between different elements, allowing teams to better understand the structure of the system or project and identify areas of potential improvement or optimisation [19].

Figure 11.10 shows the flow of values and possible revenues as a DSM. The cells along the diagonal of the matrix represent the main logical entities of the AMoD system. Each off-diagonal cells are diving into two halves: upper half

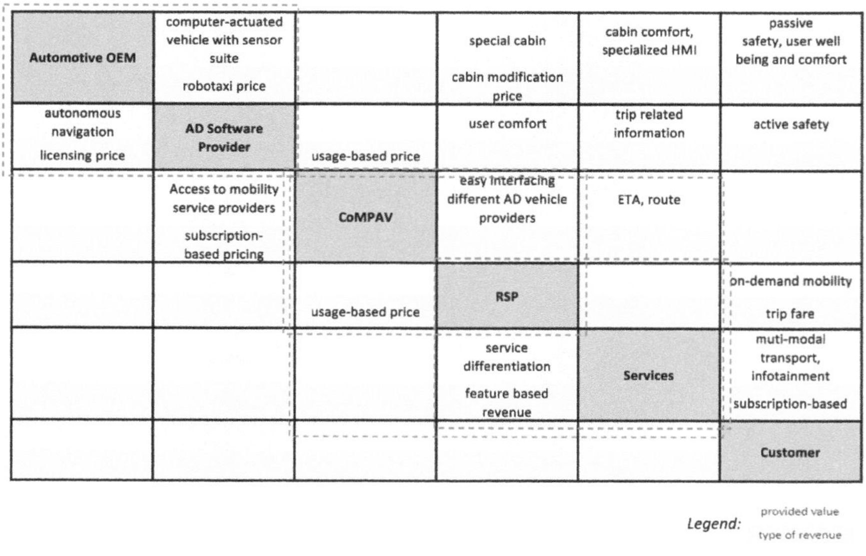

Figure 11.10   Design structure matrix of provided value and revenue types of AMoD components

represents the value additions and lower half represents the revenue type. Examining any row in the matrix reveals all the value additions and revenue type from the element in that row. Looking down any column of the matrix shows all of the value received and cost incurred to the element in that column. For example, the upper half of cell[row:1, column:2] indicates that Automotive OEM adds value to the AMoD system by providing computer-actuated vehicle with sensor suite to AD software provider. The lower half of the cell indicates the cost incurred by the OEM. Cell[row:2, column:1] indicates that AD software provider adds value to the AMoD system by providing autonomous navigation to automotive OEM and can get licensing price for its AD technology from the OEM.

The other cells in the matrix indicate the value added by each component in the AMoD system. The value added by CoMPAV is to easily access the multiple mobility service providers in the view of AD software provider while in the view of RSP, it provides easy interfacing with different AV vehicle/software providers. RSP provides on-demand autonomous mobility to the services and by including services, RSP will be able to differentiate among other competitors. There are more values that can be found while more features are integrated by the AMoD components. For example, if robo-vehicles that are wheel-chair friendly are provided by an automotive OEM, it will be beneficial for certain RSPs and service providers as a differentiating factor. The DSM is a tool used as canvas and it is not possible to discuss all possible revenue types as it depends on the contracts between entities. Each AMoD component can choose its stakeholder company based on the added value. For example, RSP will prefer AD software provider that provides passenger comfort while navigation. Similarly, RSP will choose automotive OEMs that provide customised cabin for automated vehicles with advanced HMI, better music system, etc.

The adoption of automated vehicles hinges not just on technological advancements and regulatory frameworks but also on the development of business models aimed at generating and harnessing value. These models evolve alongside technologies, user needs, infrastructure, and regulations, shaping a fresh socio-technical mobility landscape [20]. In addition to value addition and revenue source, Figure 11.10 shows possible entity aggregations as dotted lines. For example in one business model, an automotive OEM can have its own AD software technology and act as a AD mobility provider. This new integrated entity can have contractual agreement with CoMPAV and receives recurrent revenues instead of just one time revenue of robotaxi price otherwise. In another model, a mobility service provider can take the responsibility for CoMPAV and RSP. It can even be extended by including the services as well. Different cost/benefit analysis can be performed to arrive at the best possible model with minimum disturbance to the overall architecture.

11.4 Conclusion

In this chapter, we have applied a systematic SE approach based on the Arcadia method for designing an autonomous mobility on-demand system. Key results of the component and functional architecture at each levels are presented. A coherent reference architecture with clear functional boundaries is designed with careful consideration of both technological requirements and business constraints. The system as a whole does not prescribe a singular operational model and remains adaptable to exploring diverse partnership arrangements. The resultant architecture emerges from numerous iterations and has been effectively deployed across various experimental services.

The cloud-based mobility platform, CoMPAV, oversees AV fleets sourced from various providers. The supplier's cloud component gathers data from the AV fleet and shares it with CoMPAV to enhance safety and streamline route planning of RSP. To ensure seamless integration, AD suppliers are required to furnish a contractual interface with CoMPAV. Within CoMPAV, each AD supplier offers a cloud support component to facilitate integration within the broader mobility service ecosystem.

Some of the core features of the AMoDS reference architecture are:

- It is scalable in the sense that it facilitates a mobility service provider to handle AD vehicles from different providers.
- It is modular and clearly defines the functional boundaries of each actor in the system.
- It is flexible to centralised and decentralised fleet management and is configurable to the operating market needs.
- It is agnostic to business models.

The AD functionalities are integrated, and the architecture has been used for several national and international collaboration projects and deployed in urban and rural service areas. These different instances of the reference architecture demonstrate the values during the initial analysis that allow their implementation using reduced resources. In future, the use of data-driven methods shall provide more benefits; however, the underlying frameworks will remain consistent.

Acknowledgements

The authors express their thanks to our colleagues at the Renault Group that participated in the various deployments of the proposed architecture without their contribution it would have been difficult for it to evolve.

References

[1]   Shaheen SA, and Cohen AP. The impacts of shared and automated mobility. In: *Shared Mobility and Automated Vehicles: Responding to Socio-Technical Changes and Pandemics*. Khan AM and Shaheen SA, eds. 2021;20:363. IET London.

[2]   Zardini G, Lanzetti N, Pavone M, *et al.* Analysis and control of autonomous mobility-on-demand systems. *Annual Review of Control, Robotics, and Autonomous Systems*. 2022;5:633–658.

[3]   Bai J, and Tang CS. Can two competing on-demand service platforms be profitable? *International Journal of Production Economics*. 2022;250:108672.

[4]   Ibañez-Guzman J, Laugier C, Yoder JD, *et al.* Autonomous driving: context and state-of-the-art. *Handbook of Intelligent Vehicles*. 2012;2:1271–1310.

[5]   Heineke K, Heuss R, Kampshoff P, *et al. The road to affordable autonomous mobility.* McKinsey and Company; 2022. Available from: https://www.mckinsey.com/industries/automotive-and-assembly/our-insights/the-road-to-affordable-autonomous-mobility. Accessed 16 August 2024.

[6]   Ambadipudi A, Heineke K, Kampshoff P, *et al.* Gauging the disruptive power of robo-taxis in autonomous driving. *Automotive & Assembly*. 2017. www.mckinsey.com/~/media/McKinsey/Industries/Automotive%20and%20Assembly/Our%20Insights/Gauging%20the%20disruptive%20power%20of%20robo%20taxis%20in%20autonomous%20driving/Gauging-the-disruptive-power-of-robo-taxis-in-autonomous-driving.pdf. Accessed on 01 September 2024.

[7]   Pavone M. Autonomous mobility-on-demand systems for future urban mobility. In: *Autonomes Fahren: Technische, Rechtliche und Gesellschaftliche Aspekte*. Eds M. Maurer *et al.*, Springer Nature, Berlin. 2015:399–416.

[8]   Milanés V, González D, Navas F, *et al.* The tornado project: an automated driving demonstration in peri-urban and rural areas. *IEEE Intelligent Transportation Systems Magazine*. 2021;14(4):20–36.

[9]   Sifakis J, and Harel D. Trustworthy autonomous system development. *ACM Transactions on Embedded Computing Systems*. 2023;22(3):1–24.

[10]   Kacperski C, Vogel T, and Kutzner F. Ambivalence in stakeholders' views on connected and autonomous vehicles. In: *HCI in Mobility, Transport, and Automotive Systems. Automated Driving and In-Vehicle Experience Design: Second International Conference, MobiTAS 2020, Held as Part of the 22nd HCI International Conference, HCII 2020*, Copenhagen, Denmark, July 19–24, 2020, Proceedings, Part I 22. Springer; 2020. pp. 46–57.

[11]   Martin E, Cohen A, and Shaheen S Synthesis Report: Findings and Lessons Learned from the Independent Evaluation of the Mobility on Demand (MOD) Sandbox Demonstrations. FTA Report; 2023.

[12]   Roques P MBSE with the ARCADIA Method and the Capella Tool. In: 8th European Congress on Embedded Real Time Software and Systems (ERTS 2016); January 2016, Toulouse, France. hal-01258014 https://hal.science/hal-01258014/document. Accessed on 01 September 2024.

[13]   Roques P, (ed.) Systems architecture modeling with the Arcadia method. (Vol 2). In: *Implementation of Model Based System Engineering Set*. Ed (Roques P.) London: ISTE Press Ltd; 2018.

[14]   Spieser K, Samaranayake S, Gruel W, *et al.* Shared-vehicle mobility-on-demand systems: A fleet operator's guide to rebalancing empty vehicles. In: Transportation Research Board 95th Annual Meeting. 16-5987. Transportation Research Board; 2016.

[15]   Cloutier R, Muller G, Verma D, *et al.* The concept of reference architectures. *Systems Engineering*. 2010;13(1):14–27.

[16]   Wallar A, Van Der Zee M, Alonso-Mora J, *et al.* Vehicle rebalancing for mobility-on-demand systems with ride-sharing. In: *2018 IEEE/RSJ*

International Conference on Intelligent Robots and Systems (IROS). IEEE; 2018. p. 4539–4546.

[17]   Zhao X, Darwish R, and Pernestål A. Automated vehicle traffic control tower: A solution to support the next level automation. *International Journal of Transport and Vehicle Engineering*. 2020;14(7):283–293.

[18]   Cantamessa M, Milanesio M, and Operti E. Value chain structure and correlation between design structure matrices. In *Advances in Design*. Eds; ElMaraghy HA, ElMaraghy WH Springer-Verlag, London, 2006:303–313.

[19]   Steward DV. The design structure system: a method for managing the design of complex systems. *IEEE Transactions on Engineering Management*. 1981;(3):71–74.

[20]   Göcke L, and Meier P. Business model development for autonomous electric vehicles in shared fleets in multinational original equipment manufacturers (OEMs). In: *Electric Vehicles in Shared Fleets: Mobility Management, Business Models, and Decision Support Systems*. Eds. Degirmenci K, Cerbe TM and Pfau WE. Singapore: World Scientific; 2022. pp. 125–144.

Chapter 12

Robust cooperative perception for intelligent transport systems

Fuxi Wen[1], Zhiying Song[2], Tenghui Xie[2], Jiaxin Gao[2], Bolin Gao[1] and Shengbo Eben Li[1]

12.1 Introduction

12.1.1 What is cooperative perception?

Intelligent transportation system (ITS) is a term that describes various technologies that facilitate the movement of people and goods. These technologies include advanced traffic management systems, intelligent sensors, and vehicular communications systems. ITS can help make transportation more efficient and safe.

Automated driving is a prominent technology for ITS [1,2], which comprises perception, planning, and control modules [3,4]. The perception module relies on sensors to continually scan and monitor the vehicle's surroundings, which enables automated vehicle (AV) to comprehend its environments. AV perception can be categorized into individual perception and cooperative perception. Despite significant advancements driven by deep learning in individual perception [5,6], several challenges hinder its progress. First, individual perception often faces occlusion issues when operating in complex environments. Second, onboard sensors have inherent limitations in detecting distant objects. Moreover, sensor noise adversely affects the performance of the perception system.

Addressing the limitations of individual perception, cooperative perception has garnered significant attention [7,8], which leverages interactions among multiple agents. Cooperative perception operates as a multi-agent system [9], where agents share perceptual data to overcome visual constraints within the ego vehicle. As illustrated in Figure 12.1, in an individual perception setting, the ego AV detects only a portion of nearby objects due to occlusion in distant regions. In contrast, in a cooperative perception setup, the ego AV broadens its field of view by integrating information from other agents. This cooperative effort enables the ego AV to detect distant and occluded objects while enhancing detection accuracy in dense areas.

[1]School of Vehicle and Mobility and State Key Lab of Intelligent Green Vehicle and Mobility, Tsinghua University, China
[2]School of Vehicle and Mobility, Tsinghua University, China

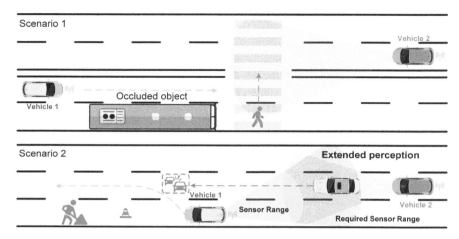

*Figure 12.1   Two scenarios where cooperative perception helps. The onboard
sensors of Vehicle One miss detect a critical object due to occlusion
(top), and Vehicle Two extends Vehicle One's perception range,
which is insufficient to complete a safe lane change*

The emerging cooperative perception aims to enhance the awareness of connected and automated vehicles in the surrounding environment. A vehicle's onboard sensors can miss the detection of critical objects moving into its path due to occlusion caused by other obstacles on the road. The occluded object may be detected too late to avoid a collision. Figure 12.1 shows such a scenario: with sensor data shared from Vehicle Two, the perception of vehicle one can cover the crossing pedestrian initially occluded by the bus from the sensors have a predefined perception range due to technological limitations. In corner cases, the effective perception range is insufficient for safe maneuvers. In Scenario 2 of Figure 12.1, Vehicle One intends to change to an adjacent lane with high-speed traffic. Factors such as lane-change time and acceleration time dictate the required perception range behind Vehicle One to complete the lane-change maneuver safely. By sharing sensor data, approaching vehicles like Vehicle Two can benefit Vehicle One with an extended perception range.

Cooperative perception is a promising technique for ITS through Vehicle-to-Everything (V2X) communications, provided that accurate relative pose information is available between the connected vehicles and the roadside unit or other connected ITS agents [10].

Figure 12.2 shows how the fusion center jointly processes the remote Collective Perception Message (CPM) and the local perception data. The updated object tracks resulting from data fusion are then encoded into CPM before being transmitted to other ITS agents [11].

12.1.2   Cooperative perception methods

The overall perception system of a connected and automated vehicle (CAV) comprises an input, base network, feature, perception head, and output stages. Based on

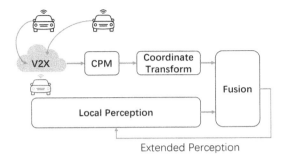

Figure 12.2 An illustration of the cooperative perception scheme for connected vehicles with the help of V2X communications

the data sharing and cooperative stage, the cooperative perception scheme can be broadly categorized into early, intermediate, and late cooperation. As depicted in Figure 12.3, each is distinguished by the nature of shared data [12]. Certain studies have embraced hybrid cooperative approaches for enhanced multi-agent cooperation, as demonstrated in [13]. These approaches amalgamate early and late cooperative strategies to integrate information from diverse infrastructures.

12.1.2.1 Early cooperation

The early cooperation approach involves the fusion of raw data at the network's input, also referred to as data-level fusion or low-level fusion [14,15]. In the context of automated driving, the ego vehicle acquires and processes raw sensor data from other agents, consolidating the transformed data onboard. Raw data encompasses the most comprehensive information and provides a detailed description of agents. As a result, early cooperation serves as a fundamental solution to overcome occlusion and long-range issues encountered in individual perception, significantly enhancing performance. However, it is essential to note that early cooperation relies on high data bandwidth, presenting a challenge in achieving real-time edge computing.

12.1.2.2 Intermediate cooperation

Considering the high bandwidth requirements of early cooperation, certain studies suggest intermediate cooperative perception methods to strike a balance between performance and bandwidth considerations. In intermediate cooperation, other agents typically transmit deep semantic features to the ego vehicle, which then integrates these features to make the final prediction. Due to its flexibility, intermediate cooperation is the most widely adopted multi-agent cooperative perception approach. However, the feature extraction process can lead to information loss and unnecessary redundancy, prompting researchers to explore appropriate feature selection and fusion strategies.

12.1.2.3 Late cooperation

Late cooperation, also known as an object-level cooperative, involves the fusion of predictions at the network output, as illustrated in Figure 12.3. In this approach,

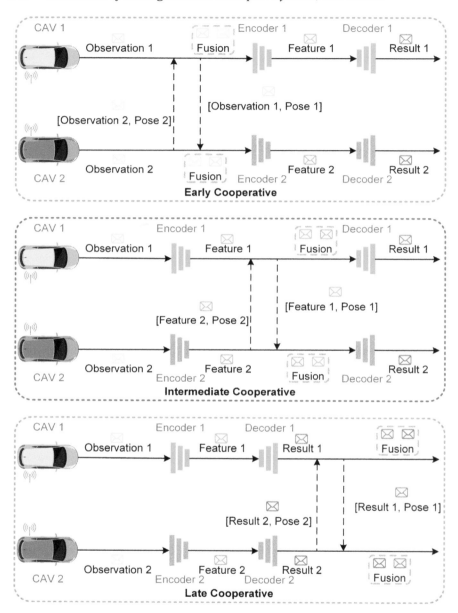

Figure 12.3   Overview of V2X cooperative schemes for automated driving

each agent independently trains the network and exchanges outputs with other agents. The ego vehicle then spatially transforms these outputs and combines them after post-processing. Late cooperation is more bandwidth-efficient and straightforward than early and intermediate cooperation. However, late cooperation has limitations as individual outputs may be noisy and incomplete, and late cooperation exhibits the lowest perception performance.

As depicted in Figure 12.3, early cooperation consolidates raw sensor observations from all participating agents, providing a comprehensive perspective [16]. This method effectively tackles occlusion and long-range obstacles in single-agent perception. However, its practical application requires significant communication bandwidth. On the other hand, late cooperation focuses on exchanging perceptual outcomes, demanding minimal data transmission [17]. While efficient in communication resource usage, late cooperation may introduce inherent noise and incompleteness into individual perceptual outputs, potentially leading to suboptimal fusion results. Intermediate cooperation has emerged as a viable solution for addressing the trade-off between performance and bandwidth constraints. This approach aggregates intermediate features across multiple agents, encompassing representative information within compressed features. Such integration holds promise for enhancing both communication efficiency and perception capabilities.

Traditional cooperative perception approaches have traditionally favored late cooperation strategies, driven by feature extraction complexity and communication capabilities constraints. These strategies typically operate at the object or track level. However, the rapid advancements in deep learning have brought intermediate cooperation to the forefront in multi-agent cooperation, primarily due to its capacity to alleviate the strain on communication resources. Introducing extraneous and unnecessary information is risky, especially in dynamic networking conditions. Addressing this challenge underscores the importance of optimizing cooperative information and selecting collaborators judiciously within a cooperative perception system. Furthermore, the deep representation introduces higher levels of abstraction and implicitness in the spatial interactions between agents, necessitating appropriate data alignment and information fusion algorithms. Developing a unified cooperative perception framework to tackle these challenges is crucial for advancing cooperative driving automation.

12.2 Robust cooperative perception

Cooperative perception has long been a subject of interest. Earlier studies [18–20] concentrated on constructing cooperative perception systems to assess the viability of this technology. However, the absence of large public datasets has hindered its effective progression. In recent years, there has been a surge in interest and research, fueled by the advancements in deep learning and the public release of extensive cooperative perception datasets [21–23].

Considering communication bandwidth constraints, most research [24] is dedicated to devising innovative cooperative modules to balance accuracy and bandwidth. However, these prior works often assume an ideal cooperative scenario. To address practical issues in automated driving applications, such as localization errors, communication latency, and model discrepancies, recent studies [25] propose corresponding solutions to ensure the robustness and safety of cooperative

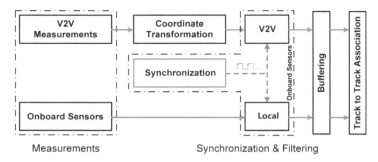

Figure 12.4   Robust cooperative perception framework

systems. Figure 12.4 shows one of the typical object-level robust cooperative perception frameworks.

12.2.1   Synchronization and coordinate transformation

Indeed, synchronization and coordinate transformation play crucial roles in cooperative perception, especially in multi-agent or distributed systems [26]. Cooperative perception involves multiple sensors or agents working jointly to build a shared understanding of the environment. Here's how synchronization and coordinate transformation are essential in this context: Synchronization ensures that the perception information is consistent across all agents simultaneously. It helps to prevent discrepancies caused by delays or differences in sensor sampling rates. Meanwhile, coordinated transformation ensures that the spatial information is consistent across all agents, enabling a shared global understanding of the environment.

When we consider sensor fusion and data integration problems, different sensors or agents might operate with their internal clocks. Synchronization ensures that the timestamps of the collected data align correctly to fuse data from various sources accurately. This is important for associating measurements from different sensors at the same point in time. Furthermore, agents may have different coordinate systems. Transforming sensor data into a common reference frame allows for a consistent and unified environment representation.

In summary, synchronization and coordinated transformation are foundational elements in cooperative perception systems, enabling seamless cooperation, consistency, and accurate understanding of the shared environment among multiple agents or sensors.

12.2.1.1   Illustration of cooperative perception with pose errors

The illustration of cooperative perception with pose errors is shown in Figure 12.5. It is a cooperative perception scenario where the ego vehicle (yellow) and the CAV vehicle (green) cooperate to detect pedestrian *M*. When the position of the green vehicle is inaccurate, its perception results cannot align with the ego vehicle's, causing the ego vehicle to identify the pedestrian at a different location incorrectly,

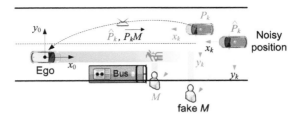

Figure 12.5 Illustration of cooperative perception with pose errors

thus making incorrect decisions, ultimately resulting in a collision with the pedestrian.

12.2.2 Sources of temporal and spatial errors

12.2.2.1 Synchronization errors

Given the significant disparities across different vehicles and the myriad factors influencing inter-vehicle communication, synchronization among multiple vehicles is essential in cooperative perception. Here are some familiar sources of synchronization errors:

Clock drift. The clocks in different devices may not be perfectly synchronized, leading to time discrepancies.

Network delays. In distributed systems, communication across a network can introduce latency, causing data to arrive at different times at different locations.

Communication failures. Issues in communication channels, such as dropped or delayed messages, can lead to discrepancies in the state of different components.

Heterogeneous environments. Systems that run on different hardware architectures, operating systems, or software versions may have difficulties achieving synchronization.

12.2.2.2 Relative pose errors

Several local pose errors can affect the estimated position and orientation accuracy when using global positioning system (GPS) and inertial measurement unit (IMU) data for pose estimation. Here are some common sources of GPS errors:

Satellite signal obstruction. Buildings, trees, or other obstacles can block or reflect GPS signals, leading to multipath errors and degraded accuracy. Temporary blockage of GPS signals due to tunnels, dense urban environments, or natural obstructions can result in temporary signal loss and pose estimation errors.

Multipath interference. Reflections of GPS signals off surfaces, such as buildings or large vehicles, can lead to multipath interference, causing errors in position estimates.

Here are some common sources of errors in IMU-based pose estimation:

Sensor noise. Inherent noise in IMU sensors, including accelerometers and gyroscopes, can lead to inaccuracies in measuring acceleration and angular rates. Bias in IMU sensors can result in systematic errors over time, causing a drift in the estimated pose. Inaccuracies in the scaling factor of sensors can lead to measurement discrepancies. Interference between different axes of the IMU sensors can introduce measurement errors.

Vibration and shock. Vibrations or shocks experienced by the vehicle or device carrying the sensors can introduce noise and affect the accuracy of IMU readings. Rapid changes in acceleration, such as during sudden stops or accelerations, can challenge the accuracy of IMU-based pose estimation.

12.3   Robust cooperative perception with temporal synchronization

12.3.1   Introduction

Although individual car temporal synchronizations can be executed effectively, in the context of cooperative perception, new factors emerge leading to temporal synchronizations.

Vehicles perceive and understand the surrounding world through onboard sensors such as light detection and ranging (LiDAR) and cameras. Initially, they collect data using these sensors, for instance, by recording real-time environmental videos with cameras, resulting in capture delays. The collected data is then fed into an object detection module, typically based on deep neural networks, to detect traffic participants in the environment. However, this process inevitably takes some time, introducing detection delays. Subsequently, vehicles share sensor perception results with each other, exchanging information over the network, akin to humans sending messages to friends. Upon receiving shared information from other vehicles, they must align temporally and spatially and fuse it with their sensor perceptions. This fusion process, known as fusion delay, incurs a certain amount of time. In summary, from the moment one vehicle perceives the current environmental state until another vehicle outputs the shared and fused results, this process involves four distinct types of delay, as shown in Figure 12.6.

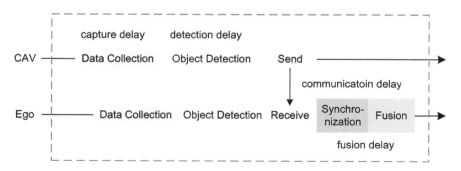

Figure 12.6   Time delay in a cooperative perception system

For example, Moradi-Pari *et al.* proposed an object-level cooperative perception solution tailored explicitly for intersections [27]. According to their findings, the overall system latency observed from a receiver onboard unit (OBU) is less than 400 ms. This latency can be broken down into the components in Table 12.1.

In achieving cooperative perception, vehicles receiving information must align the received data temporally with their context to ensure that they are processing information from the exact moment. This alignment is crucial for integrating the received information into their system effectively. In the existing literature, the standard approach involves the receiving vehicles compensating for time and inferring the state of this information at the current time point. Below are two typical examples demonstrating how models are established to compensate for time delays in feature—level and object-level cooperative perceptions.

12.3.2 Object-level temporal synchronization

A typical workflow of object-level cooperative perception is illustrated in Figure 12.7. Upon receiving perception results from the CAV, the ego vehicle first performs delay compensation. Then, the ego vehicle tracks the trajectories of objects in the scene by data association, Kalman filter, and birth and death memory modules. It is similar to the multi-object tracking methods [28].

Table 12.1   The latencies of various components comprising the OBU

Components	Latency/ms
Camera system	300
Image processing unit (IPU) detection and processing	20–50
IPU to road side unit (RSU) network	5–10
RSU processing and broadcast	<20

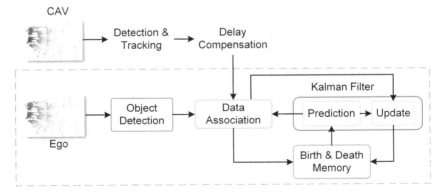

Figure 12.7   A simple architecture of object-level cooperative perception

Delay compensation is typically achieved by establishing a hypothetical motion model for each target, and predictions are made regarding the target's movement during the delay period. For each object i, a motion model is employed to derive the recursive relationship of its state at adjacent time instants

$$x_{t+1}^i = f(x_t^i, \omega_t) \tag{12.1}$$

here x^i denotes the state of object i and ω represents the error in the motion model.

Various motion models can be used to model the movement of a road user. For example, a simple linear motion model can be applied to a vehicle driving straight. This model assumes constant velocity and predicts the vehicle's position at the next step based on its current position and velocity. Other models can capture more complex behaviors, such as acceleration, deceleration, and changes in direction, depending on the characteristics of the target and the environment. These models play a crucial role in estimating the future states of objects and compensating for delays in perception data.

Inevitably, the motion model contains errors. To address this, fusion with ego's tracking results is performed using a Kalman filter, as shown in the right part of Figure 12.7. This filter combines information from the motion model predictions and the ego vehicle's tracking results to estimate the object's state accurately. By incorporating both sources of information, the Kalman filter mitigates the impact of errors in the motion model and improves the target's state estimation accuracy.

However, the method mentioned above is only effective for handling time delays that are not excessively large. When the delay exceeds a certain threshold, estimation errors in the motion model become significant and challenging to manage, weakening the cooperative perception's effectiveness.

12.3.3   Feature-level temporal synchronization

In contrast to transmitting bounding boxes in object-level cooperative perception among collaborating agents, feature-level methods involve transmitting inter-mediate features from deep neural networks. Figure 12.8 shows three pictures of deep neural network features; unlike the bounding boxes output in object-level methods, which contain only object positions, sizes, and orientations, these features are less abstract and have richer semantic information.

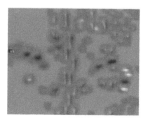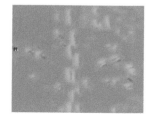

Figure 12.8   Features of deep neural network

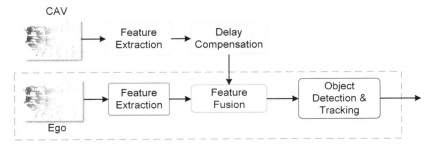

Figure 12.9   A basic framework of feature-level cooperative perception

Figure 12.9 illustrates the fundamental components of a typical feature-level cooperative perception system. In contrast to the object-level architecture depicted in Figure 12.7, this architecture is more straightforward, requiring vehicles only to extract features from deep neural networks. These features can then be shared between vehicles and ultimately fused by the receiving vehicle for unified object detection and tracking.

To achieve delay compensation, feature-level methods need to predict the motion of the features, similar to (12.1). Due to the intricate operational mechanisms within neural networks, it's challenging to establish their internal motion equations explicitly. Therefore, their motion is typically approximated using another neural network. By inputting features into a neural network module and training it along with the entire framework, the network learns to perform time compensation automatically.

Deep neural networks have demonstrated powerful capabilities in many aspects, and numerous learning-based cooperative perception methods have been proposed recently. However, there is still a lack of highly effective time compensation solutions. This is also a challenge that feature-level cooperative perception must address to be implemented in the real world, and further research is awaiting to produce more effective solutions.

12.4   Robust cooperative perception with pose calibration

Cooperative perception outcomes often hinge on the positioning accuracy of multiple agents, where inaccuracies can result in significant spatial discrepancies, especially in complex traffic environments. Therefore, achieving robust cooperative perception necessitates coordinate transformation among multiple agents. Figure 12.10 illustrates the workflow of cooperative perception methods considering pose calibration. Pose calibration methods in cooperative perception can be categorized into three levels based on the desired data types: data-level, feature-level, and object-level. At the data-level, raw data is output, at the feature-level, deep neural network intermediate features are output, and at the object-level, object-level bounding boxes are output. This section will introduce a representative example for each of these three methods.

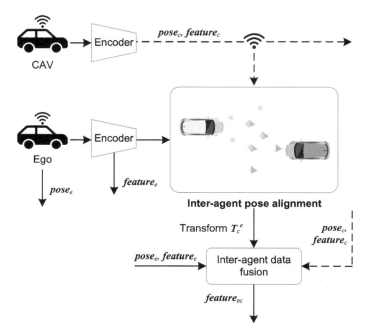

Figure 12.10   Workflow of object-level cooperative perception methods considering pose alignment

12.4.1   Data-level pose calibration

Fang *et al.* proposed an accurate and robust iterated split covariance intersection filter (Iterated Split CIF) for data-level pose calibration [29], consisting of single vehicle localization, inter-vehicle pose estimation, and Iterated Split CIF for state estimation, as shown in Figure 12.11.

12.4.1.1   Single vehicle localization

Each agent utilizes LiDAR to establish a point cloud map and derive its pose information through simultaneous localization and mapping (SLAM) and any representative LiDAR SLAM framework may be employed. The method employs the lightweight and ground-optimized LiDAR odometry and mapping (LeGO-LOAM) [30] within a single-vehicle localization module, comprising segmentation, feature extraction, LiDAR odometry, LiDAR mapping, and transform integration sub-modules. LeGO-LOAM balances localization and mapping accuracy and computational resource consumption well. The single-vehicle localization module processes 3D LiDAR point cloud data, generating a corresponding 3D point cloud map and a six-degree freedoms vehicle pose estimation.

12.4.1.2   Inter-vehicle pose estimation

The inter-vehicle relative pose estimation module leverages local maps and shared pose estimations from neighboring vehicles to generate relative pose estimations

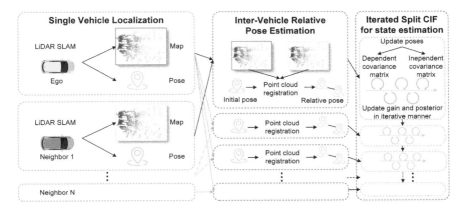

Figure 12.11   Overall system of the data-level pose calibration using Iterated Split CIF for each vehicle

via point cloud registration. It comprises two main components: the initial and continuous inter-vehicle relative pose estimation.

During the initial pose estimation phase, each vehicle independently executes its single-vehicle localization module to acquire the ego vehicle's point cloud map and the neighbor vehicles' point cloud scans. Sample consensus initial alignment (SAC-IA) [31] is employed, using the shared point cloud scans of neighbor vehicles as the source point cloud and the ego vehicle's point cloud map as the target point cloud for the initial iterative closest point (ICP) registration. SAC-IA enhances computational efficiency and feature extraction accuracy through down-sampling, leading to more precise feature correspondences.

Initial estimates of neighbor vehicles' positions in the ego vehicle's coordinate system are computed during the continuous relative pose estimation phase. Subsequently, these estimates are refined by registering the source and target point clouds using the ICP algorithm, resulting in precise poses for neighboring vehicles and inter-vehicle pose estimation. Throughout the procedure, the ongoing exchange of point cloud data and pose information between neighbor vehicles and the ego-vehicle is imperative for accurate relative pose estimation.

12.4.1.3   Iterated split CIF for state estimation

To achieve pose calibration, the iterated split CIF method further integrates shared data estimation updates of the ego-vehicle and neighboring vehicles' relative poses. The vehicle state estimation relies on a simplified kinematic bicycle model. The update process entails ego-vehicle pose updates from ego-vehicles' LiDAR SLAM and relative pose updates obtained through point cloud registration of neighbor vehicles. To address potential correlations, the iterated split CIF divides the covariance of state estimation into dependent and independent components. Additionally, it iteratively computes filter gains and posterior estimates to manage outliers, thereby resolving the outlier problem and improving the robustness of pose calibration.

12.4.2   *Feature-level pose calibration*

Given that feature-level pose calibration methods are not as straightforward to comprehend as data-level methods, we provide further detailed explanations. Unlike data-level methods, feature-level methods involve transmitting information between neighbor vehicles and the ego-vehicle including the neighbor vehicles' pose. However, unlike the transmission of raw perception information such as point cloud maps in data-level methods, feature-level methods transmit features extracted from these raw data through deep learning networks. Feature-level pose calibration methods utilize the feature information perceived by the ego-vehicle and the feature information transmitted by neighbor vehicles to calibrate inaccurate pose information.

As shown in Figure 12.12, Vadivelu *et al.* proposed a pose error correction network for feature-level spatial calibration [25], which comprises a pose regression module, a consistency module, and an attention aggregation module, all jointly trained end-to-end.

12.4.2.1   Pose regression module

Since each vehicle captures scenes from different perspectives, a convolutional neural network (CNN) is utilized to learn the disparities between the ego vehicle's viewpoint and those from neighbor vehicles. Specifically, the CNN inputs the perceptual features from the ego vehicle and neighbor vehicles, along with their respective poses, and outputs their corrected relative pose estimates. Due to the unidirectional nature of independent predictions, the relative pose estimates obtained for the two vehicles as ego vehicles are not identical.

12.4.2.2   Consistency module

The pose regression module delivers relative pose estimates between the ego-vehicle and neighboring vehicles. These estimates refine a set of globally consistent absolute poses, ensuring unanimous agreement across all vehicles. This refinement

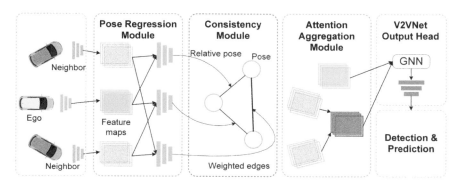

Figure 12.12   Overall system of the feature-level pose calibration using pose error correction network

process enhances accuracy by ensuring global visibility of absolute poses among vehicles.

Specifically, consistency is modeled using a Markov random field (MRF), where each vehicle pose serves as a node, conditioned on predicted relative poses. Due to outliers in predicted relative pose errors, the distribution of accurate absolute poses conditioned on these errors exhibits a heavy tail. To address this, there's a need to adjust the edge potentials to mitigate the impact of inaccurate pose regression outputs. Using weights for each term in the pairwise potential as its exponent term ensures that lower-weight terms have less influence on the estimates. Each weight is associated with a prior distribution, where the mean represents the spatial overlap fraction between the shared information passed when two vehicles perceive each other as ego vehicles. Inference on the MRF involves estimating absolute poses, scale parameters, and weights to maximize the product of pairwise potentials. This maximization utilizes iterated conditional modes [32], executed simultaneously for all nodes through weighted expectation-maximization for the t distribution [33], computable via closed form [34].

12.4.2.3 Attention aggregation module

Following the prediction and refinement of relative transformations, residual errors may persist in specific messages. We introduce a straightforward yet efficient attention mechanism to prioritize clean messages over noisy ones. This mechanism employs a CNN to forecast a normalized weight. Afterward, shared information from neighboring vehicles and their corresponding weights are jointly fed into a graph neural network (GNN) to obtain the final aggregated result, achieving pose calibration.

Training the base intermediate cooperative network and the attention network is necessary for learning. Noisy examples are identified through supervised binary classification, where clean examples are assigned high values and noisy examples low values. In generating data and labels, larger pose noise is applied to some vehicles, while weak pose noise is used to others within a scene. Both types of noise, similar to poses, consist of two translational components and a rotational component. A fixed proportion of agents receive noise from the strong distribution, while the remaining agents receive noise from the weak distribution. A message is labeled clean when both agents have noise from the weak distribution and labeled noisy otherwise. A specific function captures this labeling, and the function smooths labels to moderate the predictions of the attention module, preventing attention weights from being strictly binary. The loss for the joint training task consists of perspective-n-point (PnP) loss and binary cross-entropy loss. Introducing the binary cross-entropy loss is crucial for effectively training the attention mechanism. Using only PnP loss for training yielded a significantly less effective model.

Then, the base intermediate cooperative network and attention mechanisms are fixed, while only the regression module is trained. During this phase, all CAVs receive noise from the strong noise distribution. The training uses a loss function: the summation of losses over each coordinate.

Finally, the entire network undergoes end-to-end fine-tuning with the combined loss, facilitated by the differentiable MRF inference algorithm via backpropagation.

12.4.3   Object-level pose calibration

For object-level pose calibration, the information exchanged between the ego-vehicle and neighboring vehicles includes their respective poses and perception results of the environment, specifically the bounding boxes obtained from their perception systems. Object-level synchronization deals with significantly less data than the data and feature-level approaches discussed earlier. Because of the huge information loss, aligning poses using this sparse data is non-trivial and poses a significant challenge.

Song *et al.* proposed a robust inter-agent object association approach for achieving object-level pose calibration [35]. The approach effectively handles perception errors and outliers by leveraging object-specific characteristics. The approach maximizes feature utilization by incorporating information from all objects into each object's local context matrix. Each context matrix exhibits significant distinctiveness built from a localized perspective, bolstering resilience against outliers. Furthermore, perception errors are mitigated by pursuing global consensus via redundancy among the context matrices of all objects. As shown in Figure 12.13, the approach consists of three components: intra-agent context construction, context similarity-based coarse matching and global consensus maximization.

12.4.3.1   Intra-agent context construction

Distinct attributes characterize each traffic participant in real-world environments, including position, direction, and appearance. This uniqueness of surrounding environments from the viewpoint of individual vehicles forms their inherent context. Essentially, context encodes the relationships between nearby objects from an object's local perspective, enabling the possibility of employing context-based comparisons to detect and pinpoint identical objects across various viewpoints. In a manner akin to the preprocessing steps outlined in [36], measurements undergo standardization by being transformed into the ego coordinate system, and the transform function can be found in [36], representing the transform calculated from

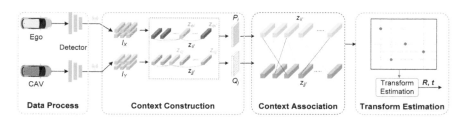

Figure 12.13   Overall system of the object-level pose calibration using the robust inter-agent object association approach

on-board localization systems. In the ego frame, the directions of objects are adopted by defining their heading toward the front in the ego frame as the forward direction.

The relative positional measurements between objects in a local coordinate system can be calculated as feature vectors based on the standardized measurements of objects' poses. Due to the substantial spatial footprint of each traffic participant and the necessity of preserving a safe distance between them, their positional vectors remain notably distinct within their respective areas, even in the presence of measurement errors. Consequently, despite such errors, feature vectors can be reliably sustained, leading us to define them as one of the object's contextual vectors in real-world traffic scenarios. Context vectors are computed for each object within the ego frame for all nearby objects. These are then aggregated together to obtain a context matrix for each object.

Given that context encapsulates object relationships and their neighboring counterparts, it inherently integrates robust spatial constraints between interconnected objects, maintaining invariance to rigid transformations such as rotation and translation. This characteristic renders context particularly well-suited for real-world driving scenarios, wherein spatial relationships are important.

12.4.3.2 Context similarity-based coarse matching

Given the context matrices of the ego-vehicle and neighbor vehicle, the similarity is computed for any arbitrary context vector within them. The similarity comprises two terms; the first term denotes the angular distance, and the second characterizes the length difference between the local context of the object in the ego vehicle's perception results and the other in the neighbor vehicle's perception results. Correspondingly, there are two parameters, with the first parameter set to tolerate angular perception errors caused by the positional error of the surrounding objects and the heading angle error of the center object, the second parameter set to handle the vector length noise caused by the positional error of both the center and surrounding objects. The use of absolute value operation is intended to avoid ambiguity caused by heading direction since detecting the direction of a road user frequently results in opposite judgments. Subsequently, the proposed algorithm selects highly similar pairs based on similarity, resulting in a global filter matrix being developed for each object pair to filter out mismatched correspondences in the preliminary correspondence matrix. Changing the two parameters mentioned can encompass many potential matches, aiming to include a significant portion of the ground truth correspondences. The solution to the object association problem can be obtained by filtering outliers from the preliminary correspondence matrix, this is achieved by maximizing the global consensus.

12.4.3.3 Global consensus maximization

A global filter matrix is developed for each object pair to filter out mismatched correspondences in the preliminary correspondence matrix. The basic idea is to assess the already matched pairs from a global perspective. Pairs accepted as matched in some objects' local frames but not embraced by all the objects are

eliminated. This operation assesses the non-zero correspondences in the preliminary correspondence matrix from each other's perspective to maximize global
consensus. Then, the improved correspondence matrix may contain one-to-many
matching correspondences, where one object in the ego-vehicle matches with
several objects in a neighbor vehicle or vice versa. After quickly obtaining the
suboptimal matching correspondences after eliminating one-to-many matching
correspondences, accurately associating the objects perceived by the ego-vehicle
with the neighbor vehicle enables subsequent calibration of relative pose estimation, thus achieving pose calibration.

Validation on real-world dataset
During the experimental validation of the approach, we first evaluated its primary
inter-agent object association performance using real-world datasets. Here, we
delineate the experimental setting and report the average precision corresponding
to the experimental findings.

Experiments setting
Dataset: Due to the nascent stage of cooperative perception technology, most of
datasets are focused on evaluating object detection performance, and very few
datasets are available for evaluating spatial robustness and object association performance. One choice is SIND [37], a real-world drone dataset captured from a
signalized intersection from a stationary aerial perspective for about 420 minutes.
The dataset includes more than 13,000 traffic participants in various types like cars,
pedestrians, and motorcycles.

   Metrics and benchmarks: Given the estimated association set $\widehat{M}$ and the
ground truth matching set M, the average precision is evaluated. Three benchmarks
are considered, including ICP [38], robust iterative closest point (RICP) [39], and
VIPS [10]. ICP is a fundamental technique for point association. Many variants
have occurred recently as a classical method, among which RICP is the latest
achievement. VIPS is the state-of-the-art method for inter-vehicle object association using graph matching techniques. VIPS achieves faster processing speed and
higher accuracy than other graph-matching-based algorithms. The proposed
method is denoted as context-based matching (CBM). In real-world traffic scenarios, non-covisible objects exist due to a limited field of view and occlusions.
These objects are outliers that severely impact matching tasks. Given the object
index set I at a single frame, The co-visible object set M is randomly sampled to
simulate cooperative perception, such that $\text{card}(M) = \eta \cdot \text{card}(I)$, $M \subseteq I$, where η
is the rate of co-visible objects. The remaining objects are evenly assigned to the
two cooperative agents, then we have $I_X \cup I_Y = I$, $I_X \cap I_Y = M$, where I_X and I_Y
denote the perceived set by the two agents.

   Perception errors: Different position and orientation angle error levels are
added to the objects in the dataset to investigate the impact of perception errors.
They are set to be Gaussian distributed as $N(0,\sigma_p)$ and $N(0,\sigma_\theta)$, respectively. For
object detection algorithms, determining the orientation of an object is a difficult
task and prone to errors. To simulate this, a direction noise is added to the orientation with a 50% probability to make it face the opposite orientation.

Localization errors: The proposed method and VIPS are not sensitive to the cooperating vehicles' initial relative pose transformation relationship, while ICP and RICP are. Since the initial relative pose transformation reflects the magnitude of the cooperating vehicles' pose error, it is set as a fixed value. In practice, the objects in I_Y are translated by 3 m in the x and y directions and rotated by 5° as a whole, i.e., the agents' relative position offset entirely based on accurate poses of the two vehicles.

12.4.3.4 Average precision of inter-agent object association

The performance of benchmarks on inter-agent object association is tested under different levels of outlier rate and perception errors (including position and orientation errors), the results are shown in Figure 12.14.

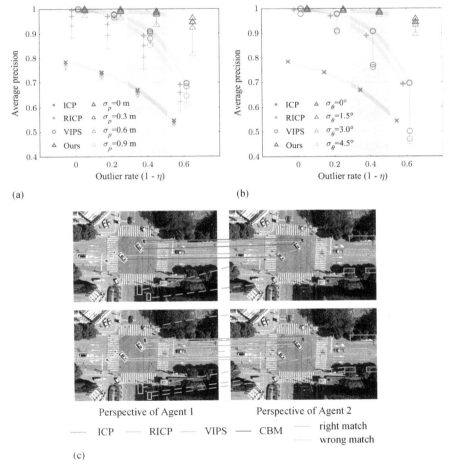

Figure 12.14 *Quantitative results and qualitative demonstration on SIND. (a) AP under different position errors, (b) AP under different heading errors, and (c) an example.*

The result shows that η has a more significant impact than standard deviations σ_p and σ_θ. RICP exhibits a higher overall average precision (AP) level than ICP, but they are both highly sensitive to η, this might be due to their use of iterative searching for the closest point in the correspondence identification step that converged to a local optimum. VIPS outperforms them in terms of AP, showing good robustness to changes in η. CBM outperforms the previous three methods' overall precision and robustness to changes in η. When considering the perception errors, it is observed that CBM achieves good robustness to position and orientation errors, with only an overall downward shift in the AP curve at $\sigma_p = 0.9$ m. For different levels of errors, the AP curve remains consistent with the zero error scenarios. RICP exhibits poor robustness to position errors, VIPS is robust against position errors but cannot handle large orientation errors. This is because VIPS uses the sine difference of the heading angles of two nodes to encode edge-to-edge similarity. This makes it fragile to errors in the objects' heading angle.

Evaluation on cooperative perception dataset
We then validate the approach's effectiveness in cooperative perception, leveraging a large-scale vehicle-to-vehicle (V2V) simulation dataset. Here, we delineate the experimental setting and analyze the performance metrics concerning object association, transform estimation, and the overarching cooperative perception efficacy.

12.4.3.5   Experiment setting

Dataset: OPV2V [22] is a large-scale dataset that contains 73 scenes for V2V-based cooperative perception, including 2170 frames for *test* subset, and 549 frames for *test culver city* (*tcc*). The latter is developed to narrow the gap between the simulated and real-world traffic scenarios, which can be used to test the adaptability and portability of the proposed algorithm. The reasons of using OPV2V are *a*) incorporation of cooperative vehicles and their locally perceived information, and *b*) provision of a wide range of scenarios, including highly complex traffic scenes with numerous participants. The first row of Figure 12.15 illustrates the distributions of co-visible object rate and absolute object counts across two test sets.

Object detection and perception errors: The object detection module provides inter-node measurements. The object detection network PointPillars is trained [40] using the *train* subset provided by OPV2V and kept the same for all benchmarks. The performance of PointPillars on the dataset is evaluated in the second row of Figure 12.15. The third subfigure of Figure 12.15 depicts the distribution of lateral (y) and longitudinal (x) position errors in the bounding boxes detected by PointPillars. It shows that the position errors in both directions approximately follow a Gaussian distribution, with more significant errors observed in the longitudinal direction. Similar error distributions are observed across both datasets. This result supports the setting of σ_p in SIND. The fourth subfigure of Figure 12.15 shows the distribution of angular errors in the network's perceived results. Specifically, we calculated the degree of deviation between the perceived bounding boxes and ground truth in the heading direction. It shows that the angular errors

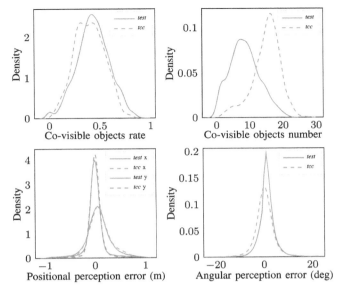

Figure 12.15   Some statistics on OPV2V

Table 12.2   Matching performance on OPV2V with $\sigma_p^L = 3$m, $\sigma_\theta^L = 5$deg. Pre.: precision, Rec.: recall, Dis.: distance

Method	tcc			test		
	Pre.	Rec.	Dis.	Pre.	Rec.	Dis.
VIPS	0.478	0.585	28.267	0.472	0.575	22.317
ICP	0.460	0.384	7.843	0.561	0.522	5.530
RICP	0.736	0.721	3.987	0.781	**0.808**	2.584
CBM	**0.995**	**0.930**	**0.317**	**0.906**	0.780	**0.498**

were distributed mainly between $-20°$ and $20°$, posing a significant challenge to association algorithms.

    Localization errors: Without loss of generality, two scenarios are considered for demonstration and comparison: the first assumes that the participating agents have no position and orientation errors. In contrast, the second one assumes that the position and orientation errors follow zero-mean Gaussian distribution with standard deviation $\sigma_p^L = 3$ m and $\sigma_\theta^L = 5°$.

12.4.3.6   Evaluation of inter-agent association performance

Precision and recall: The precision and recall results for the matching task on OPV2V are shown in Table 12.2. Note that VIPS performs significantly worse on OPV2V than

SIND, primarily due to the complex and challenging natures of the scenarios in OPV2V, such as a more significant amount of objects and perception errors.

Distance between correspondence pairs: Matching precision and recall may not be a perfect indicator of the perception performance, for example, when the two objects are close to each other, their incorrect pose estimation would not deviate significantly from the ground truth, and it would not have a significant impact on the perception results. Therefore, another metric that can assess the impact of matching performance on perception is required. A new metric average distance is defined as $d = 1/N \sum_{(i,j) \in \widehat{M}} \mathrm{d}(s_i - s_j)$ that measures the average distance between the matched object pairs, where $N = \mathrm{card}(\widehat{M})$, and operator $\mathrm{d}(\cdot)$ denotes calculating the Euclidean distance. Table 12.2 shows the performance of the methods of d on two datasets, *test* and *test culver city*. Notably, the d values of the proposed method are quite small, which means even for incorrectly associated object correspondences in the matching results, the distances between them are not too far away to cause fatal impacts on the estimation of the pose transformation in the back end.

Impact of outliers: Figure 12.16 depicts the distribution of average matching precision against the rate of non-co-visible objects on the *test culver city* and *test* datasets. The results consistently align with those obtained from the SIND dataset, as illustrated in Figure 12.14. CBM exhibits remarkable resilience to outliers.

12.4.3.7   Evaluation of transform estimation performance

There are two evaluation metrics used to measure the similarity between the estimation and ground truth: *a*) relative rotation error $\mathrm{RRE} = \arccos\left(0.5 \cdot \mathrm{Tr}(R^T \cdot \widehat{R}) - 0.5\right)$, and *b*) relative translation error $\mathrm{RTE} = \|t - \widehat{t}\|_2$ The results of RRE and RTE are shown in Figure 12.17, and different heading error levels are added to the cooperative

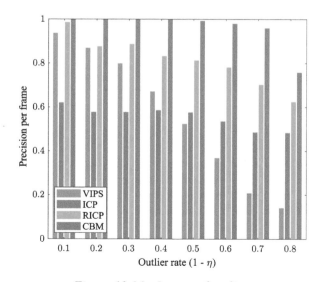

Figure 12.16   Impact of outliers

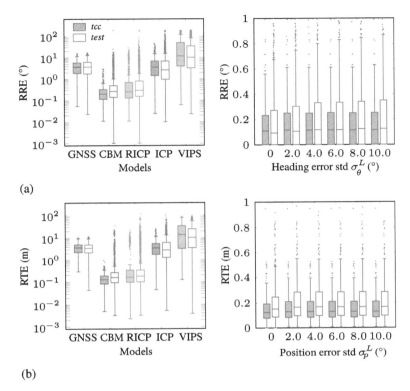

Figure 12.17 *(a) Distribution of RRE on OPV2V and (b) distribution of RTE on OPV2V*

agents for RRE estimation. Similarly, only position errors are introduced for the evaluation of RTE. When evaluating the RRE and RTE, we first compare the performance with the benchmark performance at a fixed localization error level $\sigma_p^L = 3$ m or $\sigma_\theta^L = 5$ (first column of Figure 12.17), and then conduct a detailed performance evaluation of the proposed method at different error levels (second column of Figure 12.17). The Global Navigation Satellite System (GNSS) benchmark corresponds to the results without calibration. CBM has significantly reduced the median RRE to 0.1 and the median RTE to 0.1 m, achieving an order of magnitude improvement over the GNSS solution and outperforming other benchmarks. Furthermore, the RRE and RTE distributions of the proposed method are similar at different pose error levels, indicating the robustness of the proposed method. The minor fluctuations are mainly due to the direct use of GNSS positioning information to estimate pose in the case of matching failure, which is influenced by the varying GNSS errors.

12.4.3.8 Evaluation of cooperative perception

This section evaluates the robustness of the proposed method against pose errors. The evaluation metric is mean average perception, computed by comparing the IoU

Table 12.3   Mean average precision at IoU = 0.7 on OPV2V under different σ_p^L and σ_θ^L

Dataset	Method	σ_p^L/m					
		0	**0.6**	**1.2**	**1.8**	**2.4**	**3.0**
test	Single	0.600	0.600	0.600	0.600	0.600	0.600
	GNSS	**0.805**	0.245	0.214	0.242	0.260	0.265
	ICP	0.443	0.373	0.326	0.281	0.265	0.268
	VIPS	0 .259	0.242	0.235	0.228	0.228	0.229
	RICP	0.719	**0.716**	**0.709**	**0.689**	0.674	0.643
	CBM	0.703	0.689	0.688	0.688	**0.689**	**0.689**
tcc	Single	0.470	0.470	0.470	0.470	0.470	0.470
	GNSS	**0.684**	0.288	0.258	0.253	0.255	0.259
	ICP	0.358	0.322	0.293	0.283	0.269	0.268
	VIPS	0.257	0.258	0.256	0.257	0.258	0.258
	RICP	0.623	**0.623**	**0.621**	0.597	0.578	0.517
	CBM	0.621	0.619	0.619	**0.619**	**0.619**	**0.619**

Dataset	Method	$\sigma_\theta^L/°$					
		0	**2.0**	**4.0**	**6.0**	**8.0**	**10.0**
test	Single	0.600	0.600	0.600	0.600	0.600	0.600
	GNSS	**0.805**	0.343	0.264	0.242	0.236	0.234
	ICP	0.443	0.397	0.354	0.323	0.299	0.277
	VIPS	0.259	0.242	0.234	0.234	0.231	0.230
	RICP	0.719	**0.707**	**0.697**	0.681	0.683	0.666
	CBM	0.703	0.691	0.686	**0.684**	**0.683**	**0.681**
tcc	Single	0.470	0.470	0.470	0.470	0.470	0.470
	GNSS	**0.684**	0.350	0.295	0.276	0.272	0.268
	ICP	0.358	0.330	0.302	0.296	0.282	0.270
	VIPS	0.257	0.257	0.256	0.257	0.257	0.257
	RICP	0.623	**0.623**	**0.621**	0.616	0.609	0.585
	CBM	0.621	0.620	0.619	**0.619**	**0.617**	**0.616**

of fused bounding boxes and the ground truth boxes. An IoU threshold of 0.7 is chosen. Note that the metric is quite different from the average precision. The results are shown in Table 12.3.

12.5   Conclusions

Ensuring precise perception is crucial for accelerating automated driving and addressing safety challenges in modern transportation systems. Despite the advanced object detection techniques based on artificial intelligence, existing perception techniques still encounter challenges in real-world traffic scenarios. Issues like physical occlusion and limited sensor field of view remain prevalent for individual vehicle systems.

Cooperative perception integrated with V2X technologies has arisen as a solution to overcome these obstacles and augment the capabilities of automation driving systems. Meanwhile, synchronization and coordinate transformation are crucial for cooperative perception, especially in multi-agents and distributed systems. We extensively examine two pivotal elements essential for robust cooperative perception: temporal synchronization and pose calibration. We delve into the origins and ramifications of time delays, illustrating through two prominent examples how to construct models to mitigate time delays, both in feature-level and object-level cooperative perceptions. Subsequently, we presented a detailed exposition of pose calibration in cooperative perception, accompanied by delineating three distinct methods: data-level, feature-level, and object-level pose calibration.

We have attained a nuanced comprehension of cooperative perception through thorough discussions and analyses. Our examination has yielded a comprehensive insight into the challenges confronting diverse cooperative perception methodologies as they progress toward heightened robustness and corresponding solutions. This enhanced understanding significantly enriches ongoing scholarly inquiry and exploration within cooperative perception.

References

[1] Shi Y, Liu Z, Wang Z, *et al.* An Integrated Traffic and Vehicle Co-Simulation Testing Framework for Connected and Autonomous Vehicles. *IEEE Intelligent Transportation Systems Magazine.* 2022;14(6):26–40.

[2] Zhao L, and Malikopoulos AA. Enhanced Mobility with Connectivity and Automation: A Review of Shared Autonomous Vehicle Systems. *IEEE Intelligent Transportation Systems Magazine.* 2022;14(1):87–102.

[3] Van Brummelen J, O'Brien M, Gruyer D, *et al.* Autonomous Vehicle Perception: The Technology of Today and Tomorrow. *Transportation Research Part C: Emerging Technologies.* 2018;89:384–406.

[4] Li SE. *Reinforcement Learning for Sequential Decision and Optimal Control.* Singapore: Springer; 2023.

[5] Li X, Ye P, Li J, *et al.* From Features Engineering to Scenarios Engineering for Trustworthy AI: I&I, C&C, and V&V. *IEEE Intelligent Systems.* 2022; 37(4):18–26.

[6] Zhang H, Luo G, Li J, *et al.* C2FDA: Coarse-to-Fine Domain Adaptation for Traffic Object Detection. *IEEE Transactions on Intelligent Transportation Systems.* 2022;23(8):12633–12647.

[7] Li S, Xin L, Liu C, *et al.* Promoting Connected and Automated Vehicles with Cooperative Sensing and Control Technology. In: *Cooperative Intelligent Transport Systems: Towards High-Level Automated Driving.* Stevenage: The Institution of Engineering and Technology (IET); Ed. Meng Lu, 2019. pp. 515–531.

[8] Zheng Y, Li SE, Li K, *et al.* Distributed Model Predictive Control for Heterogeneous Vehicle Platoons Under Unidirectional Topologies. *IEEE Transactions on Control Systems Technology.* 2017;25(3):899–910.

[9]   Dorri A, Kanhere SS, and Jurdak R. Multi-Agent Systems: A Survey. *IEEE Access*. 2018;6:28573–28593.

[10]   Shi S, Cui J, Jiang Z, *et al.* VIPS: Real-Time Perception Fusion for Infrastructure-Assisted Autonomous Driving. In: *Proceedings of the 28th Annual International Conference on Mobile Computing And Networking. MobiCom 22*. New York, NY: Association for Computing Machinery; 2022. pp. 133–146.

[11]   Shan M, Narula K, Worrall S, *et al.* A Novel Probabilistic V2X Data Fusion Framework for Cooperative Perception. In: *2022 IEEE 25th International Conference on Intelligent Transportation Systems (ITSC)*; 2022. pp. 2013–2020.

[12]   Han Y, Zhang H, Li H, *et al.* Collaborative Perception in Autonomous Driving: Methods, Datasets, and Challenges. *IEEE Intelligent Transportation Systems Magazine*. 2023;15(6):131–151.

[13]   Arnold E, Dianati M, de Temple R, *et al.* Cooperative Perception for 3D Object Detection in Driving Scenarios Using Infrastructure Sensors. *IEEE Transactions on Intelligent Transportation Systems*. 2022;23(3):1852–1864.

[14]   Liu C, Eben Li S, Yang D, *et al.* Distributed Bayesian Filter Using Measurement Dissemination for Multiple Unmanned Ground Vehicles with Dynamically Changing Interaction Topologies. *Journal of Dynamic Systems, Measurement, and Control*. 2017 11;140(3):030903.

[15]   Liu C, Li SE, Hedrick JK. Measurement Dissemination-Based Distributed Bayesian Filter Using the Latest-In-and-Full-Out Exchange Protocol for Networked Unmanned Vehicles. *IEEE Transactions on Industrial Electronics*. 2017;64(11):8756–8766.

[16]   Chen Q, Tang S, Yang Q, *et al.* Cooper: Cooperative Perception for Connected Autonomous Vehicles Based on 3D Point Clouds. In: *2019 IEEE 39th International Conference on Distributed Computing Systems (ICDCS)*. Los Alamitos, CA: IEEE Computer Society; 2019. pp. 514–524.

[17]   Camajori Tedeschini B, Brambilla M, Barbieri L, *et al.* Cooperative Lidar Sensing for Pedestrian Detection: Data Association Based on Message Passing Neural Networks. *IEEE Transactions on Signal Processing*. 2023; 71:3028–3042.

[18]   Kim SW, Qin B, Chong ZJ, *et al.* Multivehicle Cooperative Driving Using Cooperative Perception: Design and Experimental Validation. *IEEE Transactions on Intelligent Transportation Systems*. 2015;16(2):663–680.

[19]   Rauch A, Klanner F, Rasshofer R, *et al.* Car2X-based perception in a high-level fusion architecture for cooperative perception systems. In: *IEEE Intelligent Vehicles Symposium*; 2012. pp. 270–275.

[20]   Yee R, Chan E, Cheng B, *et al.* Collaborative Perception for Automated Vehicles Leveraging Vehicle-to-Vehicle Communications. In: *IEEE Intelligent Vehicles Symposium (IV)*; 2018. pp. 1099–1106.

[21]   Li Y, Ma D, An Z, *et al.* V2X-Sim: Multi-Agent Collaborative Perception Dataset and Benchmark for Autonomous Driving. *IEEE Robotics and Automation Letters*. 2022;7(4):10914–10921.

[22] Xu R, Xiang H, Xia X, *et al.* Opv2v: An Open Benchmark Dataset and Fusion Pipeline for Perception with Vehicle-to-Vehicle Communication. In: *2022 International Conference on Robotics and Automation (ICRA)*. IEEE; 2022. pp. 2583–2589.

[23] Yu H, Luo Y, Shu M, *et al.* DAIR-V2X: A Large-Scale Dataset for Vehicle-Infrastructure Cooperative 3D Object Detection. In: *Proceedings of the IEEE/CVF Conference on Computer Vision and Pattern Recognition (CVPR)*; 2022. pp. 21361–21370.

[24] Wang TH, Manivasagam S, Liang M, *et al.* V2vnet: Vehicle-to-Vehicle Communication for Joint Perception and Prediction. In: *Computer Vision–ECCV 2020: 16th European Conference, Glasgow, UK, August 23–28, 2020, Proceedings, Part II 16*. Springer; 2020. pp. 605–621.

[25] Vadivelu N, Ren M, Tu J, *et al.* Learning to Communicate and Correct Pose Errors. In: Kober J, Ramos F, Tomlin C, (eds.) *Proceedings of the 2020 Conference on Robot Learning. vol. 155 of Proceedings of Machine Learning Research. PMLR*; 2021. pp. 1195–1210.

[26] Lu Y, Li Q, Liu B, *et al.* Robust Collaborative 3D Object Detection in the Presence of Pose Errors. In: *IEEE International Conference on Robotics and Automation (ICRA)*; 2023. pp. 4812–4818.

[27] Moradi-Pari E, Tian D, Mahjoub HN, *et al.* The Smart Intersection: A Solution to Early-Stage Vehicle-to-Everything Deployment. *IEEE Intelligent Transportation Systems Magazine.* 2021;14(5):88–102.

[28] Weng X, Wang J, Held D, *et al.* 3d Multi-Object Tracking: A Baseline and New Evaluation Metrics. In: *2020 IEEE/RSJ International Conference on Intelligent Robots and Systems (IROS)*. IEEE; 2020. pp. 10359–10366.

[29] Fang S, Li H, and Yang M. LiDAR SLAM based multivehicle cooperative localization using iterated split CIF. *IEEE Transactions on Intelligent Transportation Systems.* 2022;23(11):21137–21147.

[30] Zhang J, and Singh S. Low-Drift and Real-Time Lidar Odometry and Mapping. *Autonomous Robots.* 2017;41:401–416.

[31] Liu L, Shi T, Liu B, *et al.* Comparison of Initial Registration Algorithms Suitable for ICP Algorithm. In: *2020 International Conference on Computer Network, Electronic and Automation (ICCNEA)*. IEEE; 2020. pp. 106–110.

[32] Besag J. On the Statistical Analysis of Dirty Pictures. *Journal of the Royal Statistical Society Series B: Statistical Methodology.* 1986;48(3):259–279.

[33] Liu C, and Rubin DB ML. Estimation of the t Distribution Using EM and Its Extensions, ECM and ECME. *Statistica Sinica.* 1995;5(1):19–39.

[34] Wang Y, Kucukelbir A, and Blei DM. Robust Probabilistic Modeling with Bayesian Data Reweighting. In: *International Conference on Machine Learning. PMLR*; 2017. pp. 3646–3655.

[35] Song Z, Xie T, Zhang H, *et al.* A Spatial Calibration Method for Robust Cooperative Perception. *IEEE Robotics and Automation Letters.* 2024;9(5): 4011–4018.

[36]   Song Z, Wen F, Zhang H, *et al.* A Cooperative Perception System Robust to Localization Errors. In: *2023 IEEE Intelligent Vehicles Symposium (IV)*. IEEE; 2023. pp. 1–6.

[37]   Xu Y, Shao W, Li J, *et al.* SIND: A Drone Dataset at Signalized Intersection in China. In: *IEEE 25th International Conference on Intelligent Transportation Systems (ITSC)*. IEEE; 2022. pp. 2471–2478.

[38]   Besl PJ, and McKay ND. A Method for Registration of 3-D Shapes. *IEEE Transactions on Pattern Analysis and Machine Intelligence*. 1992;14 (2):239–256.

[39]   Zhang J, Yao Y, and Deng B. Fast and Robust Iterative Closest Point. *IEEE Transactions on Pattern Analysis and Machine Intelligence*. 2022;44 (7):3450–3466.

[40]   Lang AH, Vora S, Caesar H, *et al.* Pointpillars: Fast Encoders for Object Detection from Point Clouds. In: *Proceedings of the IEEE/CVF Conference on Computer Vision and Pattern Recognition*; 2019. pp. 12697–12705.

Chapter 13

Cooperative architecture for transportation systems (CATS): assessment of safety and mobility in vehicular convoys

Catherine M. Elias[1,2], Elsayed I. Morgan[1,2], Christoph Stiller[3] and Omar M. Shehata[1,2]

13.1 Cooperative vehicular systems

In light of the significant development in connected and automated vehicles (CAV) and the associated V2X technology, the concept of cooperative control algorithms started to be utilized in the intelligent transportation system (ITS) to take over the control of the different driving behaviors, inspired by the swarm systems behaviors. These cooperative control algorithms are mainly using the technology of the V2V paradigm to achieve cooperative tasks among the vehicles which would potentially enhance the performance of the vehicular system.

According to the literature [1], cooperative driving is needed when a crisis occurs, and this mainly exists in two situations: intersections and highways. Moreover, a more complex situation might occur as highway joining and splitting scenarios via on- and off-ramps. These kinds of situations need a huge effort from researchers to tackle different aspects relying on pure sensor-based automated driving including the heterogeneity in vehicles' sensing capabilities, V2X communications, and vehicle control algorithms. As a result, several operational concepts are defined to provide safe driving through cooperation among the vehicles.

Based on this survey for the recently funded projects and according to a survey done in 2019 by Wang *et al.* [2], the cooperative driving of vehicles can be classified into five main operational concepts. These concepts can further be categorized into the two situations mentioned before where cooperative driving is necessary and their combination. On highways, two concepts are studied: cooperative adaptive cruise control (CACC) and platooning, and speed harmonization on highways. Meanwhile, in the intersections, we can find two additional operational concepts: cooperative

[1]Multi-Robot Systems (MRS) Research Group, Cairo, Egypt
[2]Mechatronics Department, Faculty of Engineering and Materials Science, German University in Cairo, Egypt
[3]The Institute of Measurement and Control Systems, Karlsruhe Institute of Technology (KIT), Germany

eco-driving signalized intersections, and automated coordination at non-signalized intersections. Finally, a concept is introduced to study the linkage between the two situations: cooperative merging at highway on-ramps.

In CACC systems, CAVs share their own parameters with other CAVs in the network by V2V communications, which are realized autonomously without central management. Given the fact that the communication bandwidth might become insufficient when the number of CAV increases in a CACC system, short-range wireless technologies are more accepted for V2V communications. CACC takes advantage of V2V communications to allow CAV to form platoons and be driven at harmonized speeds with shorter time headways between them. By sharing vehicle information such as acceleration, speed, and position in a distributed manner, CAV in a certain communication range can cooperate with others to obtain the following benefits: (1) driving safety is increased since actuation time is shortened compared to manually driven, and downstream traffic can be broadcasted to following vehicles in advance; (2) roadway capacity is increased due to the reduction of time/distance headways between vehicles; (3) energy consumption and pollutant emissions are reduced due to the reduction of unnecessary velocity changes and aerodynamic drag on following vehicles [3,4].

It is proven that by adopting this concept while driving, an energy saving/fuel consumption reduction is accomplished by about (15–30%) and road throughput/road density increases by (3–5) times. Yet, the platooning system causes instability of the vehicular string especially in long strings (>5), failing to solve some merging and splitting scenarios to/from the platoon due to the short-range devices and communication topologies which are mainly leader-follower-based [3–7].

Accordingly, it is noticed that all five concepts are classified as longitudinal cooperative algorithms neglecting the effect of the lateral motion of the vehicle and this is due to the fact of studying the behaviors of the vehicle in single lanes only. However, this does not match the real driving scenario where the vehicles make lateral maneuvers, and accordingly, the need for a more extended representation of the cooperative concepts became a must to include lateral cooperation as well. Therefore in this work, a sixth cooperative concept is studied as a new coordination method for the driving on highways scenario considering the cooperation in both longitudinal and lateral directions. This concept is known by **vehicular convoy system**.

13.2   Vehicular convoy system

Kato *et al.* first introduced a vehicular convoy concept [8] in 2002. It is considered as an extension of vehicular platoons. It is formally defined as:

Definition 1 *Multiple cooperative vehicles spreading over multiple lanes, traversing with a constant speed while maintaining a pre-designed formation.*

This behavior has the potential to increase the safety and mobility on the highways, in addition to having a great impact on environmental sustainability since we are now maintaining inter-vehicle spacing in the longitudinal and lateral directions. In addition to the platoon's advantages, convoys have the potential to include more vehicles, handle more complicated merging and splitting scenarios, solving communication and string instability issues. Vehicular convoys can also assist in several overtaking and cooperative lane-changing maneuvers. Moreover, it can be used for protecting high-importance cars such as the presidential car in a convoy.

13.2.1   Challenges of vehicular convoy

Comparing the vehicular convoy with the formation control in the swarm systems, it can be observed that the robotics systems are much more advanced and well-established, unlike the vehicular system which is seldom investigated. This is due to many reasons.

Convoy challenges

- **The dynamic complexity:** As in the swarm system, the particle or the kinematical model is enough to be considered for the agent to move in the different Degrees of Freedom (DOF). However, in the vehicular system, it adds more complex non-holonomic mathematical constraints which adds complexity to the control process since we can not neglect the high dynamics of the vehicle.
- **The geometric topology:** The procedure of determining a geometric topology is arbitrary. Since it is associated with a flexible configuration. This is because the vehicles are in a highly uncertain and dynamic environment, caused by the unfixed number of vehicles, the road profile, and the other road participants.
- **The communication issues:** Since the target of the convoy is to include more vehicles compared to the platoons, therefore it is important to solve the communication issue which is caused by the dependency on a single formation strategy. Accordingly, it is a challenge to select a suitable formation strategy that can solve this issue, preventing the error propagation that causes the string instability.

Based on this analysis, it is clear that the vehicular convoy is a challenge in its implementation due to the several dimensions that should be chosen carefully to get the desired behavior.

Searching for previous work that tackled the problem of formation control of multi-vehicles, we can find several pieces of literature. However, most of them are dealing with the vehicular system as a normal swarm system that only has a different kinematical model, completely ignoring the dynamics of the vehicle and the different nature of the environment. This makes these papers a bit unrelated to our problem due to the lack of practicality and not covering the different dimensions of the vehicular convoy system. However, digging deeper, we could only find very few papers that tackle some of the dimensions of the convoy in real practical scenarios related to the ITS and those papers are the ones we are interested in studying to formulate the research gap.

The multi-lane vehicular convoy concept was first introduced in 2002 by Kato *et al.* The authors studied the convoy driving behavior on highways modeled as leader-follower-based multi-lane platoons [8]. This work was limited to five cars in addition to utilizing the leader-follower strategy causes the string instability issue for longer strings.

Later in 2014, the AutoNet2030 [9] funded the research to develop a prototyped cooperative automated driving system. One of the outcomes was the work presented by Marjovi *et al.* [6] and Obst *et al.* [10] where the authors addressed the formation control of multi-lane vehicular convoys in highways using a Laplacian graph-based, distributed control law allowing the joining/splitting to/from the dynamic formation. Despite the unconstrained number of vehicles, this work only presented one pre-defined convoy with fixed parameters which makes it hard to expand.

Qian *et al.* in 2016 [7] introduced for the first time the former definition of the convoy. This work proposed a hierarchical framework to control the convoy system with the use of MPC subjected to flexible reconfiguration in case of lane-blocking or non-blocking obstacles. The limitation of this work was the very small fixed number of vehicles (to 4) in a non-realistic scenario with a single pre-defined created convoy as well without considering the joining and splitting scenarios.

Navarro *et al.* [11] addressed some convoy behaviors which are the dynamic formation and the cooperative lane changing using the curvilinear road while controlling the inter-vehicle convoy spaces via a distributed graph-based control law, allowing the convoy to adopt the shape with the road, performing sharp turns and avoiding the control competition between the Laplacian and lane keeping controller using any number of vehicles.

Extending this work, Gao *et al.* in 2019 utilized the same distributed graph theory formation control approach, enhancing the obstacle avoidance capability by controlling the modeled traffic flow using the APF [5].

However, Cai *et al.* address an ongoing project from 2019 till today [12–17] addressed a different aspect which is the formation reconfiguration and assignment processes, not the control part. These processes are used to determine the flexible configuration formation topology which will be followed by the cars after assigning each car to a specific longitudinal and lateral position. In the assignment process, they used the Hungarian algorithm developed in 1955 by Harold Kuhn [18].

In 2022, Elias *et al.* [19] proposed the cooperative architecture for transportation system (CATS) architecture that emerges the V2I and V2V technologies in the enhancement of the overall driving behavior to mimic the human-like one. However, this work is further extended [20] introducing the vehicular convoy system in the transportation system. This enables the utilization of the V2C/C2I/V2I technologies to present a platoon system that is not based on the leader-follower strategy, instead, the proposed architecture introduced a behavioral-based platoon. The use of the behavioral strategy allowed the forming of longer platoons (>5), avoiding communication issues, and also introduced the capability of creating new platoons as well as joining and splitting scenarios.

From this literature and to the best of the authors' knowledge, we can say that the vehicular convoy problem is rarely addressed and this is due to the many aspects that should be considered to achieve a comprehensive driving scenario. There exist several gaps that can be concluded as such:

Some research gaps

1. Despite the modularity of the proposed CATS architecture in [20], it is still not complete enough to tackle the convoy system (2D platoons) that has a great potential to significantly increase the number of vehicles within a single convoy on a road, hence the mobility of the overall system.
2. No criteria to define suitable formation strategies (one or more) or study the reconfigurations of the convoy subjected to changes in the road lanes number trigger.
3. There is no proper assessment of the vehicular convoy in terms of safety and mobility.

The main aim of this work is to extend the CATS architecture presented by [20]. This article studies some investigated behaviors of the vehicular convoy, especially extending the platoons to be spread over multi-lane.

Chapter highlights

The main highlight of this work is to:

1. Formulate a flexible formation reconfiguration and assignment processes with reasonable computational complexity.
2. Utilize several simultaneous formation strategies and control algorithms for achieving more behaviors and configuration spectrum.
3. Assessing the architecture performance in terms of safety and mobility.

In the paper presented by Elias *et al.* [20], the CATS architecture was proposed to have an abstracted architecture overview shown in Figure 13.1. In this work, we will mainly focus on the implementation of the convoy planning module including both the formation reconfiguration and formation assignment modules. Also, the cooperative behavioral planning and control module of the single automated vehicle will be presented.

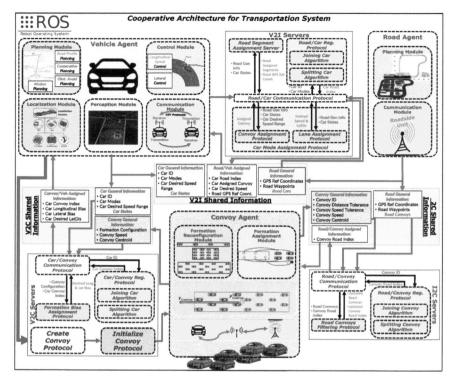

Figure 13.1   The CATS architecture modules overview [20]

13.3   Convoy formation flexible configuration

When we address the formation flexible configuration of a convoy, three main aspects should be studied. The first aspect is identifying the scenarios in which the formation reconfiguration is needed to be executed. These scenarios are known by the triggers. This will require a full understanding of the driving scenarios and the possibilities that might face and trigger our convoy to change its shape. The second aspect is identifying the possible formation of geometric topologies that the vehicles can form. This can happen by determining the longitudinal and lateral spacing between each two successive vehicles. The third and last aspect is the formation assignment which mainly drives the vehicle to be placed in a specific place within the generated topology based on predefined objectives.

13.3.1   Convoy formation reconfiguration triggers

The challenge of the flexible configuration is the hardest to handle due to the highly uncertain environment of the vehicles. There exist several other vehicles with different dynamics heterogeneity. The possibility of joining the convoy or splitting from it causes a change in the number of vehicles within the fleet, which means that the formation topology must adapt accordingly.

Furthermore, the change in the road causes a trigger for reconfiguration. This change happens in two scenarios. First, it occurs in case of changing the road width; hence the number of lances. The second case is caused by the existence of other road participants which are considered on-road obstacles. These obstacles cause the lane blockage to be either complete or partial. This blockage can be considered as a change in the number of lanes as well. Collectively, these define the different reconfiguration triggers upon which the geometric topology must be changed to a new pattern.

13.3.2   Convoy formation geometric topology

The concept of formation topology is to identify a specific geometric pattern. This pattern is composed of a number of vertices. Each of these vertices represents the desired placement of the agent. The inter-spacing between these vertices is defined to maintain the desired gaps between every two agents. In the swarm systems, this inter-spacing gap is defined as the distance between the agents. However, when we are talking about the vehicular system, it is not convenient to control the distance between the vehicles. Instead, we define the geometric pattern in terms of desired longitudinal and lateral gaps.

The formation geometry is decided based on the number of vehicles within the convoy fleet and the number of road lanes. Accordingly, four types of formation configurations are convenient with the environment considering the structure of the road which are:

The convoy formation topologies

- **String configuration:** This configuration is considered a platoon-like one where all the vehicles are assigned in a single lane keeping only a longitudinal spacing. The configuration is illustrated in Figure 13.2(a).
- **Parallel configuration:** This configuration has higher vehicle density compared to other configurations. The flexibility of formation adjustment is not as good since there is not enough spacing between the

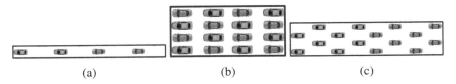

(a)	(b)	(c)

Figure 13.2   Vehicular convoy formation configurations: (a) string, (b) parallel, and (c) interlaced

vehicles to move freely which also makes it a bit dangerous if not well controlled and designed. The pattern is shown in Figure 13.2(b).

• *Equidistant configuration:* It is an upgraded version of the parallel formation, considering the length of the vehicles. This configuration is the most suitable in the case of a heterogeneous vehicular system.

• *Interlaced configuration:* This configuration provides great convenience for lane changing and formation adjusting. It is much safer than parallel formation since we are keeping a lateral spacing equal to double the desired space as in Figure 13.2(c). On the other hand, it is less dense compared to the parallel which means that to include the same number of vehicles it would require double the overall longitudinal distance. Hence, it might suffer from communication issues due to the larger range.

13.3.3   Convoy formation assignment

Another challenge in the flexible reconfiguration process is determining which vehicle shall be assigned to each vertex within the generated geometric pattern in case of triggering. This is known in the literature as an assignment/allocation process where the vehicles are allocated to the available resources which are the vertices. The assignment problem is well known in the literature in different tracks. However, a common thing is to define the objectives upon which the vehicles shall be assigned. Different objectives are possible to consider to increase mobility and safety on the highway.

Some of these objectives mainly focus on maximizing the mobility of automated vehicles by minimizing the total travel distance, minimizing the travel time, maximizing the safety measures between the vehicles by minimizing the number of lane changes, and enhancing environmental sustainability by minimizing the amount of consumed fuel, and so on.

All of these objectives can be considered on the single vehicle only or the whole fleet. Also, the objectives can be considered per trigger which means that whenever the reconfiguration occurs we neglect the history of reallocation of the vehicle, or we can consider the history which means for example if the last trigger the *i*th car changed its lane, then in the next trigger it has the priority of not being changed.

13.4   The implementation of the convoy agent

13.4.1   The convoy localization module

The localization is responsible for calculating the centroid position and actual speed of the convoy based on the positions of all vehicles within the convoy. Together, these vehicles form a formation polygon whose vertices represent a vehicle and the center represents the convoy centroid as shown in Figure 13.3.

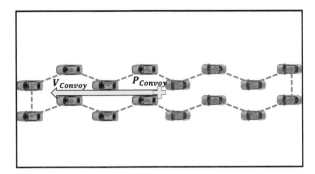

Figure 13.3 Convoy agent localization module

Accordingly, the actual convoy speed and the centroid are computed in the curvilinear coordinates relative to the Road's global axes as:

$$^{Road}P\big|^{Curv}_{Convoy} = \left(\frac{\sum_{i=1}^{m} s_{vehicle_i}}{m}, \frac{\sum_{i=1}^{m} n_{vehicle_i}}{m} \right) \tag{13.1}$$

$$^{Road}v\big|^{Curv}_{Convoy} = \frac{\sum_{i=1}^{m} V_{d_{vehicle_i}}}{m} \tag{13.2}$$

where m is the total number of joined vehicles.

13.4.2 The convoy planning module

The convoy planning is only activated whenever the convoy is triggered to change its configuration. These triggers function in the infrastructure and the vehicles. The considered triggers in this work are the environmental triggers caused by changing a number of lanes. Moreover, the vehicles can trigger changing the configuration in case of joining/splitting scenarios which caused changing the number of vehicles within the convoy.

The inputs to this module are the number of lanes, vehicle states, and the convoy centroid position. Meanwhile, the output of this module is two $m \times m$ matrices known as bias that reflect the desired signed spacing between vehicles in the longitudinal B_{Long} and lateral directions B_{Lat}. These matrices include the relative longitudinal and lateral spacing between the vehicles. Moreover, it assigns each vehicle to a place within the convoy configuration in the form of an array *Formation_Assignment*.

This can be achieved by creating three sub-modules that execute sequentially. These sub-modules are **formation reconfiguration planning**, **virtual-structure-based strategy planning**, and **formation assignment planning**.

13.4.2.1 Formation reconfiguration planning

In this work, the formation reconfiguration sub-module is designed with the capability of achieving the three introduced configurations in [12–17]. In this work, we adopted the layers and sub-layers concept introduced by [12–17] to generate the three formation configurations. However, we developed a more generalized algorithm that can work in all cases resulting in the desired bias matrices.

The inputs to this sub-module are the number of cars in the convoy and the number of lanes. Moreover, the desired longitudinal and lateral gap d_{Long} and d_{Lat} between two points in the configuration should be set. In our work, the d_{Lat} is set to be equal to the lane width.

It is noticed that tuning this selected d_{Long} will have a significant impact on the safety of the vehicles while driving in a convoy and this will be illustrated in the results.

In this planning sub-module, the developed algorithm goes through two steps. First, the **geometric topology rules** are defined by dividing the configuration into a number of layers and sub-layers, and the number of vertices that represent the number of cars in each of them shall be determined. the second step is **desired spacing generation** assigning the longitudinal and lateral gaps to each vertex within the layer and sub-layers. Finally, the bias matrices B_{Long} and B_{Lat} are generated. Here, we will mainly focus on presenting the interlaced topology, yet it can be easily extended to define the parallel topology.

Geometric topology rules

The first step for generating the interlaced configuration is to calculate the number of layers using (13.3).

$$Num\_Layers = ceil\left(\frac{Num\_Cars}{Num\_Lanes}\right) \tag{13.3}$$

Then, the number of sub-layers per layer is determined and for this topology it is set to 2.

$$Num\_subLayers\_Per\_Layer = 2 \tag{13.4}$$

The layers and sub-layers distribution is presented in Figure 13.4.

The next step is to calculate the number of cars per layer and sub-layer. Therefore, a general formula is developed that calculates the number of cars in layer i:

$$Num\_Cars\_Layer_i = Num\_Lanes - mod(mod(i \times Num\_Lanes, Num\_Cars), Num\_Lanes) \tag{13.5}$$

However, the number of cars per sub-layer is also calculated differently as such:

$$Num\_Cars\_Per\_SubLayer_j\_Per\_Layer_i = \begin{cases} min\left(Num\_Cars\_Layer_i, ceil\left(\frac{Num_Lanes}{2}\right)\right) & j = 1 \\ Num\_Cars\_Layer_i - Num\_Cars\_Per\_SubLayer_1\_Per\_Layer_i & j = 2 \end{cases} \tag{13.6}$$

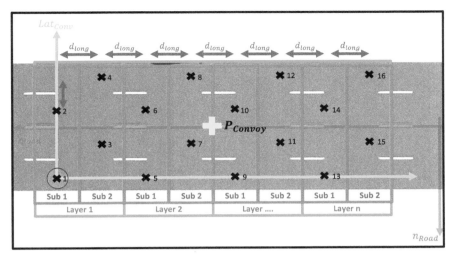

Figure 13.4   Convoy agent formation reconfiguration planning: interlaced configuration generation

Desired spacing generation

Knowing the number of cars per layer and in the sub-layers as well, the desired spacing can be generated by mapping the cars' vertices index to a corresponding layer and sublayer.

For a convoy composed of *m* number of vehicles, the car vertices index shown in Figure 13.4 can be described as an array as:

$$Car\_Vertices\_Index = \{k | k \in \mathbb{N}, \quad k \in [1,m]\} \tag{13.7}$$

The layer in which the kth vertex belongs can be computed using:

$$Vertex_k\_Layer = ceil\left(\frac{k}{Num\_Lanes}\right) \tag{13.8}$$

Moreover, the kth index can be mapped to a number between 1 and the number of lanes within the layer using the formula:

$$k\_Mapped\_Layer = mod(k - 1, Num\_Lanes) + 1 \tag{13.9}$$

Afterward, the generation of the desired spacing starts based on the desired topology.

In the interlaced topology, the sub-layer in which the mapped kth vertex belongs within the layer can be computed using:

$$Vertex_k\_SubLayer = ceil\left(\frac{k\_Mapped\_Layer}{ceil\left(\frac{Num\_Lanes}{2}\right)}\right) \tag{13.10}$$

Then, the mapped kth vertex is remapped within the sub-layer to a number between 1 and $ceil\left(\frac{Num\_Lanes}{2}\right)$ using the formula:

$$k\_Mapped\_SubLayer = mod\left(k\_Mapped\_Layer - 1, ceil\left(\frac{Num\_Lanes}{2}\right)\right) + 1 \quad (13.11)$$

Finally, the desired longitudinal and lateral positions are calculated relative to the car with vertex index 1 which is at the front of the convoy in the most left lane and highlighted in Figure 13.4. The mathematical formulas are:

$$\begin{aligned} &Vertex_k\_Desired\_Long = \\ &(2 \times Vertex_k\_Layer + Vertex_k\_SubLayer - 3) \times d_{Long} \end{aligned} \quad (13.12)$$

$$\begin{aligned} &Vertex_k\_Desired\_Lat = \\ &(2 \times k\_Mapped\_SubLayer + Vertex_k\_SubLayer - 2) \times d_{Lat} \end{aligned} \quad (13.13)$$

Having both *Vertex_Desired_Long* and *Vertex_Desired_Lat* for the m vertices, the bias matrices can be calculated by calculating the relative error in the longitudinal and lateral directions between every two successive vertices i and j.

$$B_{Long_{i,j}} = Vertex_i\_Desired\_Long - Vertex_j\_Desired\_Long \quad (13.14)$$

$$B_{Lat_{i,j}} = Vertex_i\_Desired\_Lat - Vertex_j\_Desired\_Lat \quad (13.15)$$

The overall algorithm can be summarized in [1]

Algorithm 13.1 Formation reconfiguration planning

Initialize: d_{Long} & d_{Lat}
Receive(*Num_Cars, Num_Lanes*)
Step 1: Geometric Topology Rules
Get *Num_Layers, Num_SubLayers_Per_Layer*
Get *Num_Cars_Layer, Num_Cars_Per_SubLayer_j_Per_Layer_i*
Step 2: Desired Spacing Generation
Vertex_Desired_Long = [] , Vertex_Desired_Lat = []
for (k = 1:*Num_Cars*) **do**
Get *Vertex_k_Layer*
Map index k & **Get** *k_Mapped_Layer*
Get *Vertex_k_SubLayer*
Remap index k & **Get** *k_Mapped_SubLayer*
Calculate *Vertex_k_Desired_Long* & *Vertex_k_Desired_Lat*
Vertex_Desired_Long.add(*Vertex_k_Desired_Long*)
Vertex_Desired_Lat.add(*Vertex_k_Desired_Lat*)
end
Calculate B_{Long} & B_{Lat}
Send(B_{Long} & B_{Lat})

13.4.2.2 Virtual-structure-based strategy planning

In this sub-module, the virtual-structure-based strategy is used to generate a virtual polygon subjected to the spacing matrices B_{Long} and B_{Lat} generated by the reconfiguration stage. In order to generate the virtual structure, the convoy centroid $^{Road}P|^{Curv}_{Convoy}$ should also be considered to know the exact positions of each vertex relative to the road profile.

The position of the kth vertex can be generated as:

$$s_{vertex_k} = \frac{s_{Convoy} * m - \sum_{i=1}^{m} B_{Long_{k,i}}}{m} \tag{13.16}$$

$$n_{vertex_k} = \frac{n_{Convoy} * m - \sum_{i=1}^{m} B_{Lat_{k,i}}}{m} \tag{13.17}$$

$$^{Road}P|^{Curv}_{vertex_k} = [s_{vertex_k}, n_{vertex_k}] \tag{13.18}$$

where m is the total number of cars/vertices within the convoy.

The virtual structure is finally represented as a set of points where each represents a desired position of a vertex complying with the spacing constraints.

$$^{Road}P|^{Curv}_{Virtual\_Structure} = \left\{ ^{Road}P|^{Curv}_{vertex_i} | i \in \mathbb{N}, \ i \in [1,m] \right\} \tag{13.19}$$

13.4.2.3 Formation assignment planning

The target of the assignment process is to determine which car to assign to the vertices generated from the virtual-structure-based strategy planning phase. In the literature, the formation assignment process is solved using a determined algorithm which is the Hungarian algorithm as in [12–17]. However, one limitation in this work was it is limited to only one objective which is number of lane changes and it is caused computational complexity of $O(N^3)$. It is observed that this method is not convenient or computationally efficient to solve this problem especially when we consider multiple objectives simultaneously which is completely ignored in [12–17].

Therefore, in this work, it is decided to utilize a meta-heuristic optimization algorithm to find a solution for the assignment process considering the multi-objective functions subjected to a set of constraints which would make our solution more practical and in much less computational time. It is important to take into consideration that the optimization algorithm will be activated only if a trigger occurs which means that it will not affect the overall computational time of the system.

Optimization problem formulation

In order to well formulate the addressed optimization problem, the set of optimization function(s) should be defined in terms of the decision variables, subject to all desired constraints.

In this problem, the desired objectives needed to be optimized aim to increase the mobility of the automated vehicles within the convoy by minimizing the total travel distance by all vehicles in addition to minimizing the total number of lane changes performed by the cars. Moreover, the combination of these objectives is selected to minimize the performed maneuvers by the cars which would reflect in reducing the possibility of inter-vehicular crashes hence increasing the safety within the convoy.

- **Decision variables and solution representation**
  The decision variable X of the addressed problem is assignment of Car_i to $Vert\_k$. Accordingly, the solution of this optimization problem is represented in the form of an arithmetic array. The index of each cell within the array represents a vertex index within the generated virtual structure while what is inside the cell itself is the generated solution from the optimization problem representing a specific actual car as shown in Table 13.1. An example of the generated solution is illustrated in Figure 13.5. This solution will be sent as an array called *Formation_Assignment* to the convoy agent along with the bias matrices.
- **Objective functions**
  Two objective functions are considered in this work. The first objective function is to minimize the total traveled distance by the whole convoy cars fleet. This objective function is mathematically described in terms of the cars

Table 13.1   Convoy agent formation assignment process planning: general solution representation

Array index	1	2	3	4	...	m
Solution X	Car_4	Car_m	Car_1	Car_3	...	Car_2

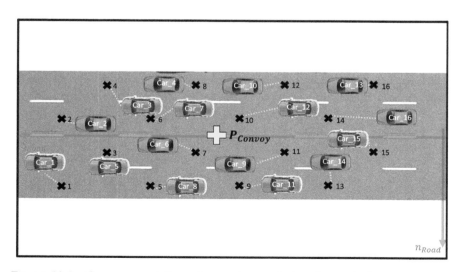

Figure 13.5   Convoy agent formation assignment planning: solution representation example

and vertices positions as:

$$f_1({}^{Road}P|^{Curv}_{Virtual\_Structure}, Convoy\_vehicles\_States) =$$
$$\sum_{i=1}^{m} \sqrt{(s_{Car_i} - s_{Vertex_k})^2 + (n_{Car_i} - n_{Vertex_k})^2} \quad (13.20)$$

where k represents the index of the vertex assigned to the corresponding car generated by the optimization algorithm and varies from one car to another.

The second objective function is minimizing the total number of lane changes done by the cars within the convoy fleet. This function can be described in terms of the lateral distance of cars and vertices and the lane width as:

$$f_2({}^{Road}P|^{Curv}_{Virtual\_Structure}, Convoy\_vehicles\_States) =$$
$$\sum_{i=1}^{m} \frac{|n_{Car_i} - n_{Vertex_k}|}{Lane\_Wid} \quad (13.21)$$

- **Constraints:**
While generating a solution, it must satisfy some constraints to ensure its feasibility. In our formulation, two constraints are considered. The first constraint is to ensure that all cars are assigned to vertices. This can be mathematically described as:

$$g_1 = length(X) \quad (13.22)$$

The second constraint is to ensure that each vertex is assigned to a unique car with no duplication.

$$g_2 = count(Car\_i) \quad (13.23)$$

Both constraints in this problem are equality where the first constraint must be equal to the total number of vertices which is the same as the number of cars. The second constraint shall be satisfied when it is equal to 1.

- **Mathematical representation:** Knowing the current position of all cars within the convoy fleet and the virtual structure position array, the final mathematical representation of the addressed optimization problem is formulated in terms of the aforementioned decision variables, and objective functions, and subjected to the constraints.

Accordingly, the overall mathematical representation can be formulated as a weighted function to find the optimal solution X^*:

$$2X^* = arg\,min_X W_1 \times f_1 + W_2 \times f_2 \quad (13.24)$$

subjected to:

$$g_1(X) == m$$
$$g_2(X) == 1$$

where W_1 and W_2 are the weights that determine the importance of each objective function.

Genetic algorithm-based formation assignment

Genetic algorithm (GA) is one of the well-known meta-heuristic optimization algorithms with the capability of solving the assignment processes. It is known for finding a near-optimal solution with high accuracy and low computational complexity, in addition to its simplicity, robustness, and independence on the problem model [21]. The GA algorithm is categorized under an evolutionary population-based optimization technique. It is designed to mimic the biological evolution process over the generations.

The basic concept of biological evolution is the existence of a generation that is composed of a number of individuals, where each individual is represented by a chromosome. The evolution of these individuals is done using three processes which are **elitism, crossing-over**, and **mutation**. Using these processes causes the existence of new better generations over time.

The analogy between the biological evolution and the GA is that the generation represents a set of N feasible solutions which is known by a population size *Pop_Size* that are generated simultaneously at the same iteration. The solution itself is a chromosome or individual that takes the form of a number of genes where each gene represents a cell in the solution representation as in Figure 13.1. GA operators are defined to mimic similar evolution processes with specific percentages that will result in generating a new set of N feasible solutions. With iterations, the solutions start to converge to an optimal solution guided by evaluating the fitness of each solution by calculating the objective function caused by each solution till reaching the optimal or near-optimal solution by the end of the number of generations generated which represents the total number of iterations i_{max}.

- **GA operators:**
  - Firstly, the Survivor Selection used in this work is the Fitness Based Selection which is the *Elitism*. The chromosomes with the best fitness value are automatically survived to the following generation. The used percentage in this work is P_e of the population size.
  - The second and third operators start by defining a *Parent Selection Criteria* upon which we decide on which parents the operator shall be applied. In this work, the parents' selection criteria is the *Rank Selection* using a *single fixed point*.
  - A defined percentage of chromosomes are subjected to the *crossover* operator, to produce new combinations of different chromosomes, which reinforces the exploitation process. Using the rank selection, the fittest chromosomes are selected alternatively, then using *random one point* crossover to generate new two children using *Davis' Order* to fulfill the crossing-over process as shown in Figure 13.6(a). The used percentage in this operator is P_c of the population size.
  - The *mutation* operator is performed also using the rank selection criteria. However, this operator chooses the worst members with a percentage P_m

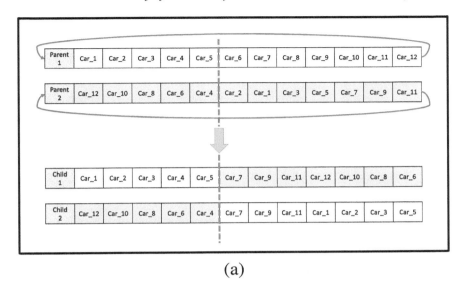

(a)

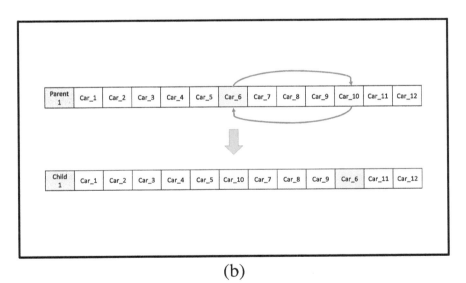

(b)

*Figure 13.6   Convoy agent formation assignment process planning GA operators:
(a) crossover and (b) mutation*

of the population size to mutate. The objective of this operator is to aid the exploration of the solution space to avoid being stuck in a local minima. The *Swapping* between two genes in the chromosome to generate new solution as shown in Figure 13.6(b).

- **GA algorithm parameters and implementation:**
  In the implementation of the GA, a number of parameters are initialized before starting the algorithm which are Pop_Size, i_{max}, P_e, P_c, and P_m. Once the parameters are initialized, the algorithm will generate initial generation Gen_0. While generating this initial generation, we have to ensure its feasibility in terms of satisfying all constraints presented in (13.22) and (13.23). It is also important to generate initial unique solutions to start the iterations with varieties that shall prevent any early convergence. The next step is to evaluate the fitness values of Gen_0, generating $Fitness\_Val_0$ by computing the objective function (13.24). Then, the population is ranked ascendingly based on the evaluated fitness values to be able to start the iterations, performing the GA operators based on the rank selection criteria.

  In each iteration i, the operators are performed to generate the next generation Gen_{i+1}. Where at first, the elite members are survived and added to Gen_{i+1}. Afterward, the crossover operator is performed between the elite members till the generation of new members with the size of $P_c\%$ of the population size. Afterward, the swapping of genes as a mutation process is performed on the worst members.

  While crossing over or mutating the parents, it is important to mention that the operator is repeated in two cases. The first case is generating already existent children where it should be ensured that unique children are generated that are different than their parents to avoid fast convergence to local minima as well. The second case is unsatisfying the constraints resulting in an unfeasible solution. In both cases, the process shall be repeated either by selecting another cut-off point in case of crossing over or selecting a random gene for swapping in case of mutation or we can choose completely different parents in case of not finding any unique children by repeating the process on the same parent for m time which is set to 3 in our algorithm.

  By the end of each generation, there is a new generation Gen_{i+1} which will be evaluated, producing $Fitness\_Val_{i+1}$ upon which the generation shall be ranked to start the next iteration in the algorithm.

  Once the number of iterations i_{max} is reached, the algorithm shall be stopped and the elite of the last generated generation which is the optimal solution X^* is returned to be the formation assignment process planning output which shall be sent to the convoy fleet to follow. The algorithm is summarized in Figure 13.2.

  By that, the planning module of the convoy agent is performed by generating the bias matrices B_{Long}, and B_{Lat}, in addition to the assignment array $Formation\_Assignment$ which indicates which vehicle to be assigned to which vertex within the convoy.

Algorithm 13.2 Genetic algorithm-based formation assignment planning

---

Initialize: $GA_{Param} = [i_{max}, Pop\_Size, P_e, P_c, P_m]$, *Generation* = []
Step 1: Initial Generation
for (j = 1:Pop_Size) **do**
$Gen_0 = []$, $Fitness\_Val_0 = []$, $X_{0,j} = []$

1 Generate Feasible Solution
while $X_{0,j}$ *is not feasible* **do**
Generate solution $X_{0,j}$
Gen_0.add($X_{0,j}$)
2 Evaluate Solution
Evaluate fitness value using (13.24) $Fitness\_Val_{0,j}$
$Fitness\_Val_0$.add($Fitness\_Val_{0,j}$)
end

3 Rank Generation
$Fitness\_rank\_ind$ = index(rank($Fitness\_Val_0$))
$Generation$ = $Gen_0[Fitness_rank_ind]$
Step 2: Start GA
for ($i = 1:i_{max}$) **do**
for ($j = 1:Pop\_Size$) **do**
Gen_i = [], $Fitness\_Val_i$ = []

1 Generate Feasible Solution
1a- Elite Operator
$Gen_{i_{Elite}}$ = $Generation[1:int(P_e\% \times Pop\_Size)]$
1b- Crossover Operator
$Gen_{i_{CO}}$ = []
while $length(Gen_{i_{CO}}) < int(P_c\% \times Pop\_Size)$ **do**
Select Parents, **Perform** Crossover, **Check** Feasibility
$Gen_{i_{CO}}$.add([Child_1,Child_2]])
end

1c- Mutation Operator
$Gen_{i_{Mut}}$ = []
while $length(Gen_{i_{Mut}}) < int(P_m\% \times Pop\_Size)$ **do**
Select Parent, **Perform** Mutation, **Check** Feasibility
$Gen_{i_{Mut}}$.add(Child_1]
end
Gen_i.add([$Gen_{i_{Elite}}$, $Gen_{i_{CO}}$, $Gen_{i_{Mut}}$])
2 Evaluate Solution
Evaluate fitness value using (13.24) $Fitness\_Val_{i,j}$
$Fitness\_Val_i$.add($Fitness\_Val_{i,j}$)
end

3 Rank Generation
$Fitness\_rank\_ind$ = index(rank($Fitness\_Val_i$))
$Generation$ = $Gen_i[Fitness\_rank\_ind]$
end
X^* = $Generation[0]$ $Formation\_Assignment$ = X^*

13.5   The implementation of the vehicle agent

13.5.1   Vehicle localization module

Each vehicle is equipped with a number of sensors that provide information about the vehicle's states of interest. These sensors are encoders, GPS, digital compass, and inertial measurement unit (IMU). This module collects information about the vehicle's position and speed in the geodetic coordinates, afterward the position is transformed to be described in the curvilinear coordinates for a unified description with the road. Therefore, the states of the vehicle can be described as $P_{Car} = \left[s_{car}, n_{car}, \psi_{car} \right]^T$ representing the curvilinear abscissa, signed lateral distance, and the relative heading, respectively. The control measurable inputs are $u_{Car} = \left[V_{d_{car}}, \delta_{car} \right]^T$ as the driving motor speed and steering angle.

13.5.2   Vehicle communication module

In this extended CATS architecture, we rely on the communication module using the communication devices to achieve inter-vehicle communication (V2V). This is due to the need for high speed of sending important messages between the vehicles for urgent situations such as achieving cooperative behavior. The utilized communication devices are emitters and receivers. The type of signals sent are radio signals through a set channel within a predefined range and baud rate. It is assumed in this work that the communication among the vehicles is ideal which means that there is no delay or data losses in the sent signals (Figure 13.7).

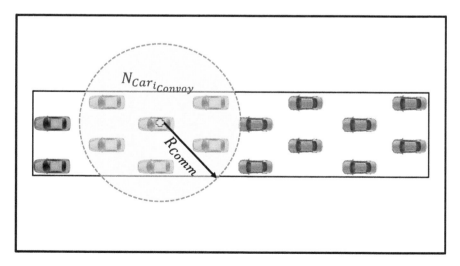

Figure 13.7   The vehicle communication range

Accordingly, the neighboring vehicles set N_{Car_i} for the ith car is mathematically represented as:

$$N_{Car_i} = \{j: ||^{Road}P|_{Car_i}^{Curv} - ^{Road}P|_{Car_j}^{Curv}|| \leq R_{Comm}\} \tag{13.25}$$

where R_{Comm} is the maximum communication range of the utilized communication devices as illustrated in Figure 13.7.

13.5.3 Vehicle planning module

In this work, the behavioral-based planning strategy is used where we defined a number of behaviors that should be considered while planning the vehicle's motion. The main behaviors included in this study are mainly divided into two categories.

The first category includes behaviors that are responsible for ensuring the safe driving of the vehicle in a non-cooperative mode. These behaviors are mainly:

• **Road profile planning** where the vehicle speed is determined based on the road speed limits as well as the assigned lane for the vehicle to follow based on an objective of distributing all vehicles within the road on the lanes to avoid congestion in one lane over the other,

• **Obstacle and collision avoidance behavior** where the vehicle uses the Additive Artificial Potential Field method to avoid any collision with environmental obstacles and the surrounding vehicles, and

• **Motion planning** including the **velocity maintaining Behavior** where the vehicle speed should be planned to ensure the passenger comfort avoiding any high jerks and the **Lane keeping/changing behavior**.

The second category includes behaviors that are essential to achieving convoy cooperative behavior between the vehicles. The cooperative behaviors include:

• **Formation maintenance** where the vehicle is required to maintain the predefined spacing with its neighbors within the same convoy

• **Convoy velocity matching** is responsible for the speed harmonization of all vehicles with the convoy speed.

These planning behaviors are only activated in the case of existing neighbors within the same convoy. This means that the net of neighbors should be defined as:

$$N_{Car_{i_{Convoy}}} \subseteq N_{Car_i} = \{Veh_j\_ID :$$
$$Veh_j\_Assigned\_Convoy = Veh_i\_Assigned\_Convoy\} \tag{13.26}$$

In order to perform both behaviors, the **graph theory** approach is utilized to plan the longitudinal and lateral speeds of the vehicle.

13.5.3.1 Formation maintenance planning

In order to plan this behavior, the overall convoy fleet is represented in a directed weighted graph $G_{Convoy} = (V_{Convoy}, E_{Convoy})$ as shown in Figure 13.8. It is assumed

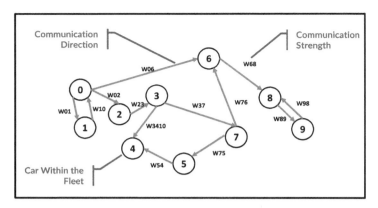

Figure 13.8   Graph representation of vehicular convoy: formation maintenance planning

that the position of each vehicle within the convoy represents a vertex in the vertex set V_{Convoy}. The communication among the vehicles is represented via a directed edge within the edge set E_{Convoy}. This directed edge indicates which vehicle is the transmitter and which one is the receiver for this particular connection. Moreover, the weight of each edge w_i indicates the communication signal strength which decays with the increase of distance between the vehicles.

Therefore, for the ith vehicle within the convoy, it is required to maintain longitudinal $B_{Long_{i,j}}$ and lateral $B_{Lat_{i,j}}$ gaps with the jth neighboring vehicle. Accordingly, the planned speed profiles shall be computed as:

$$\dot{s}_{Car_{Form_{i,j}}} = L_{i,j_{Long}} \times (B_{Long_{i,j}} - \Delta s_{i,j}), \quad \Delta s_{i,j} = s_{Car_j} - s_{Car_i} \tag{13.27}$$

$$\dot{n}_{Car_{Form_{i,j}}} = L_{i,j_{Lat}} \times (B_{Lat_{i,j}} - \Delta n_{i,j}), \quad \Delta n_{i,j} = n_{Car_j} - n_{Car_i} \tag{13.28}$$

where $L_{i,j_{Long}}$ and $L_{i,j_{Lat}}$ correspond to the Laplacian matrices subjected to weights $w_{i_{Long}}$, and $w_{i_{Lat}}$, respectively.

Accordingly, the collective formation maintenance planned speeds subjected to the whole desired spacing from all neighbors within the same convoy is expressed as:

$$\dot{s}_{Car_{Form_i}} = \sum_{j|j \in N_{Car_{i_{Convoy}}}} \dot{s}_{Car_{Form_{i,j}}} \tag{13.29}$$

$$\dot{n}_{Car_{Form_i}} = \sum_{j|j \in N_{Car_{i_{Convoy}}}} \dot{n}_{Car_{Form_{i,j}}} \tag{13.30}$$

13.5.3.2   Convoy velocity matching planning

The convoy velocity matching behavior is similar to the formation maintenance. However, it is desired to keep the velocity of the vehicles equivalent to each other and to the convoy speed as well.

In order to match the speeds of all vehicles, the corresponding desired acceleration is calculated as:

$$\ddot{s}_{Car_{Vel_{1_{i,j}}}} = L_{i,jVel} \times (0 - \Delta\dot{s}_{i,j}), \quad \Delta\dot{s}_{i,j} = \dot{s}_{Car_j} - \dot{s}_{Car_i} \tag{13.31}$$

$$\ddot{s}_{Car_{Vel_{1_i}}} = \sum_{j|j \in N_{Car_{i_{Convoy}}}} \ddot{s}_{Car_{Vel_{1_{i,j}}}} \tag{13.32}$$

where $L_{i,jVel}$ corresponds to the Laplacian matrices subjected to weights $w_{i_{Vel}}$.

Meanwhile, in order to maintain the speed of the vehicle to match the desired speed by the convoy, the desired acceleration is generated via:

$$\ddot{s}_{Car_{Vel_{2_i}}} = K_{vel_2} \times (Convoy\_Des\_speed - \Delta\dot{s}_{Car_i}) \tag{13.33}$$

where K_{vel_2} is a control gain.

13.5.4   Vehicle control module

While designing the control module, it is important to consider the behaviors that need to be controlled. Also, it is essential to consider that the controllers should fit the heterogeneous fleet of vehicles, which means that the controllers should consider the different parameters or/and be model-free as much as possible. Also, the controllers should consider the possibility of the existence of an unstructured environment; hence no lanes are available.

Since we are trying to design a driving experience that mimics human driving, thus the behaviors that should be considered in the control are longitudinal velocity control and lateral distance control. Including both control laws in the architecture would achieve controlling the driving speed while maintaining the desired lane. These two control laws are implemented in two sub-modules implemented beneath the main control module.

13.5.4.1   Longitudinal driving velocity control law

The longitudinal controller is designed to control the motion of the car in forward and backward directions maintaining a constant desired driving speed. In this work, a simple discrete Proportional Integral (PI) controller is chosen since it is convenient for our objective since it is independent of the car type. The controller is designed to generate the desired acceleration to control the car in terms of the control law proportional gain K_p and integral gain K_i.

$$e_{v_k} = V_{d_{car_{des_k}}} - V_{d_{car_k}}$$

$$A_{car_{Long}} = K_p(e_{v_k} + T_i \sum_{n=0}^{k} \Delta t \times e_n), \quad T_i = \frac{K_i}{K_p} \tag{13.34}$$

$$V_{d_{car_{k+1}}} = V_{d_{car_k}} + \Delta t \times A_{car_{Long}}$$

After generating the control action in the form of a desired longitudinal acceleration $A_{car_{Long}}$, it is important to normalize it to satisfy predefined vehicle's speeds, acceleration, and steering constraints, ensuring ride comfort.

The resulting $V_{d_{car_{k+1}}}$ is normalized to satisfy the car speed limits initialized in the car parameters.

13.5.4.2   Lateral control law

The car keeping/changing lane behavior is controlled via a control method proposed by Linderoth in 2008 [22]. This control law uses the curvilinear representation of the car position. It can be obtained in terms of the error in lateral distance $e_n = n_{car_{des_k}} - n_{Vehicle}$ and the relative error relative to the road profile which is the $e_{th} = \psi_{Vehicle}$.

The controlling steering angle is calculated by:

$$\delta_{Vehicle} = atan \left(\frac{-cos(e_{th})e_n - (k_1 + k_2)sin(e_{th})}{k_1 - (k_1 + k_2)cos(e_{th}) + sin(e_{th})e_n} \right) \tag{13.35}$$

where k_1 and k_2 are two control parameters to ensure the stability of the car's performance. The control steering action $\delta_{Vehicle}$ is normalized to satisfy the car steering limits and its rate of change as initialized in the car parameters.

13.6   Convoy results

In order to validate the CATS performance, a simulation environment is designed using Webots [23], a high-fidelity submicroscopic simulator where the simulated vehicle models incorporate basic rigid dynamics and properties of real cars. The simulated environments have been running in a machine with the following specification: ROS melodic installed on Ubuntu 18.04 LTS 64-bit operated machine of 15.5 GB memory with core i7-8750H CPU @2.2 GHz × 12 processor power.

In this section, an experiment is demonstrated as a showcase for a fixed-size convoy of 15 cars in City Map 1 as shown in Figure 13.9 (three times the maximum

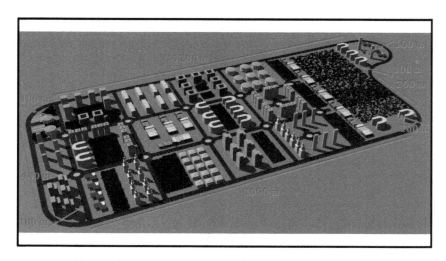

Figure 13.9   Webots simulated City Map 1 dimensions

number within the platoons). This experiment aims to show the results of the formation reconfiguration and assignment processes under the trigger of creating a
new convoy.

In this experiment, 15 vehicles are deployed in City Map 1 13.9 starting from
very close initial conditions to put the vehicles in a critical scenario. Figure 13.10
shows the initial condition of the deployed vehicles in the map at $T = 0$ and the
vehicles' positions after the formation in convoy at $T = 45 \, sec$ in an interlaced
configuration where $d_{long} = 7m$ and $d_{lat} = 2 \times lane\_width$.

To prove the validity of the mobility objective function (13.24) in the assignment algorithm, Figure 13.11 illustrates that at the beginning of the optimization, the
fitness value started at a high value and decreased within the generations, minimizing
the total objective function (collective fleet travel time and the number of lane
changes). Figure 13.12 shows the generated virtual structure, the initial condition of
the vehicles, and their assignment to the optimized vertex. It can be observed that
during the assignment, out of the 15 vehicles, only 2 changed their lanes. This proves

(a)   (b)

*Figure 13.10   Experiment 1: (a) vehicles initial condition $(T = 0)$ and (b) vehicles
in convoy formation $(T = 45$ s)*

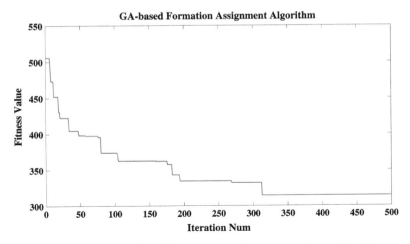

Figure 13.11   GA-based formation assignment – fitness function

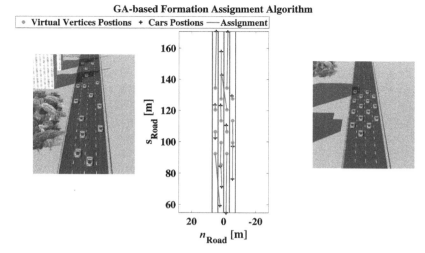

Figure 13.12   GA-based formation assignment – generated optimized solution

the increase in the mobility of the fleet on the road as well as increasing the safety by eliminating the lane changes on the road in the convoy mode.

To further prove the high safety impact of the vehicular convoy, the 15 vehicles are tested in the same environment but once in a non-cooperative driving mode as per the protocols presented in [19] while the second time they were driving the vehicles in cooperative convoy mode based on the CATS architecture presented in [20] and the extended modules presented in this work.

The performance of both driving modes are compared and assessed in terms of safety. The considered KPI is the IEEE Safety Standard presented in [24]; **the Minimum Safe Distance Violation (MSDV).**

The MSDV is defined as an instance in which the actions of the ego vehicle result in encroaching upon its safe boundaries with another (safety-relevant) entity within the scenario environment, as defined by the current velocities and acceleration capabilities of both entities.

The safety boundaries are defined by clear lateral and longitudinal distances defined by the RSS model [25] that the vehicle should maintain towards surrounding road users in order to prevent the ego-vehicle from being the cause of a road accident.

Several of the parameters involved in the responsibility-sensitive safety (RSS) minimum safe distance calculation are defined following reasonable, but subjective, assumptions on the behavior of other entities (e.g., the maximum longitudinal braking, $a_{max\_accel}^{long}$).

To calculate the MSDV, the longitudinal distance between two vehicles, d^{long} [m], the lateral distance between two vehicles, d^{lat} [m], the longitudinal velocity, v^{long} $\left[\frac{m}{s}\right]$, and the lateral velocity, v^{lat} $\left[\frac{m}{s}\right]$, of the vehicles must be obtained.

Moreover, it is important to obtain the response time ρ (the time $[s]$ it takes the AV or person to react to a situation, including reaction and actuation time), the minimum and maximum decelerations the entities could apply $a_{max\_decel}^{long,lat}$, and $a_{min\_decel}^{long,lat}$ $[\frac{m}{s^2}]$, respectively, and the maximum acceleration an entity could apply during the response time $a_{max\_accel}^{long,lat}$ $[\frac{m}{s^2}]$, must be assumed and defined to calculate the minimum safe distances.

The minimum allowable longitudinal distance d_{min}^{long} between the ego vehicle (the rear one subscript r) and the neighboring vehicle (the front one subscript f) moving in the same direction can be computed by:

$$d_{min}^{long} = \left[v_r^{long}\rho_r + \frac{1}{2}a_{r_{max\_accel}}^{long}\rho_r^2 + \frac{(v_r^{long} + a_{r_{max\_accel}}^{long}\rho_r)^2}{2a_{r_{min\_brake}}^{long}} - \frac{v_f^{long^2}}{2a_{f_{min\_brake}}^{long}} \right]_+$$

(13.36)

where $[x]_+ := max\{x, 0\}$.

The minimum allowable lateral distance d_{min}^{lat} between the ego vehicle (on the left one subscript l) and the neighboring vehicle (on the right one subscript r) moving in the same direction can be computed by:

$$d_{min}^{lat} = \mu + \left[\frac{(2v_l^{lat} + a_{r_{max\_accel}}^{lat}\rho_l)\rho_l}{2} + \frac{(v_l^{lat} + a_{r_{max\_accel}}^{lat}\rho_l)^2}{2a_{r_{min\_brake}}^{lat}} \right. $$
$$\left. - \frac{(2v_r^{lat} - a_{r_{max\_accel}}^{lat}\rho_r)\rho_r}{2} + \frac{(v_r^{lat} - a_{r_{max\_accel}}^{lat}\rho_r)^2}{2a_{r_{min\_brake}}^{lat}} \right]_+$$

(13.37)

where where $[x]_+ := max\{x, 0\}$ and μ is the lateral fluctuation margin.

The MSDV is to be formulated as:

$$MSDV = \begin{cases} 2 & d^{lat} \leq 0 \quad \& \quad d^{long} \leq 0 \quad \rightarrow \text{Collision Occurrence} \\ 1 & d^{lat} < d_{min}^{lat} \quad \& \quad d^{long} < d_{min}^{long} \quad \rightarrow \text{Dangerous Zone} \\ 0 & \qquad\qquad Otherwise \quad \rightarrow \text{Safety} \end{cases}$$

Figure 13.13 illustrates the computed MSDV index comparing the 15 vehicle performance once in the case of the non-cooperative free driving mode and the second case in the Convoy driving mode. It is clear that the vehicles in both cases did not suffer any crashes which proves the validity of the obstacle and collision avoidance behavior planning. However, it is clear that in case of the free driving the vehicles are always in a dangerous zone, unlike the case of convoy driving, and because of controlling the spacing between the vehicles in both longitudinal and lateral directions, the vehicles are actually always in a safe zone with zero violation except in the interval between 15 and 30 s where the convoy is actually formed and the vehicles are in motion to form the topology.

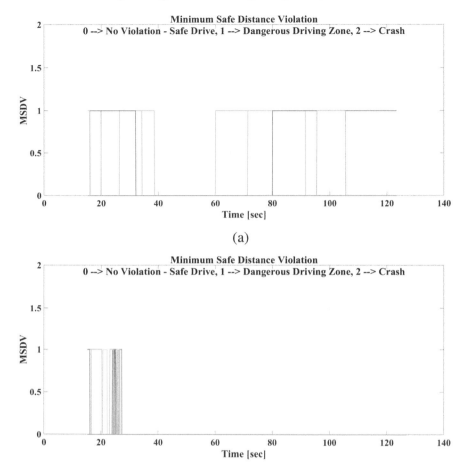

Figure 13.13   MSDV index: (a) free driving mode, and (b) convoy driving mode

13.7   Conclusion

The convoy system is rarely studied when comes to the vehicular system despite the great potential impact on safety, mobility, and environmental sustainability. In this chapter, we have demonstrated some important behaviors for the vehicular convoy, in particular the formation reconfiguration and assignment processes.

Also, we have shown that defining a set of objective functions that guarantee the optimal assignment of the vehicles will also have an impact on some important metrics and there is no limit over selecting these objective functions. Also, the motion coordination of these vehicles by maintaining both desired longitudinal and lateral spacing (unlike the platoon), improved significantly the safety of the whole fleet and this was proven by measuring the IEEE Safety Standard metric (the MSVD).

Finally, the importance of structuring and designing a unified and modular framework for the transportation system as the CATS architecture is very important when comes to investigating several integrated and complicated application in the transportation system and also ease the assessment of these applications and its modularity makes it easy to tune and select the suitable parameter that helps in improving the overall performance of ITS.

References

[1] Watzenig D, and Horn M. *Automated Driving: Safer and More Efficient Future Driving*. Berlin: Springer; 2016.

[2] Wang Z, Bian Y, Shladover SE, *et al.* A survey on cooperative longitudinal motion control of multiple connected and automated vehicles. *IEEE Intelligent Transportation Systems Magazine*. 2019;12(1):4–24.

[3] Wang Z, Wu G, and Barth MJ. A review on cooperative adaptive cruise control (CACC) systems: Architectures, controls, and applications. In: *2018 21st International Conference on Intelligent Transportation Systems (ITSC)*. IEEE; 2018. pp. 2884–2891.

[4] Dey KC, Yan L, Wang X, *et al.* A review of communication, driver characteristics, and controls aspects of cooperative adaptive cruise control (CACC). *IEEE Transactions on Intelligent Transportation Systems*. 2015; 17(2):491–509.

[5] Gao L, Chu D, Cao Y, *et al.* Multi-lane convoy control for autonomous vehicles based on distributed graph and potential field. In: *2019 IEEE Intelligent Transportation Systems Conference (ITSC)*. IEEE; 2019. pp. 2463–2469.

[6] Marjovi A, Vasic M, Lemaitre J, *et al.* Distributed graph-based convoy control for networked intelligent vehicles. In: *2015 IEEE Intelligent Vehicles Symposium (IV)*. IEEE; 2015. pp. 138–143.

[7] Qian X, De La Fortelle A, and Moutarde F. A hierarchical model predictive control framework for on-road formation control of autonomous vehicles. In: *2016 IEEE Intelligent Vehicles Symposium (IV)*. IEEE; 2016. pp. 376–381.

[8] Kato S, Tsugawa S, Tokuda K, *et al.* Vehicle control algorithms for cooperative driving with automated vehicles and intervehicle communications. *IEEE Transactions on Intelligent Transportation Systems*. 2002;3(3): 155–161.

[9] De La Fortelle A, Qian X, Diemer S, *et al.* Network of automated vehicles: the autonet2030 vision. In: *21st World Congress on Intelligent Transport Systems Conference*; 2014.

[10] Obst M, Marjovi A, Vasic M, *et al.* Challenges for automated cooperative driving: The autonet2030 approach. In: *Automated Driving*. Berlin: Springer; 2017. pp. 561–570.

[11] Navarro I, Zimmermann F, Vasic M, *et al.* Distributed graph-based control of convoys of heterogeneous vehicles using curvilinear road coordinates. In: *2016 IEEE 19th International Conference on Intelligent Transportation Systems (ITSC)*. IEEE; 2016. pp. 879–886.

[12] Cai M, Xu Q, Li K, *et al.* Multi-lane formation assignment and control for connected vehicles. In: *2019 IEEE Intelligent Vehicles Symposium (IV)*. IEEE; 2019. pp. 1968–1973.

[13] Xu Q, Cai M, Li K, *et al.* Coordinated formation control for intelligent and connected vehicles in multiple traffic scenarios. *IET Intelligent Transport Systems*. 2021;15(1):159–173.

[14] Cai M, Xu Q, Chen C, *et al.* Formation control for multiple connected and automated vehicles on multi-lane roads. In: *2021 IEEE International Intelligent Transportation Systems Conference (ITSC)*. IEEE; 2021. pp. 1940–1945.

[15] Cai M, Xu Q, Chen C, *et al.* Formation control for connected and automated vehicles on multi-lane roads: relative motion planning and conflict resolution. *IET Intelligent Transport Systems*. 2023;17(1):211–226.

[16] Cai M, Xu Q, Chen C, *et al.* Formation control with lane preference for connected and automated vehicles in multi-lane scenarios. *Transportation Research Part C: Emerging Technologies*. 2022;136:103513.

[17] Cai M, Xu Q, Chen C, *et al.* Multi-lane unsignalized intersection cooperation with flexible lane direction based on multi-vehicle formation control. *IEEE Transactions on Vehicular Technology*. 2022;71(6):5787–5798.

[18] Kuhn HW. The Hungarian method for the assignment problem. *Naval Research Logistics Quarterly*. 1955;2(1–2):83–97.

[19] Elias CM, Shehata OM, Morgan EI, *et al.* Emerging of V2X paradigm in the development of a ROS-based cooperative architecture for transportation system agents. In: *2022 IEEE Intelligent Vehicles Symposium (IV)*. IEEE; 2022. pp. 1303–1308.

[20] Elias CM, Shehata OM, Morgan EI, *et al.* Cooperative architecture for transportation system (CATS): development of a convoy agent for (V/I) 2C applications. In: *2022 IEEE 25th International Conference on Intelligent Transportation Systems (ITSC)*. IEEE; 2022. pp. 2335–2340.

[21] Katoch S, Chauhan SS, and Kumar V. A review on genetic algorithm: past, present, and future. *Multimedia Tools and Applications*. 2021;80(5):8091–8126.

[22] Linderoth M, Soltesz K, and Murray RM. Nonlinear lateral control strategy for nonholonomic vehicles. In: *2008 American Control Conference*. IEEE; 2008. pp. 3219–3224.

[23] Michel O. Cyberbotics Ltd. Webots^{TM}: professional mobile robot simulation. *International Journal of Advanced Robotic Systems*. 2004;1(1):5.

[24] Wishart J, Como S, Elli M, *et al.* Driving safety performance assessment metrics for ads-equipped vehicles. *SAE International Journal of Advances and Current Practices in Mobility*. 2020;2(2020-01-1206):2881–2899.

[25] Shalev-Shwartz S, Shammah S, and Shashua A. On a formal model of safe and scalable self-driving cars. arXiv preprint arXiv:170806374. 2017.

Chapter 14

Traffic management using floating car data in low- and middle-income countries

Megan M. Bruwer[1], Johann Andersen[1] and Kia Eisinga[2]

Traditional traffic engineering tools have been dominated by volume-centric traffic data. For example, level of service (LOS) measures, including density (freeways), delay (intersections), and time spent following (two-lane roads) are all estimated from traffic volume input. This approach to traffic engineering has been heavily influenced by the fact that traffic volume data were the primary traffic state variable collected through manual counts and long-trusted static (Eulerian) traffic sensors.

Recently, alternative traffic state data (particularly speed, travel time and vehicle routing data) have become widely available through the development of global positioning systems, enabling the observation of floating car data (FCD) collected along the trajectory of vehicles within the traffic stream. FCD application in transportation engineering and planning disciplines has surged with increasing FCD availability, plausibly stimulating a paradigm shift toward speed based and Lagrangian traffic state observation.

This chapter considers the impact that FCD are having on digitalization of traffic systems. Importantly, this chapter will examine the appropriate application of FCD in traffic planning and management, ensuring maximum benefit from this new data source, while maintaining the integrity of traditional traffic engineering methods. Particular cognizance is given to the low- and middle-income country (LMIC) context, where limited sensor distribution over extensive road networks hamper traffic studies. FCD are of immediate importance to LMICs, where digitalization promotes leapfrogging of legacy traffic sensor systems.

The content of this chapter focuses primarily on research and industry implementation of FCD and intelligent transport systems (ITS) in the sub-Saharan African context, as this is the area of application in which the authors mostly work. The discourse has, however, been structured to ensure wider applicability to the majority of LMIC contexts.

[1]Department of Civil Engineering, Stellenbosch University, South Africa
[2]TomTom International BV, Amsterdam, The Netherlands

14.1   Measuring traffic state from FCD

14.1.1   Coordinate systems for traffic state measurement

Traffic state attributes are the primary input to all traffic planning and management tasks. Traffic state describes traffic conditions according to quantifiable characteristics, including flow, average speed, and density, as well as measures of quality such as LOS.

Traffic state data are grouped by the coordinate system in which traffic state variables are measured, namely Eulerian data and Lagrangian data. Eulerian data are collected at fixed geographic locations by static sensors [1], while Lagrangian data are collected along the trajectory of individual vehicles by mobile sensors.

14.1.1.1   Eulerian data

Static Eulerian sensors include pneumatic tubes, piezoelectric sensors, magnetic induction loops, roadside infrared and microwave radar sensors, and video image detection technologies [2]. These are typically mature technologies, that have been widely deployed over many decades for observation of traffic streams. Eulerian observations include traffic volume measures (hourly flow and average daily traffic), occupancy (a measure of traffic density), and vehicle speed [1,2]. These data can be indicated per individual vehicle or averaged to provide arithmetic mean volume, speed, and occupancy for the traffic stream at a singular location.

The benefits of Eulerian data are plentiful: the mature technologies are trusted and provide excellent accuracy of traffic state measurement. Furthermore, data on all vehicles in the traffic stream are observed. There are, however, also important disadvantages to static traffic sensors that need consideration, particularly that these systems require the installation of sensors, ancillary communications, and electricity supply. Static traffic monitoring is, therefore, accompanied by high installation, operational, and maintenance costs, which significantly limits the area of observation to the highest order roads, where the distribution of sensors may still be sparse [3,4]. This limited device deployment is particularly pertinent in LMICs where the extensive road network and limited fiscal resources result in minimal traffic monitoring [5]. Constrained sensor distribution requires the extrapolation of traffic state data from incomplete and sparse observations [6].

14.1.1.2   Lagrangian data

In contrast with Eulerian data, Lagrangian traffic state variables are measured along the trajectory of vehicles by mobile sensors. Lagrangian data are grouped according to three main sources: test vehicles, reidentification methods, and FCD.

Test vehicle Lagrangian data comprise short-term data collection strategies using a single or limited number of specifically tracked vehicles along specified routes [2] in which study participants are aware that they are reporting positional data. This data is not collected or distributed widely, and so has limited application potential for comprehensive deployment.

Vehicle reidentification methods use static traffic sensors to develop Lagrangian data by identifying vehicles through Automatic Number Plate

Recognition, electronic toll tags, or Bluetooth sensors at set locations along a device monitored route [2].

The final form of Lagrangian data, **FCD**, are the main focus of this chapter. Vehicle traces are collected by mobile global navigation satellite system (GNSS) sensors, usually through GNSS-enabled mobile phones, onboard navigation systems, and vehicle tracking telematics systems widely used in the logistics industry [7]. FCD-reporting vehicles travel freely, essentially *floating* through the road network, as suggested by the term "floating car data".

FCD overcome key challenges of Eulerian, test vehicle, and vehicle reidentification data: first, FCD are widely available over the entire road network; and second, FCD systems do not require the extensive deployment of additional sensors and communications. FCD sensors are already extensively deployed for other use cases and FCD can take advantage of existing mobile communications infrastructure and GNSS [3].

14.1.2   Floating car data systems

There are a few key elements required for FCD systems to operate. These include: (1) probe devices, (2) communications, (3) processing, and (4) user platforms, as visualized in Figure 14.1.

The devices that report FCD are called probes, with FCD sometimes referred to as *probe data*. The probe devices report the timestamped location of the vehicle that carries the probe, usually recorded as a GNSS pin. Existing communications, including mobile internet coverage, are typically used in FCD systems to transfer positional data to a processing platform. Processing comprises the collection and

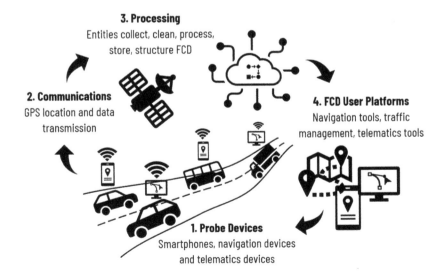

Figure 14.1   FCD system components (adjusted from Bruwer, Behrens, and Andersen (2023)) [25]

storage of the raw GNSS trace data, as well as the cleaning and handling of the data from GNSS traces to useful traffic state data. Once the data has been processed to usable traffic information, it may be distributed and sold to users through various user platforms. Users of FCD may be individual consumers (typically travelers using navigation tools), logistics companies with linked commercial drivers using commercial telematics systems, and entities such as roads authorities and transport practitioners using FCD for transport planning and traffic management systems.

14.1.3   Categories of FCD

Sources of FCD are categorized according to the nature of the data processing, the users, and the FCD owner, which may be a research or transport planning entity, a commercial FCD provider, or a telematics company. There are four broad categories of FCD: *commercial FCD*, *privately developed FCD systems*, *telematics generated FCD*, and *Mobile Network Operator (MNO) generated FCD*.

Commercial FCD: *FCD provided by third-party traffic data entities who collect, store and process widespread FCD to sell to consumers.*

Commercial FCD sources are available to various commercial traffic data providers such as TomTom, HERE, Google, and INRIX. These entities obtain vehicle tracking data from commercial fleet telematics companies or through transfer agreements with mobile phone providers and mobile networks often via navigation apps [8]. Commercial FCD entities encourage travelers to share their anonymized trip information in exchange for navigation assistance augmented with actual traffic conditions. Commercial FCD providers have sophisticated processing methods that ensure that good quality aggregated traffic data are generated from vehicle trajectory data, which can then be purchased for use by transport practitioners and planning authorities. In this way, good quality traffic data can be obtained without the need for FCD collection platforms and extensive systems to clean, collate, and interpret unformatted probe data [9] because the data are already processed and aggregated.

Privately developed FCD systems: *FCD obtained from a local, discrete vehicle fleet for particular use by a single, or restricted number of entities.*

Privately developed FCD systems are *fit-for-purpose* FCD created by private entities for research or private fleet management. A specific set of vehicles are equipped with GNSS devices providing the input data. Data analytics are specific for the requirement of the entity that collect, own, manage, and use the data. One such FCD system is FLEET, developed by the Austrian Institute of Technology. FLEET collects trajectory data from taxi fleets, commercial delivery and service vehicles, and public transport vehicles in certain cities where the institute is active for research purposes [10].

Telematics generated FCD: *FCD obtained as a by-product from commercial vehicle tracking by vehicular fleet management systems.*

Fleet management systems are developed by telematics companies for use by logistics businesses, public transport operators, and government entities to track their vehicle fleets. Ride-hailing apps also generate FCD in this category. Fleet management systems track vehicles, evaluate driver behavior, and provide dashboard operational reports. These systems improve driver behavior, control of assets, track delivery indicators, and assess delivery and time-table adherence. FCD generated by telematics systems are usually a by-product of the tracking capabilities of telematics systems.

MNO generated FCD: *positional data generated by mobile phones as a by-product of providing cellular network coverage to customers.*

MNO generated FCD are based on positional data generated from mobile phone network operations [11]. Locational information from mobile networks is established through interaction of mobile phones with cellular base stations. A call detail record (CDR) is generated whenever a call or internet connection is made, or a mobile phone connection transfers from one base station to another [11]. MNO CDR data are not strictly trajectory-based; they do not provide continuous positional data because CDRs are only generated when a mobile phone is in use and have poor positional accuracy. MNO CDR data, therefore, falls somewhat out of the scope of true FCD.

The four categories of FCD can be allocated to quadrants defined by the ease with which the generated data can be used for traffic analysis, and the geographic dispersal of the collected data, according to Figure 14.2. FCD generated by telematics companies and MNOs are typically unstructured and not readily applied to traffic studies as they are the by-product of tracking capabilities instead of built-for-purpose systems for traffic analysis. MNO FCD and commercial FCD are both widespread as the data are collected from widespread probes, while privately developed FCD and Telematics Generated FCD are limited to the distribution of the privately tracked probes. The benefits of commercial FCD are clear from Figure 14.2, being both widespread and specifically developed to provide input to traffic studies and traffic management.

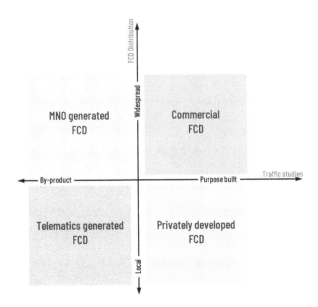

Figure 14.2   Categorization of FCD sources

14.1.4   Traffic state variables reported by FCD

Probe devices transmit spatiotemporal information to entities that collect and process the data [12]. This raw spatiotemporal information is processed to extract useful traffic information, including speed, travel time, and trip routing information. This chapter refers to this processed data as FCD. During the processing of incoming data from probes, commercial FCD are aggregated per road segment. FCD aggregation and the aggregated traffic state variables made available in commercial FCD are described below.

14.1.4.1   Segment aggregation of commercial FCD

FCD may be both aggregated and disaggregated. Disaggregated FCD provide movement characteristics (travel time, speed, and routing information) of individual vehicles. Aggregated FCD provide characteristics of the traffic stream averaged over a defined time period per road segment, such as space mean speed and average travel time. Aggregation times vary from short intervals, for example 60 s [12], to longer time periods (15 min, an hour, or even a full day). Aggregated FCD are the most common form of commercial FCD, because the aggregation process anonymizes the data, which is important for privacy concerns (refer to Section 14.1.6).

Aggregated FCD reports on the average traffic characteristics of the traffic stream observed as a fluid continuum. Commercial FCD sources aggregate spatiotemporal probe input by route segment, dividing roads into discrete segments which remain fixed over time. Historically, FCD aggregation segments were defined by traffic message channel (TMC) codes [13], which are universally applied for mapping and data evaluation of numerous commercial FCD providers. TMC codes define segments with end nodes positioned at significant points, such as intersections, speed limit changes, and changes in road geometry. TMCs however, are not coded in certain countries, particularly many LMICs. For example, TMCs are coded for South Africa, but no other African countries.

TomTom has subsequently redefined FCD aggregation segments through a new map encoding procedure, OpenLR, that is widely applied by other mapping companies and commercial FCD providers [13]. OpenLR segments are shorter, providing more granular traffic data, but fit within original TMC defined segments (where defined), allowing for continuous comparison of traffic and mapping characteristics. FCD aggregation segment lengths depend on road characteristics and are, therefore, not of a consistent length. Shorter segments are typically defined on roads where speeds are more variable.

14.1.4.2   Typical aggregated traffic stream characteristics

Typical aggregated traffic stream characteristics reported per road segment by commercial FCD include the travel time per segment, harmonic (space) mean speed, median speed, and percentile speeds (5th, 25th, 75th, and 95th percentile) per route segment. This information is utilized for navigation by routing engines to calculate the fastest routes and estimate travel times. This information is then presented to road users through navigation applications. Traffic stream speed information is also provided on maps with different speeds indicated in a color array. FCD-reported speed data can be used to detect anomalies like road closures, lane blockages, traffic accidents, and zones of reduced speed limits.

Aggregated traffic stream variables are made available for traffic studies and traffic management though various commercial application programming interfaces (APIs). Some sources of commercial FCD report the number of probes that have reported data on each route segment during the aggregation period. Number of probes, when reported, assist in estimating probe penetration rate (the proportion of the traffic stream reporting FCD), which is a good measure of the level of representativeness of the FCD, discussed further in Section 14.1.5.

Behavioral traffic data are also frequently reported by commercial FCD. These aggregated data report the routing of traffic through origin-destination matrices between traffic analysis zones [8], the percentage of vehicles using a particular route, and maps which show the likelihood of trips moving from one zone to another. Similar to speed data, the routing data are aggregated, indicating number of trips starting and ending in traffic analysis zones (for example a neighborhood or school district), rather than the start and end points of individual trips.

The one traffic state component that has not previously been readily reported by FCD is traffic volume because FCD are reported by a sample of vehicles and not the full population, making vehicle counting a complex task. Recently, with increasing FCD probe penetration levels, the FCD-based estimation of traffic volumes has come into focus [14]. The first FCD-reported traffic volume products are just starting to enter the market, primarily addressing road network planning use cases rather than real-time traffic management applications at this stage.

FCD-based traffic observation allows for more cost efficient, network-wide traffic state monitoring when compared to static Eulerian sensors. In the long term, it is anticipated that FCD-based traffic volume, speed, travel time, and routing data will widely replace the classic detector-based traffic monitoring approach due to the widespread data availability and cost effectiveness.

14.1.5   *Accuracy of FCD for traffic state evaluation*

FCD probe penetration rates are increasing steadily as more probe sources come online and are shared, increasing FCD accuracy annually. It is, however, still important to report on the accuracy of FCD. FCD-reported speed accuracy is evaluated by comparing FCD speeds to benchmark data to calculate a *speed bias* according to various statistical measures, including absolute average speed error (AASE), speed error bias (SEB) and mean absolute percentage error (MAPE) [4]. Speed bias is allowed to fall within predefined accuracy bands to indicate adequate FCD accuracy, for example an error within 5 miles per hour, or 10% of average speed. Global studies generally report good FCD speed accuracy. Only a few FCD accuracy studies have been carried out in LMICs, but these too have shown adequate speed bias levels [15].

The accuracy of FCD varies according to probe penetration rate, operating speed, time of day, congestion level, and road class as detailed below.

14.1.5.1   Penetration rate and accuracy

Not all vehicles are equipped with a probe device and so FCD are reported by a sample of the traffic stream. The accuracy of FCD for speed measurement, as measured by speed bias, improves as probe penetration increases [4,16]. This is because the FCD sample becomes more representative of the full driver population at higher penetration rates.

Relatively low probe penetration rates are required to ensure adequate accuracy of FCD-reported speeds. Penetration as low as 1 to 3% along uninterrupted freeway segments have been shown to yield adequate speed accuracy [1]. In fact, studies have shown that penetration rates higher than 5% do not produce significantly better speed-reporting accuracy [17], meaning that most FCD systems will have adequate penetration rates to ensure reasonable accuracy of FCD-reported speed and travel time information, even in LMICs. 2020 penetration rates for a particularly commercial FCD provider were found to be between 8% and 11% on urban freeways in various cities around South Africa [15].

Higher penetration rates are required for behavioral traffic observations of routing and origin-destination information. Low penetration rates may generate some level of bias when trip routing is reported, especially in LMICs, where traffic is often heterogenous and lower income groups are not as likely to report trip data, further discussed in Section 14.2.

14.1.5.2   Operating speed and accuracy

Speed bias decreases as operating speeds increase. The range of speed bias observations are also found to be less at higher operating speeds along major roads [4]. This indicates that there is lower variability of FCD-reported speeds to population speeds at higher operational speeds. This is likely because lower operational speeds result in a situation where vehicles are more likely to be speeding up or slowing down (for example in zones close to intersections or interchanges). Where speeds are consistently high, there is lower variability in the speeds of all vehicles, and so FCD-reported speeds are more likely to provide a good representation of overall average speed.

14.1.5.3   Temporal variation of accuracy

FCD accuracy tends to be better during the day than at night [4,18]. During the night, traffic volumes are generally lower, allowing for greater variability in speed selection by individual drivers, increasing speed variability, and reducing FCD speed accuracy. The time of the day with the best FCD accuracy is during recurrent peak traffic hours in the morning and afternoon [4]. During these high flow, high density periods, overtaking slower vehicles becomes increasingly difficult, and so the speeds of all vehicles tend toward the average speed of the traffic stream.

14.1.5.4   Accuracy and road class

Finally, the accuracy of FCD differs according to road class: FCD-reported speeds are more accurate on higher order mobility roads. For example, FCD along urban freeways (interstates) may exhibit accuracy levels up to two times higher than FCD-reported speeds on lower order arterial roads. Research has indicated that the reason for this is due to the different traffic control methods used for different road classes [16]. Freeways maintain uninterrupted flow through grade separation at interchanges, while at-grade intersections along arterials interrupt and control flow through traffic signal, stop or yield control. Some vehicles can travel through intersections at high speeds, while others are required to stop, increasing variability in speeds, resulting in higher speed bias along control interrupted arterial roads.

These findings reiterate the importance of variability in the speed of the general vehicle population in FCD speed accuracy: FCD accuracy increases as speed variability decreases (in higher speed areas and times of high flow), and as more variability is captured by the FCD sample (with higher penetration rates).

14.1.6   Privacy concerns

International privacy laws require that commercial FCD be aggregated to protect the personal information of the people providing data [3,6]. No personal identifiers may be linked to data distributed by users. This is particularly important when considering vehicle routing information and the observation of the start and end points of trips, which point to personal addresses, workplaces, and schools. Furthermore, this data may not be used to identify speed transgressions.

Aggregation of commercial FCD is one of the best ways to ensure that personal information is removed from reported FCD because individual vehicle traces and speeds are not reported. The following typical data processes are applied to commercial FCD to further mitigate data privacy concerns:

- Incoming data from individual probe devices are assigned a unique code which is not linked to the IP address of the device. This code changes per trip and per day to ensure that the same vehicle or person cannot be tracked past a single reported trip.
- The beginning and end points of trip traces are trimmed to prevent the identification of people reporting the data based on their place of residence.
- FCD data providers do not offer access to unprocessed FCD to external FCD users. Only data which is fused and aggregated into traffic state products that are not subject to privacy concerns are made available.
- Internally, access to unprocessed FCD are restricted, and the retention time for storing the data is limited.

14.2   The application environment of FCD in LMICs

14.2.1   The value of FCD for LMICs

The high installation, operational, and maintenance costs associated with traditional static traffic sensors results in a "traffic data desert" in many LMICs. Traffic sensors that are installed are generally limited to main routes and important urban centers, leaving vast swathes of road networks in LMICs unmonitored.

FCD obtains widespread traffic data without extensive traffic sensor installation. Numerous studies have indicated that commercial FCD are a particularly suitable source of traffic data in LMICs due to wide network coverage with limited infrastructure and communications requirements [5]. FCD can position LMICs to leapfrog static traffic sensor systems. Leapfrogging is the process whereby regions bypass technology stages to better technology saturation.

Commercial FCD are readily usable by industry practitioners, transport planning authorities, and road authorities because these FCD sources are pre-formatted and aggregated, unlike unformatted FCD sources common of systems not developed for transportation studies (MNO and telematics generated FCD). Furthermore, privately developed FCD necessitates extensive Big Data analytics capacity to convert individual probe spatiotemporal data into usable traffic information.

Another important role of FCD in the LMIC context is the promotion of transport system digitalization. Digitalization is the application of digital technologies to improve efficiency and management of various systems. Examples of transportation sector digitalization include smartphone accessed shared mobility and ride-hailing [19], connected and autonomous vehicles, and automated traffic control systems such as adaptive traffic signals. Digitalization technologies also generate data, further enriching FCD.

Examples transport sector digitalization in Africa

Internet connectivity, smartphone use, and tech startups are growing rapidly in Africa, allowing for increasing digitalization of the transportation sector. Ride hailing services like Uber, Lyft, Bolt, Careem (Egypt), Yego (Rwanda), and Little Cab (Kenya) are quickly becoming entrenched as a useful mode of transport on the continent. Uber announced that they had served 1-billion trips in Africa by May 2022, operating in eight African countries (Egypt, Nigeria, Ghana, Kenya, Tanzania, Uganda, Ivory Coast, and South Africa) [20].

Many ride hailing startups are emerging in Africa, taking into account the specific mobility needs and existing travel behaviors of the region. SafeBoda, an e-hailing platform from Uganda, connects passengers and *boda-boda* (motorcycle taxis) services. *Boda-boda* are one of the most used public transport modes in the country, but are known for poor safety, accounting for nearly a quarter of all road traffic injuries in Uganda [21]. SafeBoda drivers have been proven to show improved safe driving behavior than other *boda-boda* drivers. This is attributed to training provided by SafeBoda, penalties for unsafe driving imposed by the company, better bike maintenance, and increased likelihood of SafeBoda drivers to wear helmets [21]. Motorbike taxi apps such as SafeBoda are also creating a dataset of movements by a transport mode that would previously not have been obtained due to the informal nature of the mode.

14.2.2   FCD availability in LMICs

Commercial FCD are widely available and used in high income countries (HICs) and are poised to become an important source of traffic data in LMICs, however, coverage of FCD in LMICs lags behind more industrialized countries. Figure 14.3 presents the global coverage of FCD procured, processed, and distributed by TomTom by the first quarter of 2024. North America, Europe, Southeast Asia, and Australia have extensive FCD coverage. South and Central America have relatively good coverage with seven of the largest countries well accounted for. Africa, the Middle East and much of Asia are, however, not as well represented by FCD.

In HICs, FCD is collected from a variety of sources, including fleet management systems, mobile phone apps, and connected private vehicles. FCD collection in LMICs is dominated by mobile phone data, collected by smartphone operators and navigation app providers (note that this is not MNO-generated FCD, but rather commercial FCD).

FCD coverage expressed by penetration rate in many middle-income countries is equivalent to many HICs. For example, penetration rates of up to 14% are observed in South Africa and Egypt, providing substantial input for all types of FCD applications. Penetration rates are often significantly lower in low-income countries. The lower penetration in low-income countries with less mobile phone coverage still allow good quality travel speeds to be estimated.

14.2.3   LMIC-specific aspects impacting FCD

There are two key issues which need to be taken into account when using any source of FCD in the LMIC context as these may impact the representativeness of FCD and its accuracy. First, where penetration rates are low, the sample of FCD may not be fully representative of the full traffic stream population; and second, there are urban and mobility system differences between LMICs and HICs which may affect FCD usage and trip representation. The impact of these issues on FCD applicability are discussed in this section.

Figure 14.3   FCD coverage by TomTom up to the first quarter of 2024

14.2.3.1   Sampling issues related to users reporting FCD in LMICs

Geographic location and demographic profile of the person carrying a probe device will impact the sample providing FCD, making certain groups of the population less likely to be equivalently represented in the data. Mobile phones are a considerable source of non-commercial probe input to FCD, and so mobile phone distribution can be used as a proxy for FCD sampling in this discussion. Mobile phone ownership, particularly in LMICs, is linked to age, gender, income level, and domicile [22,23]. For example, educated men from larger households are more likely to own a mobile phone than other demographic groups in many LMICs. Rural populations also present lower mobile phone ownership than urbanized groups [24]. Reporting of FCD from rural groups may be further impaired by irregular or limited mobile network coverage [22,24].

The sample reporting FCD may, therefore, contain a less than representative proportion of rural, poor, and female-owned probes, resulting in somewhat biased mobility data against these groups [22,25]. The level of bias, however, cannot be quantified because the data must remain anonymous due to privacy conditions.

The bias in FCD movement patterns is compounded because mobility is also impacted by demographics [23]. Wealthier individuals with higher educational levels and greater access to private vehicles travel further and more frequently. Men make more trips by private vehicle than women in Sub-Saharan Africa – 22% of women in South Africa hold a driver's license, compared to 40% of men [26]. Urban communities are more likely to travel more often than rural groups [23]. These mobility traits of LMICs compound the bias created by the demographic of FCD sources: the population sample most likely to report FCD are also making the most trips.

One further sampling issue related to FCD in LMICs relates to the high proportion of digitalized paratransit operations within the traffic stream (refer to the discussion panel on SafeBoda). This likely leads to FCD being more frequently reported by certain vehicle fleets, particularly ride hailing and micro-courier services. FCD with a high proportion of data input from one type of entity might misrepresent the general traveling population.

14.2.3.2   Urban from differences in LMICs

The mobility systems of HICs vary significantly from LMICs. LMIC cities often have limited formal public transport services, which have led to intensive growth in private vehicle usage. This, coupled with underdeveloped road infrastructure, result in excessively high congestion levels in LMIC cities [27].

Land use distribution in LMICs also differs from HICs. LMIC cities are often segregated by demographics, with poorer communities located on the outskirts of cities, exacerbated by rampant urbanization and poor planning. Further land use segregation results in residential areas located far from commercial centers and services [27]. The urban poor (who are less likely to be reporting FCD to the same levels as higher income travelers) are therefore required to travel further distances within cities.

The urban form and transport infrastructure differences observed in LMICs may, therefore, impact the density of FCD originating in different zones within cities in LMICs. This implies that, while FCD-reported speeds are usually of a good quality in LMICs [15], the use of FCD for observation of trip routing behavior and traffic volumes are likely not yet suitable. Significantly more research is needed to determine how the LMIC specific context impacts FCD [25].

14.3 Enabling intelligent transport systems application through FCD

FCD can be applied in ITS as an enabler to, *inter alia*, optimize transportation systems, enhance traffic management, and improve road safety. Application areas described in this section include freeway management systems, public transport systems, parking management, and intersection control systems.

14.3.1 Freeway management systems

Freeway management systems aim to ensure more efficient urban travel, improve road safety, and maximize throughput. These systems can be significantly enhanced by providing real-time information on prevailing traffic conditions, allowing transportation agencies to make data-driven decisions as traffic conditions change. Examples of the elements of FCD application in freeway management systems that are deployed in South Africa on Freeways in major cities are described below.

Real-time traffic monitoring: FCD enables transportation authorities to monitor traffic conditions in real-time providing insights into traffic patterns, congestion levels, travel demand, and incidents on different freeway segments. This is often integrated into advanced traveler information system (ATIS) software in the traffic management center and is used in conjunction with other traffic managements devices to manage freeway traffic.

In Figure 14.4, images of the roadway network covered by the freeway management system in Cape Town, South Africa, are shown. The FCD visualization is integrated into the ATIS software and displayed both on the desktop of the traffic management operator, as well on the video wall.

Travel time estimation: By integrating FCD into the freeway management systems, travel times between different points on the freeway can be calculated. Accurate and up-to-date information on travel times helps drivers make informed decisions about route choice and necessary departure timing in both the trip planning stage (usually through traveler information websites and navigation apps), and while traveling using variable message signs (VMS). Figure 14.5 shows a VMS on a freeway in Gauteng, South Africa. VMS may display information about traffic movement, potential incidents, or display travel time to a known point on the freeway network, assisting road users by providing meaningful traveler information.

Incident detection and management: Transportation agencies can quickly identify incidents by considering changes in vehicle speeds and movement patterns reported by FCD. Rapid identification of incidents allows response teams to be dispatched to the

Figure 14.4   SANRAL Traffic Management Centre Training Room, Cape Town

Figure 14.5   VMS on Freeway in Gauteng (photo: Paul Vorster)

affected areas more quickly, minimizing disruptions to traffic, improving the outcome of injured travelers, and reducing the risk of secondary incidents. Figure 14.6 describes non-recurrent congestion related to a traffic crash that occurred on a freeway in Cape Town, where a vehicle overturned, landing in the median. Congestion, evaluated from FCD

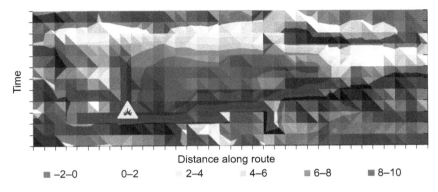

Figure 14.6   Incident detection from FCD

speeds, is reported as an index from 0 to 10, where 0 indicates typical traffic state, and 10 indicates severe non-recurrent congestion. The spatiotemporal extent of the non-recurrent congestion (reported by FCD) is clearly observed in Figure 14.6, according to both the distance and time that traffic state was impacted.

Dynamic traffic management: Freeway management systems can automatically adjust variable message signs, variable speed limits, and lane control systems to optimize traffic flow and safety with input from real-time traffic data provided by FCD.

14.3.2   Public transport management

Public transport management can be enhanced with FCD by utilizing real-time information about traffic conditions to improve the efficiency, reliability, and quality of public transportation services. Some examples include:

Real-time traffic monitoring: FCD allows transportation authorities to monitor traffic conditions in real-time by tracking movement and speed of vehicles on the road network, providing insight to traffic congestion and other factors that may affect bus operations.

Dynamic route planning: Transit agencies can automatically adjust bus routes to avoid congested areas or minimize delays by integrating FCD with public transport management systems.

Real-time passenger information: FCD from bus and rail fleet management systems are integrated with passenger information systems to provide passengers with real-time information about bus and train location, expected arrival times, and service disruptions. Passengers can access this information through mobile apps or electronic displays at bus stops.

14.3.3   Parking management

The overall traveler experience can be significantly enhanced with the use of FCD enhanced parking management applications, such as:

Real-time parking availability: FCD enables transportation authorities to monitor parking availability by tracking the movement and parking behavior of vehicles equipped with probe devices. This data provides insights into parking occupancy levels, available parking spaces, and parking duration.

Dynamic parking guidance systems: Integrating FCD with parking guidance systems can provide drivers with real-time information about parking availability. As an example, mobile apps or electronic signs can display the number of available parking spaces in nearby parking lots or on-street parking areas, helping drivers locate parking more quickly and efficiently.

Traffic flow management: By providing drivers with real-time information about parking availability and directing them to available parking spaces efficiently, FCD helps reduce traffic congestion and improve traffic flow in downtown areas. Drivers will spend less time searching for parking, leading to smoother traffic flow and reduced emissions.

14.3.4   Intersection control systems

FCD can enhance the efficiency of intersection control by providing real-time information on traffic conditions, allowing signal timing to be adjusted dynamically based on actual traffic demand.

Real-time traffic monitoring: FCD can be used to monitor the movement of vehicles approaching an intersection in real time providing insights to traffic patterns, congestion levels, and travel demand at various intersections through the network. This application can also offer opportunities to enhance **maintenance** systems, an important use-case in LMICs. Through FCD, the transportation authority can determine when traffic signals are not functioning correctly, facilitating timely response.

Dynamic signal timing adjustments: Traffic signal parameters (such as green time and cycle length) can be adjusted dynamically to respond to changes in traffic demand reported by FCD. For example, if FCD indicates a buildup of traffic on one approach, the signal timing can be modified to allocate more green time to that direction, improving traffic movement. This allows traffic signals to respond adaptively to changes in traffic flow, optimizing signal timing to minimize delays and improve efficiency. Research at the University of Michigan has shown that traffic signals can be recalibrated in real-time to reduce congestion and delays at intersections using FCD reported with penetration rate as low as 6% [28]. This application offers a beneficial opportunity in the LMIC environment for reducing costs in managing traffic control systems.

14.4   Case study: pothole detection

Potholes cause extensive damage to vehicles and are a significant safety issue. Road safety would be greatly improved if pothole location could be readily determined, allowing roads authorities to plan and conduct urgent road maintenance and giving drivers the opportunity to select routes that avoid potholes or anticipate potholes when alternative routes are unavailable.

Data needed for automatic pothole detection should be able to be remotely obtained, without the need for specifically equipped test vehicles and trained technicians to report potholes. Furthermore, appropriate data should be passively reported, not requiring active reporting by the public, for example through apps such as Waze. This will ensure that pothole reporting is not limited by the participation of active users, which may not always be forthcoming, especially in rural areas. Lastly, adequate data for remote pothole detection needs to be available over the vast rural road network, and so cannot rely on static sensor equipment.

Recent work by the authors of this chapter has looked into the applicability of commercial FCD for detecting potholes. Initial results have proven positive, indicating the potential for the use of FCD to detect potholes on rural roads in LMICs.

Pothole detection from commercial FCD was developed and tested on 250 km of rural routes in the Free State Province of South Africa. Figure 14.7 gives an indication of the condition of some of the roads included in this study. Potholes were plentiful, but mostly grouped into short zones, separated by long stretches with decent pavement quality. Some potholes extended over the full road width, example in Figure 14.7(a), while other potholes were limited to the edge of one side of the road (Figure 14.7(b)). The need to slow down significantly at the type of potholes indicated in Figure 14.7(a) is obvious, and slower speeds are clearly reported by FCD. The study found that FCD-reported speeds were also significantly impacted by more moderate potholes, such as those indicated in Figure 14.7(b), which bodes well for pothole detection from FCD.

Figure 14.8 indicates the impact of the location of potholes (indicated by the orange vertical lines) on the harmonic mean speed profile reported by TomTom FCD. It is clear that vehicle operating speed reduced more in zones where there were a number of potholes. Further analysis indicated that the relationship between the number of potholes per kilometer and the speed reduction from the speed limit (dashed red line) is statistically significant at the $p < .05$ level ($p < .001$). The more potholes there are per segment, the more operational speed is impacted, and this relationship was clearly evident from commercial FCD.

Figure 14.7   Severe (a) and more moderate potholes (b) along rural roads in South Africa

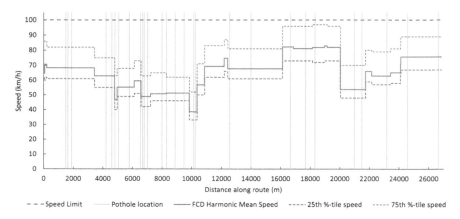

Figure 14.8   FCD speed variation compared to pothole occurrence

A simple model was developed to detect severe and moderate (or widely spaced) potholes using commercial FCD speeds as the only input. Initial results were promising: the model was able to correctly identify 96% of potholed road segments and 85% of the routes were correctly categorized as either having or not having potholes.

Further research is needed to test the applicability of the model over a wider area, and to include machine learning to improve pothole detection. The application of FCD to detect potholes proves the extensive benefits that can be derived from commercial FCD in the LMIC context, which can be developed specifically for the user needs in these countries.

14.5   The future of commercial FCD in LMICs

14.5.1   Increasing FCD penetration rates

The availability of FCD has grown substantially in recent years. Figure 14.9 describes the increasing availability of FCD by presenting the number of kilometers (in billions of kilometers) of probe reported data that are obtained by TomTom per week. The average kilometers per week in each month is indicated in the figure. In January 2017, an average of 8.8 billion kilometers of probe data were reported per week, growing to nearly 44.6 billion kilometers per week by January 2024 (an average growth of approximately 30% per annum). The cyclical nature of trip generation is evident in Figure 14.9 – decreased kilometers of FCD are generated in the winter months of the Northern Hemisphere due to lower numbers of trips made and distances travelled in colder months. The impact of COVID-19 on travel behavior throughout 2020 is also clearly visible in Figure 14.9.

The growing availability of FCD means that FCD penetration rates are increasing rapidly in large parts of the world, especially in Europe and the United States. For example, typical penetration rates in the Netherlands for TomTom are measured to be around 26% on highways and 18% on other roads when compared

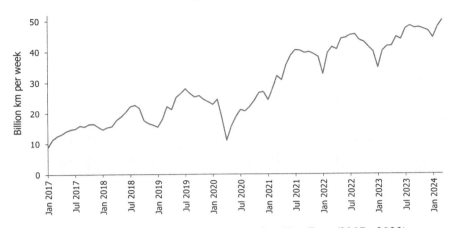

Figure 14.9   Growth in FCD reported to TomTom (2017–2023)

to loop detector data [14]. In the United States, penetration rates have been estimated to be as high as 40% to 50% on certain highways (estimated from traffic volumes calculated using the fundamental diagram and FCD-reported speed). These high penetration rates enable FCD-based volume estimation to be realized. FCD-based volume estimation is a significant use case of FCD because of the importance of traffic volumes for transportation planning, infrastructure planning, and traffic impact analyses on topics such as emissions and economic evaluation of transport projects.

In addition to the growth in the number of kilometers of FCD that are being reported, the geographical coverage of FCD is increasing steadily. As the majority of HICs have good FCD coverage, commercial FCD entities are now focusing on growth in the LMIC sector. Penetration rates are not yet at levels observed in the global north (for example, 2022 FCD penetration varied between 8% and 11% on urban freeways in South Africa [15]), however the FCD available in LMICs is of sufficient quality for accurate speed and travel time estimation. The growth of additional FCD penetration in these areas will continue to unlock better traffic state estimations in LMICs.

14.5.2   Extended FCD

Additional to the growth in the availability of commercial FCD, in recent years commercial parties have started to collect so-called extended FCD (xFCD). xFCD entails any data being collected in addition to standard FCD reported by a GNSS probe. Examples of xFCD include: (1) usage information from navigation applications, such as routing requests and application usage; (2) active community input, for instance the reporting of traffic events and infrastructure damage such as potholes; (3) sensor-derived observations from modern cars equipped with road sensing, including distance to lead vehicle, accelerations and sudden braking events, and traffic sign and lane marking observation data.

When combined with GNSS probe based FCD, the distance to lead vehicle measurements can assist in improving the positioning estimations of the floating vehicle, allowing for more accurate inferences of various traffic state characteristics, such as speed estimation, map making, and real-time vehicle positioning. This accurate positioning will enable traffic state estimations on lane-level rather than road-level, which is currently the level of accuracy that GNSS FCD can deliver, such as lane-level speed estimations and lane closure detections.

14.5.3   Stimulating FCD availability and usage in LMICs

Growth in the use of commercial FCD in LMICs will benefit from a wider distribution of FCD throughout LMICs. It is also necessary to conduct research and commercial projects using FCD, and then disseminate information about successful examples of the implementation of this data in transport planning and traffic engineering applications. This concluding section considers three aspects that will stimulate the generation and use of commercial FCD in LMICs: embracing digitalization of the transport sector, methods to encourage commercial FCD providers to invest in expanding their FCD sources in LMICs, and research into commercial FCD use-cases in LMICs.

Digitalization of the transport sector is accelerating; however, the process is often organic and undirected, resulting from external pressures from commercial digital role players, particularly in LMICs. For optimal digitalization that is planned and well implemented, it is important to have governmental input and oversight of digitalization programs [19]. A top-down approach, with governmental facilitation, would encourage development of the necessary platforms to support digitalization through the centralization of planning for digital architecture. Planning authorities should establish digitalization administration departments that coordinate digitalization processes [19]. Protection of personal information and data sharing policies are also required, which need to be overseen by governmental digitalization officers.

Geographical coverage of FCD in LMICs is steadily increasing, however FCD is still not available in many LMICs, particularly throughout Africa, the Middle and Far East, as well as less developed countries in South America. FCD in the majority of LMICs where FCD are available may be sparse and typically have lower penetration rates than FCD in HICs. A number of commercial FCD entities are promoting the growth of their data through the LMIC sector. Public private participation (PPP) may offer opportunities for expedience of digitalization and expansion of FCD probe sources of commercial FCD in LMICs.

The usefulness and viability of commercial FCD in LMICs should be extensively researched and examples of use cases disseminated through the academic and commercial sector. Insufficient studies of the accuracy of commercial FCD in LMICs exist, in comparison to extensive research in HICs [25]. Additionally, studies of use cases of traffic state estimation, and particularly reporting of movement behavior using FCD have been lagging in LMICs. Further research will assist transport practitioners to understand the application environment of commercial FCD, giving ideas for effective application, and allowing confidence of applying commercial FCD in appropriate application areas.

14.5.4 Future FCD use cases

FCD applications form part of a new generation of data-driven software solutions for traffic management, monitoring and transport planning. Artificial intelligence (AI) platforms with input from FCD will, in the future, be able to generate, analyze, predict, and optimize traffic flows across entire city regions in seconds. These goals are achieved by using data from floating probes and traffic networks, with no need for expensive hardware or infrastructure modifications.

AI platforms, predominantly empowered by FCD, will present a cost-effective avenue for integrating data, viewing network analytics, monitoring traffic flows, optimizing multi-modal traffic signals, and conducting on-demand transportation project evaluations.

14.6 Conclusion

FCD are collected along the trajectory of vehicles within the traffic stream. FCD are widely available because data collection is not limited by static sensor deployment. FCD are collected by devices already widely carried by the public while traveling (including smartphones and onboard navigation devices) and make use of existing communications and internet connectivity. This chapter described the impact that FCD is having on traffic state reporting, particularly in the LMIC context, where digitalization is promoting leapfrogging of legacy traffic sensor systems, which are limited in distribution over the extensive road networks of LMICs.

Traffic state information reported by commercial FCD are aggregated over pre-defined road segments and for specified timeframes to ensure that the data are anonymized according to international privacy protection laws. FCD-reported traffic state data include travel time, harmonic mean speed, and percentile speeds per route segment. Additionally, behavioral traffic data are provided by commercial FCD, indicating traffic routing reported as origin-destination matrices and the percentage of vehicles using a particular route between two zones. FCD are widely used to enable Intelligent Transport Systems in LMICs. Examples were provided of the importance of FCD in freeway managements systems, incident detection, public transport management, parking management, dynamic intersection control, and detection of damage to road surfacing.

FCD are becoming more widespread in LMICs due to inclusion by multinational, commercial FCD providers. FCD collection in LMICs is dominated by GNSS-based mobile phone data. FCD coverage expressed by penetration rate in many middle-income countries is equivalent to many HICs. For example, penetration rates of up to 14% are observed in South Africa and Egypt, providing adequate quality of both speed and trip routing data. Penetration rates are often lower in low-income countries, but are still able to provide good quality speed data.

Growth in the use of commercial FCD in LMICs will benefit from top-down governmental input and oversight of digitalization programs. It is also necessary to continue research efforts and commercial projects using FCD, and then disseminate information about successful examples of the implementation of this data in transport planning and traffic engineering applications in LMICs to ensure the widespread benefits that can be drawn from this passively and widely collected source of traffic data.

References

[1]   Herrera J., and Bayen, A. 'Incorporation of Lagrangian Measurements in Freeway Traffic State Estimation'. *Transportation Research Part B.* 2010;**44**:460–481.

[2]   Garber, N., and Hoel, L. *Traffic and Highway Engineering.* 5th Edition, Enhanced SI Edition. Boston, Massachusetts: Cengage Learning.

[3]   Lovisari, E., Canudas De Wit, C., and Kibangou, A. 'Density/Flow Reconstruction via Heterogeneous Sources and Optimal Sensor Placement in Road Networks'. *Transportation Research Part C.* 2016;**69**:451–476.

[4]   Ahsani, V., Amin-Naseri, M., Knickerbocker, S., and Sharma, A. 'Quantitative Analysis of Probe Data Characteristics: Coverage, Speed Bias and Congestion Detection Precision'. *Journal of Intelligent Transportation Systems.* 2019;**23**(2):103–119.

[5]   Di Lorenzo, G., Sbodio, M., Calabrese, F., Berlingerio, M., Pinelli F., and Nair, R. 'AllAboard: Vision Exploration of Cellphone Mobility Data to Optimise Public Transport'. *IEEE Transactions on Visualization and Computer Graphics.* 2016;**22**(2):1036–1050.

[6]   Van Erp, P. 'Relative Flow Data: New Opportunities for Traffic State Estimation'. PhD Thesis: TRAIL Research School, Delft University of Technology, Delft, 2020.

[7]   Verendel, V., and Yeh, S. 'Measuring Traffic in Cities through a Large-Scale Online Platform'. *Journal of Big Data Analytics in Transportation.* 2019; **1**:161–173.

[8]   Van der Loop, H., Kouwenhoven, M., Van Bekkum, P., Meurs, H., and Kijk in de Vegte, N. 'Validation and usability of floating car data for transportation policy research' *Proceedings of the World Conference on Transport Research*; Mumbai, India, May 2019.

[9]   Braun, M., Kunkler, J., and Kellner, F. 'Towards Sustainable Cities: Utilizing Floating Car Data to Support Location-Based Road Network Performance Measurements'. *Sustainability.* 2020;**12**(8145).

[10]   Graser, A., Dragaschnig, M., Ponweiser, W., Koller, H., Marcinek, M., and Widhalm, P. 'FCD in the real world – system capabilities and applications'. *Proceedings of the 19th ITS World Congress*; Vienna, Austria, 2012.

[11]   Calabrese, F., Ferrari, L., and Blondel, V. 'Urban Sensing Using Mobile Phone Network Data: A Survey of Research'. *ACM Computing Surveys.* 2014;**47**(2):1–21.

[12]   Treiber M., and Kesting, A. *Traffic Flow Dynamics: Data, Models and Simulation.* Heidelberg: Springer; 2013.

[13]   Young, S., Juster, R., and Kaushnik, K. 'Traffic Message Channel Codes: Impact and Use within the I-95 Corridor Coalition's Vehicle Probe Project'. I-95 Corridor Coalition, 2015.

[14]   Eisinga, K., and Lorkowski, S. 'Network-Wide Traffic Volume Estimation Based on Probe Vehicle Data'. *Proceedings of the 103rd Transportation Research Board 103rd Annual Meeting*; Washington, DC, January 2024.

[15] Bruwer, M.M., Walker, I., and Andersen, S.J. 'The Impact of Probe sample bias on the accuracy of commercial floating car data speeds'. *Transportation Planning and Technology*. 2022;**45**(8):611–628.

[16] Uong, L., Adu-Gyamfi, Y., and Zhao, M. 'Machine Learning Framework for Improving Accuracy of Probe Speed Data'. *ASCE-ASME Journal of Risk and Uncertainty in Engineering Systems, Part A: Civil Engineering*. 2021;**7**(2).

[17] Dai, X., Ferman, M. and Roesser, R. 'A simulation evaluation of a real-time traffic information system using probe vehicles'. *Proceedings of the 2003 IEEE International Conference on Intelligent Transportation Systems*, Shanghai, China, 2003.

[18] Van Erp, P., Knoop, V., and Hoogendoorn, S. 'Macroscopic Traffic State Estimation: Understanding Traffic Sensing Data-Based Estimation Errors'. *Journal of Advanced Transportation*. 2017;**2017**:1–11.

[19] Creutzig, F., Franzen, M., Moeckel, R., *et al.* 'Leveraging digitalization for sustainability in urban transport'. *Global Sustainability*. 2019;**2**:1–6.

[20] Kene-Okafor, T. 'Uber hits 1 billion rides in Africa'. 2022. [Online]. Available: https://techcrunch.com/2022/05/24/uber-hits1-billion-rides-in-africa//. [Accessed 9 October 2022].

[21] Muni, K., Kobusingye, O., Mock, C., Hughes, J., Hurvitz, P., and Guthrie, B. 'Motorcycle taxi programme is associated with reduced risk of road traffic crash among motorcycle taxi drivers in Kampala, Uganda'. *International Journal of Injury Control and Safety Promotion*. 2019;**26**(3):294–301.

[22] Milusheva, S., Lewin, A., Begazo Gomez, T., Matekenya, D., and Reid, K. 'Challenges and opportunities in accessing mobile phone data for COVID-19 response in developing countries'. *Data and Policy*. 2021;**3**(e20):1–20.

[23] Wesolowski, A., Eagle, N., Noor, A., Snow, R., and Buckee, C. 'The impact of biases in mobile phone ownership on estimates of human mobility'. *Journal of the Royal Society Interface*. 2013;**10**:1–8.

[24] Erikson, S. 'Cell Phones ≠ Self and other problems with Big Data detection and containment during epidemics'. *Medical Anthropology Quarterly*. 2018; **32**(3):315–339.

[25] Bruwer, M.M., Behrens, R., and Andersen, S.J. 'Commercial floating car data application in sub-Saharan African transport planning contexts: a critical review and research agenda'. *Scientific African*. 2023;20.

[26] Stats S.A. 'Gender Series Volume VIII: Gender patterns in Transport, 2013–2020'. Statistics South Africa, Pretoria, 2021.

[27] Cervero, R. 'Linking urban transport and land use in developing countries'. *The Journal of Transport and Land Use*. 2013;**6**(1):7–24.

[28] Wang, X., Jerome, Z., Wang, Z., *et al.* 'Traffic light optimization with low penetration rate vehicle trajectory data'. *Nature Communications*, 2024; **15**:1–14.

Chapter 15

European ITS Platform: evaluating the benefits and impacts of ITS corridors

Orestis Giamarelos[1] and Daniel Cullern[2]

15.1   The European ITS Platform (EU EIP)

The European ITS Platform (EU EIP) project was a platform co-funded by the European Commission which included a broad range of study and harmonisation activities in the field of ITS. The platform, coordinated by the Italian Ministry of Infrastructure and Transport, ran from 2016 to 2021 and involved an extensive list of European ITS stakeholders, member states and ITS experts cooperating in an open forum and working towards strategy and policy objectives for the betterment of ITS services along core European road corridors. This project was the follow up of projects previously supported by the European Commission (EC) TEN-T programme, namely "European ITS Platform (EIP)" (2013–2015) and "European ITS Platform+ (EIP+)" (2014–2015).

Prior to the launch of EU EIP, DG-MOVE, CINEA and European road authorities had been working together for several years to implement ITS systems and services in response to the ITS Directive 2010/40/EU, in particular the wider deployment of Safety-Related Traffic Information (SRTI) and Real-Time Traffic Information (RTTI), better traffic management and services relating to the freight industry. The implementation of these services was primarily through five related multi-member state ITS Road Corridors: Arc Atlantique, Crocodile, MedTIS, NEXT-ITS and URSA MAJOR. These Works Projects were co-funded by the European Commission within the CEF Map ITS Call 2014 and each of them corresponded to defined stretches of the trans-European Core and Comprehensive networks which also largely coincided with the CEF Multimodal Core Network Corridors (CNCs). In the course of EU EIP, the corridor projects went through three phases with partly different objectives and different Member States involved. In total, 24 countries participated in EU EIP and/or any of the three phases of the five ITS road corridor projects, leading to a combined practically pan-European

[1]Federal Highway Research Institute (BASt), Section Traffic Management and Road Maintenance Services, Bergisch Gladbach, Germany
[2]WSP for National Highways, Department Intelligent Infrastructure, London, UK

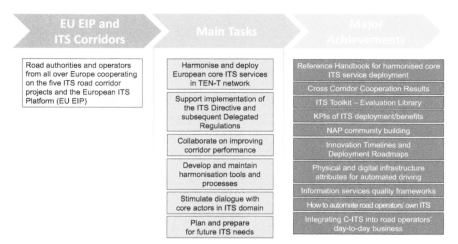

Figure 15.1   Overview of the main EU EIP tasks and achievements

coverage. By monitoring, processing, evaluating and disseminating results delivered by the ITS corridor projects, EU EIP could be considered as the technical European ITS "Knowledge Management Centre", contributing significantly to the most effective use of ITS standards and specifications.

EU EIP incorporated numerous working groups on a wide range of ITS and C-ITS related objectives. Figure 15.1 depicts its main tasks and most significant achievements. Further information on the achievements of EU EIP can be found on the EU EIP website [1]. In addition, as a result of the cooperation of EU EIP with CEDR and ASECAP, key results achieved in this action were collected and released as a comprehensive CEDR project report "Intelligent Transport Systems for Safe, Green and Efficient Traffic on the European Road Network – Findings from the European ITS Platform" [2], addressing various topics relevant for the digitalisation and management of the European road network.

The focus of this chapter is the EU EIP Evaluation Group, which undertook the task of evaluating the results provided by the five ITS Road Corridor projects, in order to consolidate a harmonised and substantiated impact evaluation of the socio-economic benefits of ITS services.

15.2   The EU EIP Evaluation Group

Over the course of the EU EIP programme, the EU EIP Evaluation Group, which comprised experts from the EU EIP Member States and CEF-funded ITS corridors, actively worked together to harmonise a common approach for the evaluation of co-funded ITS projects which could yield consistent and comparable results and assessment of benefits.

Using expertise from within the European operator and supplier communities, EU EIP developed and compiled supporting guidance and advice to assist road authorities and operators to evaluate the beneficial impacts of ITS implementation on road efficiency, safety and the environment in support of European policy objectives for transport.

To this end, the EU EIP Evaluation Group developed a fully adopted suite of tools to harmonise an approach to evaluation within the ITS corridors and, on that basis, to demonstrate the significant benefits of EU co-funded ITS projects targeting mobility inefficiencies, improved safety and reduced environmental pollution on the trans-European road network.

The EU EIP evaluation approach was developed in conjunction with the ITS corridors and subsequently adopted by them, which in turn yielded consistent results which could demonstrate the significant benefits of the ITS implementations undertaken within the ITS corridors.

15.3   The EU EIP evaluation approach

Integral to the EU EIP evaluation approach was the development and adoption of common **KPI Definitions**. These KPI Definitions were developed and agreed at the outset of the programme through extensive consultation with Member State experts, ITS corridors and DG MOVE with the intent of providing a single convenient and practical reference point for ITS evaluators. The KPI Definitions incorporated a range of pertinent Deployment and Impact KPI definitions along with suggested innovative estimation methods for the calculation of the corresponding benefits. The EU EIP KPI definitions were developed in parallel to (and were explicitly cross-referenced with) the high-level ITS KPI definitions developed by DG MOVE for the purpose of Member State reporting on the ITS Directive. The EU EIP KPI Definitions were significantly more detailed and granular, but they were nonetheless consistent with the DG MOVE KPIs.

The **EU EIP Deployment KPIs** relate to the length and percentage of road network coverage, or number of nodes enhanced by the provision of the following ITS services:

Incident Detection and Incident Management

Automated Speed Detection

Dynamic Public Transport Traveller Information

Travel Condition and Travel Time Information Service

Adaptive Traffic Control or Prioritisation

Intelligent Vehicles

Intelligent Services in accordance to Delegated Regulations under the ITS Directive

Speed Limit Information

Variable Speed Limits

Forecast and Real Time Event Information

Weather Information Service

Co-Modal Traveller Information

Dynamic Lane Management

Hard Shoulder Running

HGV Overtaking Ban

Traffic Management Plan Service for Corridors and Networks

Dynamic Information on Intelligent Truck Parking.

Ramp Metering

DATEX II Data Exchange Services

The **EU EIP benefit KPIs** represent the benefit gained in various aspects and they are described in Table 15.1.

Table 15.1   Overview of the EU EIP benefit KPIs

Benefit KPI	Description
Change in traffic flow	Change in traffic flow measured at specific locations of the road network affected by the implementation of the relevant ITS system.
Change in road traffic journey time variability	Change in journey time variability as measured coefficient of variation. Change in coefficient of variation measured along the road network affected by the implementation of the relevant ITS system.
Change in bottleneck congestion	Total delay or vehicle hours lost expressed as the difference between the total time spent and a weighted reference journey time.
Change in journey time	Indicator to determine the impact of a measure on the road users along the road network affected by the implementation of the ITS system.
Change in demand for travel	Change in vehicle-km travelled (flow) measured along the road network affected by the implementation of the relevant ITS system.
Change in mode share	Change in mode share (% mode share points) on corridors where ITS has been implemented.
Change in accident numbers and severity	Absolute and % change in number of reported accidents of all severities as well as accident rates (i.e. accidents per vehicle-km travelled) measured along the road network affected by the implementation of the relevant ITS system.
Change in CO_2 emissions	Change in annual CO_2 emissions (tons) (absolute and % difference) measured along the road network affected by the implementation of the relevant ITS system.
Public eCall timeliness	Total time taken between accident occurrence to initiation of public eCall (112), to the presentation of the content of MSD in the Public Safety Answering Point.
Change in noise pollution	Noise reduction impacts generated by ITS are considered negligible.

Further information and examples of the calculation of the benefit KPIs can be found in the ITS Deployment and Benefit KPI Definitions – Technical Reference document [3], available on the EU EIP evaluation website [4].

The EU EIP website promoted the results of the EU EIP Evaluation Group and also incorporated an extensive **Evaluation Library** [5], which served as a single access point for all EU EIP Evaluation reference and guidance materials, as well as being a repository for all Evaluation Reports from the ITS corridors and an archive of pre-CEF ITS Evaluation Reports and support materials. The Evaluation Library has remained accessible beyond the end of the project and continues to be a rich source of information on ITS evaluation, containing all related ITS evaluation reports made available before the end of the project (end of 2021) (Figure 15.2).

A common **Evaluation Report Template** [6] with integral guidance was also developed. The report template was consistent with, and structured in line with, the EU EIP KPI Definitions, both of which were promoted across the EU EIP and ITS corridor community and made publicly available from the Evaluation Library.

Moreover, the **EU EIP Evaluation Toolkit** was an extensive database of key meta-data derived from available corridor evaluation reports. The Toolkit was a "live", publicly accessible online tool for filtering ITS Evaluation results from individual ITS implementations by six key criteria: deployment KPI, benefit KPI, location, corridor, related ITS directive priority area and ITS Directive priority action. It was updated in regular intervals and was operational until the end of the project. The online Toolkit interface enabled users to directly locate relevant ITS evaluation reports and extract results for the purpose of research and supporting decision making.

Finally, another result of the EU EIP Evaluation Group was the **Report on the socio-economic assessment of the priority actions of the ITS Directive** [7]. For this purpose, a methodology and process were developed for an inventory of relevant ITS projects in Europe. This included the early establishment of an evaluation plan to assess the impacts, benefits and costs of ITS investments using European KPIs.

Figure 15.2   The EU EIP Evaluation Library

A workshop held in Stockholm presented initial results and proposals for assessing ITS Directive priority actions, gathering input to refine the assessment that followed. To facilitate the assessment, over 500 projects from 22 national progress reports were initially recorded in an Excel tool, with partners selecting 2–3 most relevant projects from each country. The feedback from the workshop guided the ongoing activities and provided clarity for completing the assessment, which focused on assessing the deployment of services related to the ITS Directive. A subsequent online workshop aimed to refine the assessment results and proposals for priority service coverage. The final socio-economic assessment report was based on the review of twelve national reports and identified deployment and coverage of five priority services and their associated benefits. In addition, the report assessed impacts and recommended KPIs for each service in the context of the operating environments defined by the EasyWay Programme, a precursor of EU EIP.

It is noteworthy that the EU EIP Evaluation Group benefited from strong, regular and direct ITS corridor engagement throughout the EU EIP programme, and the ITS corridors, in turn, were integral to the development of the common approach defined in the EU EIP support materials. This facilitated the full adoption of the harmonised EU EIP evaluation approach, as the ITS corridors aligned their evaluation reporting with the KPI definitions, adopted the use of the common report template and transferred their results to the ITS Toolkit and reports to the EU EIP evaluation library.

15.4   ITS corridor evaluation results

This section presents the five ITS corridors, including their objectives and a summary of the estimated impacts of their last two phases, based on their evaluation reports as well as their contributions and close cooperation with the EU EIP evaluation activity, facilitated by the corridor coordinators. An attempt to estimate global benefits, based on the combination of the impacts reported by each ITS corridor, is presented in the conclusion of this chapter.

The results of the ITS corridors presented in this section have collectively been reported in EU EIP project reports and deliverables: The "EU EIP Evaluation Interim Report 1" [8] included the results of the second phase of the ITS corridors. This report was complemented by the "EU EIP Evaluation Final Report" [9] published three years later, which added the results from their third phase.

Especially, the results of the second phase of the ITS corridors constituted a primary focus of the dissemination activities of EU EIP: A collection of estimated impacts of ITS deployments based on these results was presented in the book "Digitalisation of Road Transport in Europe" [10], published by EU EIP. Its official release was accompanied by a webinar [11], as part of the EU EIP Web ITS Forum [12], which consisted of presentations from the five ITS corridors and a panel discussion. Finally, a dedicated chapter on the EU EIP evaluation activities and the results of the second phase of the ITS corridors was part of the previously mentioned CEDR Project Report [2] containing the main findings from EU EIP.

15.4.1 Arc Atlantique

The Arc Atlantique ITS corridor aimed to accelerate the deployment of traditional and innovative ITS systems and services on the core and comprehensive networks with the key objectives of: (1) increasing the efficiency of the trans-European road network; (2) improving safety; and (3) improving environmental performance. Its focus was to deploy ITS-enabled traffic management and safety solutions in known problem areas, such as bottlenecks and routes with chronic and acute congestion which cause increased transport costs, pollution and were often associated with an unsatisfactory safety record.

These objectives were in line with the EU's policy objectives to reduce the overall cost of transport for the economic benefit of the Union, to reduce societal impacts by improving the safety record of the network, to improve air quality and to contribute to the implementation of the Paris Agreement on climate change. In addition, through having a positive impact on congestion, particularly at bottlenecks, the ability for the Union to deliver goods and services more effectively supported transport cohesiveness, economic vitality and wellbeing. Furthermore, it extended and built digital communications and cross-border cooperation through the implementation of harmonised systems and services such as RTTI and SRTI whilst contributing to multimodal information via National Access Points. These were implemented in accordance with applicable European Regulations and assisted Member States in meeting their obligations under the ITS Directive.

The Arc Atlantique ITS corridor network was largely aligned with the North-Sea Mediterranean and Atlantic Core Network Corridors. Together, the corridor partners worked towards improved multimodal transport links across the western reaches of the Union for which the Arc Atlantique ITS corridor deployed technology and digital services on the road network. The partnership was made up of public highway authorities or concessionaires operating on behalf of public authorities from Ireland, United Kingdom, the Netherlands, Belgium, France, Spain and Portugal.

The Arc Atlantique was implemented over three phases and completed its work in 2021. A few examples of its main achievements in each phase are provided below:

1. During Arc Atlantique Phase I, 22 RTTI schemes were implemented on the corridor, benefitting 19,000 km of network. The work included upgrades to traffic management centres and new digital communications. For the same period, 19 projects concerning SRTI were implemented, benefiting 7600 km of the network. Furthermore, the network benefited from new and improved services in co-modal information, truck-parking and the roll out of DATEX II.
2. The Arc Atlantique Phase II implemented a further 36 ITS projects designed to deliver enhancements in traffic and congestion management and safety and environmental improvement at specific locations on the network.
3. The improvements delivered to users and operators during Arc Atlantique Phase III related to significantly enhanced levels of service through the deployment of ITS on over 16,000 km of the network, with over 60 projects aimed at improving traffic management and traffic information services.

Using the available evaluation results and modelling, the impact of the new and improved ITS systems and services deployed on the completed Arc Atlantique two network over a period of five years was estimated. The expected savings were calculated as follows: (1) 236 minor injuries saved per year (1,180 minor injuries over five years); (2) 28 seriously injured saved per year (140 seriously injured over five years); (3) 11 fatalities saved per year (55 fatalities over five years).

Applying these different realistic hypotheses and taking into account the level of investment, the Arc Atlantique II Programme as a whole delivered minimum safety socio-economic savings of €36 million per year and a projected Return On Investment (ROI) of approximately three years.

In Arc Atlantique III, the operational and practical impact of most (70%) of the measures deployed was to reduce response times in the case of an event or accident, thus reducing the occurrence of secondary accidents. The evaluation built on the evaluation of the second phase and proceeded with a more conservative estimate of the impact of the measures included in Arc Atlantique III, reasonably expecting a lower impact on average. The results of the evaluation were as follows:

1. Safety benefits: Approximately 76 accidents with three fatalities, 13 serious injuries and 100 minor injuries saved per year.
2. Congestion benefits: Approximately 236,000 vehicle hours lost, 980,000 litres of fuel and 2640 tons CO_2 saved per year.

The resulting annual socio-economic savings are estimated at €25.8 million (€19.4 million in safety savings and €6.4 million in congestion savings). With a total investment cost of €65 million, the ROI is calculated at approximately 2.5 years.

More detailed information on the evaluation of the second phase and the third phase of Arc Atlantique is provided in the *EU EIP Activity 5: Evaluation – Final Report* [9].

15.4.2   MedTIS

The main objective of the MedTIS ITS corridor was to foster the implementation of ITS for better traffic management services and better traffic information services on the core and comprehensive networks of the Mediterranean Corridor. By contributing to the evolution of local traffic management modes towards coordinated and cross-border management modes, MedTIS had a key role in improving corridor efficiency in terms of road safety and capacity of the trans-European road network.

By deploying ITS on approximately 9000 km of the network and with a budget of more than €50 million, the MedTIS Programme directly addressed the objectives of the ITS Directive to smooth road traffic on major networks while improving user safety and environmental performance, in line with the EU's objectives of decreasing societal costs by reducing road casualties and improving air quality.

Bringing together almost 30 public and private road operators from the Member States of Italy, France and the Iberian Peninsula, MedTIS contributed to the improved management of a strategic axis serving several major ports such as

Venice, Genoa, Marseille, Barcelona and Valencia. As such, it addressed areas where traffic conditions are often difficult, with high traffic volumes, especially in the summer, heavy truck traffic, large international flows, but also, in some sectors, significant commuter traffic.

The main measures taken to improve control, information and traffic management were: (1) the deployment of automatic event detection equipment; (2) the deployment of the first on-board information systems; (3) the deployment of speed control systems; and (4) the display of information on truck parking occupancy.

One of the key actions of MedTIS was the *cross-border travel time* flagship project, which significantly improved the operational efficiency and the level of service on the France–Spain and France–Italy borders. Through a close collaboration between the three cross-border operators, who developed their exchanges of traffic information and traffic data, and thanks to DATEX II, new border traffic management plans and generalised travel time services were deployed between Spain, France and Italy.

Between 2014 and 2018, the MedTIS Programme deployed 85 projects, which were deployed in response to the following specific challenges encountered on MedTIS II network:

1. improvement of traffic management and reduction of congestion at critical points, such as the approach to major cities, and cross-border areas;
2. improvement of safety in and around tunnels;
3. improvement of the quality of traffic and event information to users, including travel time information.

As a result of the MedTIS Programme investments, new or improved traffic management services covered 6600 km of the corridor, while new or improved traffic information services covered 2300 km of the corridor. This extension of proven traditional systems brought significant benefits to the management and improved level of service of the MedTIS corridor and, through the use of modern telecommunications solutions, would facilitate the introduction of connected and automated vehicles in the future.

Of the 85 projects, ten were evaluated, using ex-post or ex-ante evaluations based on the EU EIP KPIs: (1) change in accident numbers and severity; (2) change in bottleneck congestion; (3) change in CO_2 emissions.

The deployments selected for individual evaluations covered a variety of road configurations (cross-border sections, interurban sections, urban sections, mountainous areas with tunnels) and technologies. As such, the impact on benefit KPIs varied from one deployment to another:

1. The benefits of these projects on congestion and environment ranged from 2% to 10% in terms of hours lost and CO_2 emissions.
2. Concerning safety, the benefits of these projects in terms of reduced accident numbers varied between 2% and 5%.
3. When comparing the costs and benefits of individual deployments, the results were equally satisfactory, with ROI ranging from 2 to 9 years.

Overall, the improvement in congestion levels and road safety was significant. The operational and practical impact of most of the projects deployed by motorway operators was to shorten response times in the event of an incident or accident, and these projects made it possible to reduce the occurrence of secondary accidents. On this basis, the evaluation methodology used by MedTIS showed that the deployments potentially reduced the number of accidents and congestion levels on the relevant sections of the motorway networks by 1.6%.

On the MedTIS II network as a whole, over a period of five years after the deployment of the programme, total projected potential savings were: eight fatalities; 53 serious injuries; 277 slight injuries; 642,000 hours lost; 2.7 million litres of fuel; and 7,200 tons of CO_2. The socio-economic benefits corresponding to these savings, calculated from the figures presented in the *Handbook on External costs of Transport on Road Safety* by the European Commission [8], amounted to €9.42 million per year. Therefore, for the investment programme deployed in MedTIS, the estimated ROI was approximately five years. It should be noted that these results were derived from the limited selection of deployment projects within the programme, and as such they should be considered as minimum results.

Finally, the evaluation of the third phase of MedTIS (MedTIS III) using the same methodology produced similar results regarding the socio-economic benefits of the programme. Specifically, the following potential annual savings were estimated: approximately 50 accidents resulting in 75 injuries and two fatalities; 130,000 hours lost; 540,000 litres of fuel; and 1,460 tons of CO_2. Using the same figures from [13], as well as market prices for fuel and CO_2, these savings account for a benefit of approximately €13.8 million. Therefore, with a total deployment cost of approximately €68 million, the ROI for MedTIS III is estimated to be slightly less than five years. Further information on the evaluation methodology can be found in the MedTIS III Corridor Evaluation Report [14].

15.4.3 NEXT-ITS

The NEXT-ITS II corridor formed the Northern part of the Scandinavian–Mediterranean Corridor. The corridor connected Northern Europe with Western and Southern European transport networks and offered the primary road transport connections between Western/Central Europe and Norway and the St. Petersburg region of Russia. Increased traffic load and the extensive presence of trucks had made the NEXT-ITS corridor and core network vulnerable to disturbances. Furthermore, the road network of the sparsely populated areas of Northern Europe offered limited possibilities for alternative routes and large parts of the network were subject to recurring hard weather conditions, particularly in wintertime.

The main objective of the NEXT-ITS II ITS corridor was to improve network performance, in terms of efficiency, reliability, safety and environmental impact, of the Northern part of the Scandinavian-Mediterranean CEF corridor from Oslo and the Finnish-Russian border in the north via Copenhagen, Hamburg and Bremen to Hannover in Germany. Cross-border continuity of traffic management services were targeted through their coordinated deployment as well as through major upgrades of traffic management centres.

The measures included in NEXT-ITS II aimed to fill gaps concerning coverage, accessibility, dissemination, quality and content of the core traffic management services, as well as to improve the cost-efficiency in the operation of traffic management. The following ITS deployment projects were completed during NEXT-ITS II: (1) implementation and upgrade of Traffic Management Centres; (2) development and implementation of traffic management plans; (3) update of roadside control software to enable service integration; (4) implementation and update of roadside information panels for driver information and control and (5) data fusion and data quality control at traffic management centres.

The measures of NEXT-ITS II mainly concerned the northern part of the Scandinavian-Mediterranean Corridor, but also influenced the adjacent road network to the corridor and, especially where general improvements and enhancements of traffic centres were carried out, larger parts of the main road network. NEXT-ITS II therefore estimated the impacts on the network affected by the services deployed. The impacts were not limited to the NEXT-ITS corridor, but were the total estimated impacts on the affected network, as the corridor included a number of deployments related to upgrades of central systems in traffic management centres, which in reality affected a larger network than just the corridor itself.

The cost calculation included all costs related to those measures that were fully deployed during NEXT-ITS II. For a five-year period, the estimated implementation costs of all NEXT-ITS II deployment measures were approximately €33 million (incl. VAT), and the annual operation and maintenance costs were approximately €3 million, resulting in €15 million for five years. Therefore, the total cost for five years was approximately €48 million.

The NEXT-ITS II evaluation focused on estimating the "average" total annual benefits. The evaluation work did not include an attempt to estimate the level of minimum benefits or a range of impacts. Instead, the work focused on carrying out a socio-economic assessment based on the "average impact per year" and subsequent sensitivity analyses. The five-year savings of NEXT-ITS II deployments were: 2,035,000 vehicle hours driven; 571,000 vehicle hours spent in congestion; 0.5 fatalities/fatal accidents; 10.8 non-fatal injuries/injury accidents; 45,600 tons CO_2 emissions.

Overall, the estimated main impacts of NEXT-ITS II measures were particularly evident in improved traffic flow, as indicated by the KPIs vehicle hours driven (reduced by 400,000 vehicle hours per year), and vehicle hours spent in congestion (reduced by 114,000 vehicle hours per year). This was a result of the deployed measures, which were mainly aimed at improving traffic and incident management, and supporting it with improved traffic information. In addition, NEXT-ITS II deployments resulted in an annual reduction of 9,000 tons of CO_2 emissions while a very conservative estimate of the safety benefits was an annual reduction of two serious accidents.

Even with these conservative estimates, the total value of the annual benefits in 2017 was calculated to be approximately €12 million. This represents a ROI of approximately four years when compared to the total cost of implementation and annual operation and maintenance for five years.

The evaluation of NEXT-ITS III proved to provide even higher socio-economic benefits. The estimated annual impacts for NEXT-ITS III on corridor level were the following: 925,000 vehicle hours driven; 166,500 vehicle hours spent in congestion; 0.4 fatalities/fatal accidents; 6.4 non-fatal injuries/injury accidents; 16,200 tons CO_2 emissions.

Based on these figures, the ROI for NEXT-ITS III is estimated to be less than three years. More information on the evaluation of NEXT-ITS III can be found in the NEXT-ITS III Evaluation Report [15].

15.4.4   URSA MAJOR

URSA MAJOR II targeted the deployment of ITS services to improve freight traffic on the TEN-T road network, primarily along the RHINE-ALPINE and the Scandinavian Mediterranean Core network corridor, linking North-Sea ports, the Rhine and Ruhr area, metropolitan areas in southern Germany and in Italy. Parts of the Rhine-Danube core network corridor were also addressed due to important freight traffic routes linking those corridors in the middle of Europe. International freight transport between EU Member States is one of the three main pillars for a Single Europe Economic Area. As such, improving services for international freight traffic along the corridors was the main European Added Value of URSA MAJOR II. Countries involved in the project were Germany, Italy and The Netherlands. Switzerland was an active partner without EU co-funding, while Austria was included in its role of transit country, based on operational agreements for cross-border traffic management plans.

The main objective of the Evaluation activity was the assessment of the overall impact of URSA MAJOR on traffic efficiency, safety and environment, based on the results from ex-post evaluation studies carried out by URSA MAJOR partners. As such, evaluation studies were based on measured real impacts on mobility. Moreover, the comprehensive usage of Floating Car Data complemented the project-wide evaluation and the study was supported by a GIS data tool.

The most significant results from the evaluation studies of the individual projects implemented in URSA MAJOR II (18 studies) are presented below:

Impact on traffic efficiency: The most remarkable impacts were the increase of traffic flow, intended as throughput, with Dynamic Lane Management, +17%/+23%), the reduction of travel time with Dynamic Re-routing and Dynamic Lane Management (770,000 hours per year and 8%/50%), a good percentage of rerouted users with dynamic re-routing (10%/43%), the reduction of vehicle hours lost thanks to traffic monitoring and management (48%/86%) and a good result in congestion cost savings with dynamic re-routing and traffic monitoring and management.

Impact on safety: The analysis on safety reported in evaluated URSA MAJOR projects showed few results related to this area, where the most relevant indicator was the change in the ratio between the number of accidents and the change in traffic flow, which resulted in 7% reduction with an implementation of Traffic Monitoring and Management. Furthermore, a safety campaign on VMS obtained 91% user satisfaction.

Impact on environment: The ITS service that presented more results within the evaluated URSA MAJOR projects is the Dynamic Lane Management, with a reduction in fuel consumption of 28%–55% and a change in fine particle emissions equal to −75%. In a dynamic re-routing application, an annual reduction of 3650 tons of CO_2 was calculated.

Other impact identified: Improvement of the event detection time, which was reduced by 93%/97% in one URSA MAJOR implementation.

The overall impact of URSA MAJOR was estimated based on a combination of the results of the evaluated ITS implementations and comparable available impact data to ensure a more solid statistical basis. The results were expressed through the KPIs defined by DG MOVE, using only those applicable and pertinent to URSA MAJOR implementations. The first step was the calculation of KPIs for each type of ITS service, using combined data from the evaluation studies and from literature. Impact results were then extended to the whole URSA MAJOR corridor using a weighted average of the indicators over the number of implemented projects for each ITS service type.

Table 15.2 represents the assessed average impact along routes where the ITS systems included in the URSA MAJOR project have been implemented.

The estimation model applied to the Italian and German projects on the Corridor allowed an estimation of the following annual savings: 71 accidents; 79 slightly injured persons; 22 seriously injured persons; 2 fatalities.

By converting these benefits into economic value, an annual gain of €11.5 million was estimated. Dividing the total project investment of approximately €46 million by the estimated annual benefit of €11.5 million, the resulting ROI is approximately four years.

The third phase of the URSA MAJOR ITS corridor, URSA MAJOR neo, had two of the projects on the corridor evaluated, producing monetary values as an outcome. A simplified methodology, agreed at EU EIP level, was used to assess the impact of the whole corridor, using indicators provided by the MedTIS corridor. The results of the evaluation were as follows:

1. Safety results: Approximately 100 accidents with 2.5 fatalities, 12 serious injuries and 150 minor injuries saved per year.
2. Congestion results: Approximately 440,000 vehicle hours lost, 95,500 litres of fuel and 617 tons CO_2 saved per year.

The resulting annual socio-economic savings are estimated at €29.4 million. With a total investment cost of €150 million, the ROI is calculated at approximately 5.1 years.

Table 15.2   Assessed average impact of ITS deployments on the URSA MAJOR corridor

Impact area	Benefit KPI along routes with implemented ITS	Change
Traffic efficiency	Change in journey time	−13%
	Change in traffic flow	+9%
Safety	Change in number of accidents	−34%
Environment	Change in annual CO_2 emissions	−22%

More detailed information on the evaluation of URSA MAJOR II and URSA MAJOR neo can be found in the *EU EIP Activity 5: Evaluation – Final Report* [9].

15.4.5   Crocodile

The CROCODILE project, in each of its three phases running from 2013 to 2022, worked on the exchange of accurate, high quality and reliable data between road operators, private stakeholders and administrations, in order to generate information services focused on road safety and truck parking information in particular. In its third phase (CROCODILE III), the project focused on organisational coordination and the implementation of technical standards, as well as the improvement of management strategies and end-user services. The aim was to improve services so that road users can access more comprehensive and reliable information through familiar channels such as websites and apps, thereby contributing to the continuity of services outlined in the EU ITS Directive.

Specifically, the main focus areas of the CROCODILE III corridor were the following:

1. Implementing European legislation, including the EU ITS Directive and its supplementary delegated regulations. This includes ensuring access to data and the set up or improvement of National Access Points (NAP) for data availability and exchange, in accordance to the relevant Delegated Regulations.
2. Implementation of DATEX II nodes: Complementing to the above, completing DATEX II nodes ensures data availability and exchange. CROCODILE III continued the work on the finalisation and harmonisation of DATEX II nodes, building on the results of the previous project phases.
3. Cross-border information services, focusing on end-users: The work continued towards the enhancement of the services (web service and application) based on previous efforts and solutions of CROCODILE. Another focus was on improving the channels to the actual end-users in order to provide them with valuable information of high quality. A prominent example in this area is the development of the cross-border traffic management plans application.
4. Strengthening cooperation among corridor projects, as well as sustaining strategic and technical efforts, with CROCODILE III serving as a prominent stakeholder forum of high relevance and significant visibility.

As the focus of CROCODILE was significantly different from the other corridors, a similar evaluation in terms of socio-economic benefits was not possible. Therefore, in the case of CROCODILE, the impact of the project can be shown indirectly through its significant achievements in the above focus areas. These are provided in published brochures for the second [16] and third [17] phases, while further information and supporting material is available on the dedicated CROCODILE website [18].

15.5   Conclusion – global benefits

In order to estimate the pan-European benefits of the ITS corridor projects, a combination of the reported results from the second and third phase of the ITS corridors is made in an attempt to estimate global results. As already mentioned, these estimated benefits can be considered as minimum global benefits, as they are based on the specific conservative assumptions and the specific aspects evaluated, as described previously in this chapter.

Based on the available results from the second phase of four ITS corridors (Arc Atlantique, MedTIS, NEXT-ITS, URSA MAJOR), pan-corridor **estimated global minimum five-year safety benefits** and **ROI** were calculated.

In terms of minimum **safety** and **socio-economic** savings as a result of combined investment in the ITS corridor programme over a five-year period, it was concluded that at least **75 lives** would be saved, and that at least **2166 injuries** would be prevented.

As a result of these safety benefits alone, it was concluded that the combined ITS corridor investments in the second phase resulted in a significant financial benefit. Based on the investment of €232 million across the four ITS corridors, the following **minimum** savings based on safety impacts alone were calculated: an annual safety benefit saving of at least **€55 million**; a minimum overall ROI of about **four years** on average; a benefit-cost ratio of approximately **3 and higher**, based on an average lifetime of ten years for the related infrastructure components.

The evaluation for the third phase of ITS corridors focused on the benefits of improved safety, such as reducing the number of lives lost and injuries, as well as reducing congestion, including vehicle lost hours, fuel over-consumption and CO_2 over-emissions. The evaluation used data from the same four ITS corridors as in the second phase, also incorporating extrapolated results for the CROCODILE corridor.

The results for the third phase of the ITS corridors over a five-year period were calculated as follows:

1. Safety benefits: At least **40 lives** saved and a minimum of **1846 injuries** prevented (195 serious injuries and 1651 slight injuries)
2. Congestion benefits: A reduction of at least **5,057,475 vehicle hours lost**, a saving of at least **11,995,830 litres of fuel** and a reduction of at least **34,000 tons of CO_2 emissions**.

These results indicate a minimum annual socio-economic benefit of €**79 million**. Considering the consolidated investment cost of €360 million for the third phase of the ITS corridors, an estimated minimum ROI of approximately **4.5 years** was calculated. Therefore, the evaluation of the third phase of the ITS corridors also confirms the significant socio-economic benefits of investing in ITS in Europe and the high ROI that these investments generate.

References

[1] EU EIP. *Achievements of the European ITS Platform* [online]. 2021. Available from https://www.its-platform.eu/achievement [Accessed 15 March 2024].

[2] EU EIP / CEDR / ASECAP. *Intelligent Transport Systems for Safe, Green and Efficient Traffic on the European Road Network – Findings from the European ITS Platform*, CEDR PR2022-01, ISBN: 9791093321615, CEDR, 2022. Available from https://cedr.eu/docs/view/6244782f39e78-en and https://www.its-platform.eu/book [Accessed 15 March 2024].

[3] EU EIP. *ITS Deployment and Benefit KPIs Definitions – Technical Reference* [online]. 2017. Available from https://www.its-platform.eu/wp-content/uploads/ITS-Platform/AchievementsDocuments/Evaluation/Evaluation%20Library/EU%20EIP_Activity%205_WP2_KPI%20Definitions_2017%2002%2008_FINAL.pdf [Accessed 15 March 2024].

[4] EU EIP. *Benefits, KPI & Evaluation* [online]. 2021. Available from https://evaluation.its-platform.eu [Accessed 15 March 2024].

[5] EU EIP. *Evaluation Library* [online]. 2021. Available from https://www.its-platform.eu/EvalLib [Accessed 15 March 2024].

[6] EU EIP. *Evaluation Report Template* [online]. 2017. Available from https://www.its-platform.eu/wp-content/uploads/ITS-Platform/AchievementsDocuments/Evaluation/Evaluation%20Library/EU%20EIP_Evaluation%20Report%20Template%20(v10.0).docx [Accessed 15 March 2024].

[7] EU EIP. *Socio-Economic Assessment of ITS Directive Priority Actions (Focus on Services and Service Deployments)* [online]. 2017. Available from https://www.its-platform.eu/wp-content/uploads/ITS-Platform/Achievements Documents/Evaluation/Evaluation%20Library/EU%20EIP%20Evaluation%20-%20Compilation%20report%20V1.pdf [Accessed 15 March 2024].

[8] EU EIP. *Activity 5: Evaluation – Interim Report 1* [online]. 2019. Available from https://www.its-platform.eu/wp-content/uploads/ITS-Platform/Achievements Documents/Evaluation/Evaluation%20Library/EU%20EIP%20Evaluation%20-%20Interim%20Report%202019%20-%20Final.pdf [Accessed 15 March 2024].

[9] EU EIP. *Activity 5: Evaluation – Final Report* [online]. 2022. Available from https://www.its-platform.eu/wp-content/uploads/ITS-Platform/Achievements Documents/Evaluation/Evaluation%20Library/EU%20EIP%20Evaluation%20-%20Final%20Report%20%28v1.0%20-%20final%29.pdf [Accessed 15 March 2024].

[10] EU EIP. *Digitalisation of Road Transport in Europe – Highlights from the Benefits of ITS Programme Co-funded by CEF*. EU EIP, 2021. ISBN: 9788897212126. Available from https://www.its-platform.eu/digitalisation-book [Accessed 15 March 2024].

[11] EU EIP. *4th Web ITS Forum – Digitalisation of Road Transport in Europe: the Great Progress under the Connecting Europe Facility* [online]. 2021. Available from https://www.its-platform.eu/communicationevents/forum-and-webinars/web-its-forum/4th-web-its-forum-webinar-digitalisation-of-road-transport-in-europe-the-great-progress-under-the-connecting-europe-facility [Accessed 15 March 2024].

[12] EU EIP. *Web ITS Forum* [online]. 2021. Available from https://www.its-platform.eu/communicationevents/forum-and-webinars/web-its-forum [Accessed 15 March 2024].

[13]   European Commission, Directorate-General for Mobility and Transport, Essen, H., Fiorello, D., El Beyrouty, K. et al., *Handbook on the External Costs of Transport – Version 2019 – 1.1* [online], Publications Office, 2020. Available from https://data.europa.eu/doi/10.2832/51388 [Accessed 15 March 2024].

[14]   MedTIS. *MedTIS III Corridor Evaluation Report* [online]. 2022. Available from https://www.its-platform.eu/wp-content/uploads/ITS-Platform/Corridor Documents/MedTis/2016-EU-TM-0588-W_MedTIS%203_Evaluation%20 Report%20VDEF.pdf [Accessed 15 March 2024].

[15]   NEXT-ITS. *NEXT-ITS 3 Corridor Evaluation Report* [online]. 2022. Available from https://www.its-platform.eu/wp-content/uploads/ITS-Platform/ CorridorDocuments/Next-ITS/NEXT-ITS%203%20Evaluation%20Report %20v1.0.pdf [Accessed 15 March 2024].

[16]   CROCODILE. *CROCODILE 2 Final Brochure* [online]. 2019. Available from https://www.its-platform.eu/wp-content/uploads/ITS-Platform/Corridor Documents/Crocodile/CROCODILE%202%20Final%20Brochure.pdf [Accessed 15 March 2024].

[17]   CROCODILE. *CROCODILE 3 Final Brochure* [online]. 2022. Available from https://www.its-platform.eu/wp-content/uploads/2022/12/CROCODILE_ 3_201x210_2022_final.pdf [Accessed 15 March 2024].

[18]   CROCODILE. *Achievements of the CROCODILE Corridor* [online]. 2021. Available from https://crocodile.its-platform.eu [Accessed 15 March 2024].

Part IV

Discussion and conclusions

Chapter 16

Key findings and the future of Intelligent Transport Systems

Meng Lu[1]

16.1 Intelligent transport systems deployment and evaluation

Intelligent transport systems (ITS), based on information and communication technologies (ICT), have been developed and deployed for around four decades. The core technologies in the ITS domain are systems and control, positioning (including relative positioning and absolute positioning), sensing, communications and information processing. Different options can be combined in different ways to create stand-alone (autonomous) in-vehicle systems and cooperative systems using communications between vehicles and with infrastructure [1]. ADAS (advanced driver assistance systems) is a collective name for a whole range of ICT based in-vehicle systems, intended to support the driver in the driving task. ADAS applications are typically meant to improve traffic safety, efficiency, network capacity and comfort of driving, and hold the promise to also improve driver performance (e.g. avoiding or correcting human error). Telematics (telecommunication and informatics) can be defined as provision of information and services, via wireless communications, to and from vehicles and their occupants. Over the past decade, enormous R&D efforts have been made in the field of high-level automated driving systems (ADS) for the future of mobility, focusing on (cooperative and) automated vehicles (AVs).

Cooperative intelligent transport systems (C-ITS) and C-ITS services have been intensively developed with an expectation to bring added value by V2X (vehicle to everything) applications [2]. For instance, in 2005 the European Commission (EC), under the FP6-IST funding scheme, launched three IPs (integrated projects) targeting cooperative systems: SAFESPOT (co-operative systems for road safety "Smart Vehicles on Smart Roads", focusing on the in-vehicle side and traffic safety) [3], CVIS (cooperative vehicle infrastructure systems, focusing on the infrastructure side and traffic efficiency) [4] and COOPERS (CO-OPerative SystEms for Intelligent Road Safety, focusing on the domain of the road operator) [5]. In 2009, the EU-funded project

[1]Aeolix ITS, Utrecht, The Netherlands

FREILOT (urban freight energy efficiency pilot) [6] was launched, which aimed to develop C-ITS services for freight transport. DRIVE C2X [7] substantially contributed to the development and evaluation of V2X communication technologies for accelerating cooperative mobility in Europe. In early 2014, the EC launched a C-ITS deployment platform, to take a more prominent role in the deployment of connected driving. After Phase I (2014–2016), the resulting shared vision on the interoperable deployment of C-ITS towards cooperative, connected and automated mobility in the European Union (EU) was further developed in Phase II (2016–2017) [8–10]. The perspective of the C-ITS platform is that ICT infrastructure-based cooperative, connected and automated transport is an option for enhancing traffic safety, traffic efficiency and energy efficiency, and for reducing fuel consumption. In 2019, the EC set up the cooperative, connected and automated mobility (CCAM) Platform to conduct pre-deployment activities [11]. In 2023, ITS America published its national V2X deployment plan [12].

ITS applications enable safer, more efficient, cleaner, more accessible and more homogenous transport systems. Enabled by the rapid development of technology, the past years saw a proliferation of consumer electronic devices, ubiquity of commercially available wireless coverage, emergence of communications and sensing technologies, and connected vehicles and automated driving going mainstream. It should be emphasised that automated driving can be implemented based on just autonomous systems, while combined autonomous and cooperative systems are an alternative option. Just like the first edition [13], the second edition of the book targets road transport – part of surface transport. The ITS applications for road transport are reaching the stage of large-scale implementation in, for instance, the domains of:

1. private vehicles, e.g. ADAS, telematics, and better information for multi-modal travellers; and vulnerable road users, e.g. two-wheeled motor vehicles, mopeds, scooters, bicycles and pedestrians;
2. public transport, e.g. e-ticketing, multi-model services, user services via data exchange and telecommunications;
3. commercial vehicles and freight transport, e.g. tracing and tracking goods, telematics and document automation in supply chain management, also called e-freight;
4. infrastructure-based traffic management and control, e.g. extended floating car data (xFCD), traffic control centre, traffic information centre, variable message signs (VMS), incident and tunnel management, road pricing and low-emission zones.

Two levels of evaluation can be distinguished: microscopic evaluation and macroscopic evaluation. The former provides inputs to the latter. Microscopic evaluation targets technical assessments and performance evaluation. Main methods at the microscopic level are: tests, (on-line or off-line) simulation, (theoretical) modelling and qualitative analysis. Macroscopic evaluation targets impact evaluation (benefits and costs) for decision making. Two categories of commonly used (macroscopic) evaluation methods can be distinguished [14]: (1) Economics-based evaluation methods, e.g. cost-benefit analysis (CBA) [15,16], cost-effectiveness analysis (CEA) [17], planning balance sheet (PBS) [18,19] and goal achievements matrix (GAM) [20]; and (2) normalisation-based evaluation methods, e.g. analytic hierarchy process (AHP) [21,22], simple additive weighting (SAW) [23], technique for order preference by similarity to ideal solutions (TOPSIS) [24], ÉLimination Et Choix Traduisant la RÉalité (ELECTRE) [25], Preference Ranking Organisation METHod for Enrichment Evaluations (PROMETHEE) [26–28], fuzzy multiple criteria decision aid [29] and grey relational analysis [30,31]. Each method from both categories in essence provides a procedure to process the evaluation matrix in order to provide a preference ranking of the alternatives.

Key performance indicators (KPIs), qualitative evaluation and educated guess can also be used for macroscopic evaluation. KPIs can only target the benefits. Proper determination of common KPIs may enable the transferability and comparison of the results of ITS implementation in different countries and regions. However, policy-makers should also identify a proper way to actually use KPIs after determining them. For instance, the EC determined some ITS KPIs for road transport, and intended to encourage Member States to use these common KPIs at European level. However, it seems that commitment of Member States to use the proposed KPIs is largely lacking. The reason could be that Member States do not want to be compared with other Member States who may have a better performance, due to the use of an absolute value of each KPI. To overcome this, an effective solution would be to use a relative value, i.e. only identify the improvement under each KPI made by each Member State itself, without comparing with the others. Such "relative value" approach may ensure the acceptance of and willingness-to-use for an evaluation approach.

16.2 Key findings in intelligent road transport

ITS applications have been booming in the past decades, with the aim to improve driving comfort, traffic safety, transport efficiency, environment and energy efficiency. However, the impact of ITS applications was not systematically evaluated, nor well documented. The contribution of the book is to fill this gap by providing an overview of ICT-based intelligent road transport systems, by presenting and discussing up-to-date methods for assessing the ITS development and deployment in different regions (Europe, North America, Asia Pacific, Middle East and Africa), and by providing and reviewing related evaluation results. There is a common understanding about the rationale for and needs of ITS impact assessment, which can be summarised as follows [32]:

1. demonstration of the benefits of ITS implementations;
2. enabling comparison of various ITS project results, even in different contexts;
3. facilitating the choice of the most appropriate ITS solution, based on previous experiences, for specific problems or contexts;
4. supporting forecasted "ex-ante" benefits of future implementations;
5. avoiding repetition of bad practices, based on lessons learned;
6. checking achievement(s) of the objectives;
7. justifying investments.

The evaluation results of intelligent road transport systems and services presented in this book are summarised in Table 16.1. Note that the results of impacts can show substantial differences in different countries and regions, due to differences in assumptions, input data, methods used and/or ways of using a same or very similar method. Therefore, evaluation results should be interpreted with great care and interpretation should be based on proper understanding of the assumptions, the available data and the method used. Appropriate comparison and analysis cannot be done by merely sticking to the numbers of benefits and costs.

Table 16.1   Summary of the results and key findings of the evaluation of intelligent road transport systems and services

No.	Chapter title	Results and key findings
1	History and deployment of (Cooperative) Intelligent Transport Systems	The development of ITS started since late 1980s, based on control systems, absolute and relative positioning, communications, sensing and information processing, with the potential to create substantial social, economic and environmental impact. A disrupting impact is forecasted for the deployment of higher levels of automated driving through different industry branches. Roadmaps and ITS action plans indicated a further deployment of (C-)ITS independent of vehicle type and manufacturer. An essential transformation is a shift in focus from hardware to software. Besides the software-defined vehicle, the role of large data sets provides new technological approaches.
2	Challenges in the evaluation of automated driving	The automation levels defined by SAE are the most widely used. There is a long tradition of evaluation of ADAS with naturalistic field operational tests (FOTs). Challenges for the evaluation of higher-level automated driving remain, due to the absence of naturalistic FOTs with vehicles equipped with L3 or L4 ADS. "The technological readiness level of the highly ADS and the safety requirements of public road tests do not yet allow for FOTs with unsupervised ordinary drivers or naturalistic driving studies" [33]. With such limited piloting of functions, evaluation has to be restricted to the technical performance of the function alongside scaling-up through simulations to examine impacts.

(Continues)

Table 16.1 (*Continued*)

No.	Chapter title	Results and key findings
3	Field operational tests – (still) the ultimate answer to impact assessment?	FOTs are still the ultimate answer for impact assessment, as they are essential for developing more in-depth knowledge of driver behaviour and of the impacts of various ITS solutions, in real environments under different conditions or circumstances. To overcome financial, operational, legal or methodological challenges, FOTs could be of smaller size and more focused, while new digital tools could play an important role in facilitating data collection and analysis. FOTs must rely on scientific, systematic and robust methods. FESTA provides a fundamental evaluation framework, but further development of a new evaluation method is required to cope with the complexity of the CCAM trials.
4	Driving the future: role of artificial intelligence in road vehicles	The AI algorithms and systems rely on data-driven methods, inferences and training. AI plays a transformative role in the development of the AV use cases. Continuous data collection and analysis by AI algorithms are expected to help with performance improvements and to drive innovation within the AV industry. Despite AI challenges (e.g. real-world uncertainties, real-time processing demands, interpretability, cybersecurity and ethical decision-making), the AV industry anticipates significant advancements towards Level 4 and Level 5 AVs. Standardisation and regulations are essential for ensuring uniformity, safety, quality and compatibility in the responsible development and deployment of AVs.
5	Assessment method for prioritising transport measures and infrastructure development	The assessment method for policy options (AMPO) is developed as a novel framework addressing decision-making processes by integrating CBA and MCA, while promoting stakeholder engagement. This method has been applied for the selection of different packages of policy measures, both similar and divergent measures, such as infrastructure projects and traffic management projects. The AMPO has a broad vision and encourages the exploration of innovative solutions for policy packages. This holistic, flexible and transparent approach fosters creativity, avoids the

(Continues)

Table 16.1   (Continued)

No.	Chapter title	Results and key findings
		limitations of a cost-centric analysis and ensures that the chosen initiatives are impactful and financially sustainable.
6	Evaluation of ITS: opportunities and challenges in the era of new pervasive technology	The advent of new pervasive technology has enabled the development of a new range of ITS schemes, based on personal devices and often using social functionality in addition to transport-focused functions. *New mobility schemes* and *social innovation schemes* are compared to illustrate how evaluation methods evolve and to highlight the interface with the common evaluation paradigm, the shift in impacts and the ways in which the traditional evaluation approach needs to be extended. The common evaluation paradigm is typically a cost-benefit approach supplemented by an environmental and safety assessment. With growing understanding of the usefulness and commodification of data, ethical considerations become essential.
7	The potential benefits of heavy goods vehicle (HGV) platooning	Based on the results of HelmUK, the impact of HGV platooning on road safety, fuel savings, emissions, as well as the effects on the road network and economic benefits are demonstrated and quantified. The future of platooning for road operators, governments and the freight industry is discussed. A series of recommendations for platooning development and deployment are provided, including consideration of regulation of low headways at junctions, and a strong recommendation to deploy the underlying platooning systems, at more typical Adaptive Cruise Control and larger headway driving conditions, where they offer safety benefits.
8	C-ITS deployment in Australia – achievements and key learnings	C-ITS services (based on ITS-G5 – an effective and reliable medium) have been successfully deployed in Australia, with strong collaboration between authorities, industry and academia, as a significant step towards safer multimodal transport. According to users' response to their driving experience, C-ITS is beneficial for road safety, and helps users comply with the rules through better situational awareness.

(Continues)

Table 16.1   (Continued)

No.	Chapter title	Results and key findings
		Positive impacts on road safety were observed, including reductions in average speed and smoother driving. A potential reduction of 13–20% serious crashes across Southeast Queensland is estimated for a wider implementation. Further C-ITS deployment is a key enabler towards sustainable mobility.
9	C-ITS evaluation on C-ROADS – findings from C-ROADS Germany	The C-Roads Platform is a collaborative initiative involving European Member States and road operators, aimed at advancing C-ITS across Europe. It includes a working group that is dedicated to assessing the effectiveness of C-ITS implementations, ensuring they meet safety, efficiency and environmental goals, and to determining the operational efficiency and policy alignment of C-ITS deployments. The C-ITS services, such as Green Light Optimal Speed Advisory (GLOSA) and Roadworks Warning (RWW), deployed in the German C-Roads Pilots have convincingly demonstrated a positive impact on road safety, traffic efficiency and the environment.
10	Impact assessment of large-scale C-ITS services in Greece	Several C-ITS services have been deployed in Greece and have demonstrated positive impacts, e.g. improving safety (reduction of accidents), ensuring traffic efficiency (smoother traffic flow, traffic jam prevention), improving driver awareness, enhancing data analysis and road operator knowledge of traffic conditions and incidents, facilitating optimal route selection, saving energy and time, and improving the comfort aspect of the driving experience. Both user acceptance and the impact at network level (regarding safety, traffic efficiency and CO_2 reduction) are positive. However, the willingness to pay for the services remained low, and it was expected that the user interface design and ease of use for certain functions would be improved.

(Continues)

Table 16.1   (Continued)

No.	Chapter title	Results and key findings
11	System architecture for the deployment of autonomous mobility on-demand (AMoD) vehicles	An AMoD system is developed by applying a systems engineering approach. A coherent reference architecture with clear functional boundaries is designed, carefully considering both technological requirements and business constraints. The system as a whole does not prescribe a singular operating model and remains adaptable to exploring diverse partnership arrangements. The cloud-based mobility platform, called CoMPAV, oversees AV fleets sourced from various providers. Within CoMPAV, each AD supplier offers a cloud support component to facilitate integration within the broader mobility services ecosystem. The architecture includes scalable, modular, flexible and agnostic features.
12	Robust cooperative perception for intelligent transport systems	To overcome the challenges of the perception techniques in real-world traffic scenarios, cooperative perception integrated with V2X technologies has been proposed. Synchronisation and coordinated transformation are crucial for cooperative perception, especially in multi-agent and distributed systems.
		Two pivotal elements essential for robust cooperative perception have been extensively examined: *temporal synchronisation* and *pose calibration*. The research has provided a comprehensive understanding of the challenges confronting diverse cooperative perception methods as they progress towards increased robustness and associated solutions.
13	Cooperative architecture for transportation systems (CATS): assessment of safety and mobility in vehicular convoys	Some key behaviours for the vehicular convoy have been demonstrated, in particular the formation reconfiguration and assignment processes. Defining a set of objective functions that guarantee the optimal assignment of the vehicles will also have an impact on some important metrics, and there is no limit to selecting these objective functions. Also, the motion coordination of these vehicles by maintaining both the longitudinal and lateral spacing as desired (unlike the platoon), significantly improved the safety of the

(Continues)

Table 16.1   (*Continued*)

No.	Chapter title	Results and key findings
		whole fleet, and this was proven by measuring the IEEE Safety Standard metric. It is essential to structure and design a unified and modular framework for the transportation system.
14	Traffic management using floating car data (FCD) in low- and middle-income countries (LMIC)	Commercial FCD is available to provide traffic state information (e.g. travel time, harmonic mean speed and percentile speeds per route segment) and behavioural data (e.g. traffic routing, origin-destination matrices and the percentage of vehicles using a particular route between two zones). FCD is widely used to enable ITS applications in LMICs, for example freeway management, incident detection, public transport management, parking management, dynamic intersection control and detection of damage to road surfacing. FCD coverage up to 14% has been observed in South Africa and Egypt, providing adequate quality of both speed and trip routing data. The FCD implementation is beneficial for the LMICs.
15	European ITS platform (EU EIP): evaluating the benefits and impacts of ITS corridors	The estimated benefits of the third phase of the ITS corridors over a five-year period are: at least 40 lives saved and 1,846 injuries prevented (195 serious injuries and 1,651 minor injuries); a reduction of at least 5,057,475 vehicle hours lost, a saving of at least 11,995,830 litres of fuel and a reduction of at least 34,000 tonnes of CO_2 emissions. The results indicate a minimum annual socio-economic benefit of €79 million. Considering the consolidated investment cost of €360 million, an estimated minimum ROI (Return On Investment) of ~4.5 years was calculated. Significant socio-economic benefits of investing in ITS in Europe and the high ROI that these investments generate have be confirmed.

16.3   Evaluation challenges and recommendations

Different evaluation methods are being used, and substantial differences may occur in the way a particular method is being used in different situations. Furthermore, data for doing the evaluation are often not well structured, and certainly not always available, or not available with sufficient detail. In addition, the assumptions made for evaluation are often not clearly and explicitly provided, and an adequate discussion of the shortcomings of an evaluation method and how to best address these is regularly lacking in evaluation reports. The various evaluation methods that are being used in practice do all have shortcomings and generally lack a strong theoretical basis. Therefore it remains a challenge to define adequate and acceptable approaches to do evaluation.

Both economics- and normalisation-based methods have their advantages and limitations, and the results often give a room for arguments. Economics-based methods express attribute values as much as possible in a monetary unit. In practice, this often appears to be costly, and sometimes inoperable. But the less stringent this condition is applied (e.g. in CEA, PBS and GAM), the less feasible it becomes to obtain a clear analytical answer. Normalisation-based methods try to remove the issue of dissimilar units, but none of them has a rigorous theoretical foundation. Each of the normalisation-based methods is, in fact, no more than an advanced calculation recipe, and some of these methods are not always able to provide an unambiguous ranking order. The presence of multiple attribute value types that cannot be expressed in monetary units inherently precludes the use of an economics-based method [14].

CBA has been widely used, especially in the United States, in the domain of ITS in the past decades to justify the investment of public money and to establish priority between projects. CBA aims to provide representative results for decision making, but there is no ground to believe that the CBA tool fits the aim, due to its deficiencies. These relate to: (1) the difficulty or impossibility to make a consistent estimation of the values of statistical life (VSL) and injury [34–36], and the arbitrary government guidelines for determining the VSL; (2) the foundation of CBA: new welfare economics, the premises of which apply only in rare circumstances; (3) the often rather arbitrary determination of the discounting rate [16].

KPIs seem very easy to use for providing an overview of the entire ITS domain, and do not require substantial knowledge on economics and mathematics. However, there are serious drawbacks. The domain of ITS for road applications is very broad and can be categorised from different perspectives and regarding different aspects by different stakeholders. Generic KPIs cannot sophistically take such dissimilar circumstances and characteristics into account. The sustainable impact of intelligent road transport systems has three main dimensions, each with several sub-dimensions:

1. economic, e.g. efficiency, cost-effectiveness, quality and responsiveness;
2. environmental, e.g. emissions, noise and natural resources utilisation; and
3. social, e.g. safety, health and employees.

Generic KPIs do not always adequately cover the (sub-)dimensions of the impacts. Moreover, very often too few KPIs were selected to enable usability, while the KPIs were expected to be used for a wide variety of situations. This substantially limits the coverage and the scope of the evaluation. In addition, the successfulness of using KPIs is strongly influenced by data availability, collection, analysis and reporting, and may be strongly dependent on the type of ITS technologies that need to be assessed, and in which context this needs to be done.

At national and international level, harmonised evaluation methods and common KPIs, at the macroscopic evaluation level, are needed, nationally and internationally. Criticism of the evaluation methods should not hamper achievement of evaluation results. On the contrary, constructive criticism may:

1. encourage and stimulate researchers to find (theoretical and practical) solutions to overcome disadvantages of the (current) methods, and to develop new methods as well as guidelines for best ways of using the methods;
2. support decision-makers in avoiding bias (by selecting independent bodies to carry out evaluation studies), in choosing appropriate methods and using them properly, in unambiguously clarifying assumptions, and in interpreting the results in a reasonable and sensible way; and
3. help all stakeholders, including industry partners, with understanding the challenges that are at stake, and in providing relevant high-quality input data for evaluation.

For the evaluation of intelligent road transport systems, the availability of complete, accurate and well-structured data is a prerequisite. To improve data availability and quality, efforts from authorities, industry and academia are needed. Moreover, the evaluation framework should be transparent, easy to be operated and widely acceptable. Concerning macroscopic evaluation methods, a harmonised approach needs to be explored. Integrated methods and/or an index could be investigated. Furthermore, the assumptions used must be clearly stated. In addition, there may be a need for a change in mindset regarding the idea that it is possible to obtain accurate results in "true" numbers. Due to the fact that all current evaluation methods have drawbacks (see above), it may in principle not be feasible to determine "true values" for benefits and costs by using any of these methods. Therefore, instead of struggling to identify "true values", an alternative approach may be necessary. Imagine that one common (harmonised and acceptable) method (even though not perfect) would be used in many instances, then, at least, comparable results would be obtained. The idea of such fundamental approach for determining relatively accurate results in many instances in the same way by using a common method may drive our thinking to remodel ITS evaluation. It would concern a common evaluation method, which should be transparent, can be easily adopted, and quickly and widely accepted by all stakeholders. This certainly deserves further discussion, and research in the domain of evaluation of intelligent road transport systems.

The ITS sector has seen substantial development and rapid deployment in the past decades. Re-thinking, in a holistic way, of approaches for future planning, modelling, design, control and comprehensive management is required. The challenges for industry, academia and administrations in the domain of evaluation of intelligent road transport systems are to enhance technical assessment and impact assessment, to overcome evaluation obstacles and to substantially reduce investment risks.

16.4   Discussion and thoughts for the future

ITS applications cover all transport modes (road, rail, waterborne and air), with the aim of improving sustainability for people and goods transport; to reduce the frequency and consequences of traffic accidents; to reduce local nuisances and greenhouse gas emissions; to reduce the need for (more) public space for transport; and to offer transport that is accessible and affordable to all.

At the beginning of the ITS development in the 1980s, the main focus was on creating substantial positive impacts on road transport, in terms of traffic safety, driver comfort, traffic efficiency, energy/fuel efficiency and emission reduction. Since the 2000s, the important role of infrastructure for the ITS deployment became internationally increasingly recognised, and huge investments have been made on intelligent infrastructure, especially in Western countries. Substantial tangible benefits are achieved from economic, social and environmental perspectives. However, the associated high costs are not negligible. The prerequisites for the development and deployment of intelligent multi-modal transport systems are sophisticated well-designed physical infrastructure and maintenance, as well as consistent long-term investments and political willingness to support them. The (re) design of the physical and digital infrastructure must balance supply and demand of the transport network, and meet the needs and requirements of transport policy and of end-users [37]. From a financial perspective, (intelligent) transport infrastructure is a huge challenge for developed countries, and even more so for developing countries.

C-ITS, also called Connected Vehicles, has a broad spectrum of applications for sustainable, secure and resilient transport (all modes) to shape the future of mobility. In the 2010s, the focus of the ITS domain has shifted from system development to broad applications of C-ITS services, for which the role of infrastructure becomes even more essential. In addition to road transport, (C-) ITS and services support other modes of transport, including multimodal travellers, intermodal transhipment (i.e. road, rail and waterborne), hubs (airports, dry ports – railway stations, sea ports, inland waterway and logistics hubs) and Urban Air Mobility (UAM) such as eVTOL (electric Vertical Take-Off and Landing) aircraft and drones.

The technical development of ITS boomed, especially due to intensive R&D in various regions, between the 1990s and 2000s. Innovative concepts and substantial technical achievements mainly arise from these decades. Since the 2010s, in addition to the change in the focus of ITS, there has also been a context shift, probably due to political misunderstanding about the technical maturity of ITS and/ or slower market take-up than expected. It seemed that in some regions innovation was sometimes no longer addressed from a traditional solid technical perspective, but moved towards "soft" or "social" discussions where lay people could easily be involved, and where technical discussions turned into political debate. Another change that occurred concerns the research method. In the past, different new methods, algorithms and mathematical models were developed, and practical and reliable solutions were tested, e.g. for the ADAS applications. However, in the recent decade, most research is largely based on the same data-driven method, in recent years hyped as AI (artificial intelligence), which has becomes a main stream of R&D for most popular applications. In general, AI is based on programming and statistics, and neither new nor in any way intelligent. For instance, *expert systems* were formally introduced around 1965, and artificial neural network (ANN), together with other classical statistical models, have been widely used since the 1990s. With the development of computer science and computer hardware, data processing has gradually become more advanced, but also increasingly extensive, with an associated substantial increase in energy use and greenhouse gas (GHG) emissions. AI is one approach but not the approach – neither an AI algorithm nor an AI system is a must for any application. It is also very questionable why it is necessary to use AI-based solutions, which may bring high risks in terms of safety, and consume much more energy than non-AI-based approaches. The AI-hype generates a high number of publications in many domains, including ITS, just because it is easy to use, if one has data, which are ubiquitous these days. As it is so popular, it is used as a means to attract public resources and private investment, and to create a popular public image, although the current AI applications are not innovative, at least from scientific and technical perspectives.

In 2021, five levels of on-road driving automation systems for motor vehicles have been technically standardised [38] and the SAE standard has been adopted internationally. Level-2 automated vehicles are available on the market. However, the further development and deployment of higher level of automated driving are treated very differently in different regions and countries. The following patterns were observed, which are open for discussion and debate: (1) in Asia Pacific, traditional technical development has been very practically continued, and the industry is looking for market take-up and innovative selling points, together with new standards development activities; (2) in the United States, based on the local legislation and regulation of some states, certain automated driving services have been successfully commercialised, despite safety concerns and debates; (3) in Europe, significant public funds have been continuously used for research and demonstrations, and practical regulatory efforts have been made. To strengthen the position of ITS, the key players in the ITS domain shall have a common goal and, with commitment, joint actions, solid technical breakthroughs and financial co-investments, and shall not be mainly dependent on public funds.

In addition, we are facing climate change, environmental degradation and an energy crisis that constitute some of the main man-made challenges of today's world. Transport is one of the major contributors to GHG emissions. Global GHG emissions are to a large extent created by energy use in the following sectors: industry (24.2%), transport (16.2%) and building (17.5%). Within transport, road transport counts for 11.9%, aviation for 1.9%, shipping for 1.7%, rail for 0.4% and pipeline for 0.3% [39]. Future mobility, if the needs of digitalisation and electrification would be much better understood, may create higher negative impact not only on the transport sector, but also other sectors. The transport sector directly linked to huge energy consumption, ICT development and political sustainability goals. The ICT-based systems and services applied in transport and other sectors account for 2–4% of global GHG emissions, and especially due to widely use of AI, could reach 14% of global GHG emissions by 2040, according to an estimation in 2018 [40]. Sustainable mobility shall no longer only refer to the operational phase, but take into account the whole lifecycle and supply chain of the entire multi-modal transport system, including infrastructure, products, services and maintenance [41,42]. The AI approach does substantially increase energy consumption due to extensive and at times quite unnecessary data collection, processing and storage. It has high negative impacts on energy consumption, land use and other resources, and therefore, on climate change. In general, the rather superficial public discussions about AI, often characterised by a lack of knowledge and understanding, and the commercial and research uses of AI, may raise further concerns about the quality of current R&D and education in each sector, as well as about the development of the intelligence of human beings. Intelligent road transport, in the context of regional development, is expected not only to improve the quality of life but also to create thriving communities, by restoring of nature and the environment, maintaining strong economic growth and increasing the potential positive social impact.

16.5   Conclusion

Although most impact assessment methods are mature for basic applications, researchers and industry partners are continuously improving these methods and developing new algorithms. Impact evaluation is still challenging. Methods for evaluation provide a recipe for analysis and ranking of different available alternatives for achieving a certain goal or objective. In general, it may be said that no algorithm can act as a complete substitute for human judgement. Due to (fundamental) drawbacks, neither common evaluation methods nor KPIs are widely used and accepted by all stakeholders. There is a need for a harmonised evaluation approach, which is a challenge for researchers, industry and authorities. Last but not least, evaluation studies for decision making should be done by independent bodies.

As the backbone of mobility, further digitalisation and electrification of ITS deployment shall continue to focus on creating positive social, economic and environmental impact, while it is more important and urgent to mitigate the negative impact for planet earth in terms of excessive and unsustainable use of scarce resources (such as fossil fuels, water and land). To make future ITS and services sustainable, a change in mentality is required.

Acknowledgements

The author sincerely thanks the Past Chairs and VPs of IBEC (International Benefits Evaluation Society for ITS) and other colleagues for their leadership of this international forum (launched in 2002), for their substantial contribution to the ITS impact assessment and for providing essential support in the past decades: Jane

Lappin, Eric Sampson, Tom Kern, Andrew Somers, Risto Kulmala, Torsten Geißler, Alan Stevens, Jennie Martin, Eric Kenis, Keith Keen, András Varhelyi, Martin Böhm, Caroline Visser, Dick Mudge, Jianping Wu, Luca Studer, Paul Vorster and Stephane Dreher.

References

[1]   Lu, M., Wevers, K., and Van der Heijden, R. (2005). Technical feasibility of advanced driver assistance systems (ADAS) for road traffic safety. *Transportation Planning and Technology*. 28(3):167–187.

[2]   Lu, M. (ed.) (2019). *Cooperative Intelligent Transport Systems: Towards High-Level Automated Driving*. Institution of Engineering and Technology (IET), London. DOI:10.1049/PBTR025E.

[3]   SAFESPOT Consortium (2005). SAFESPOT (Co-operative Systems for Road Safety "Smart Vehicles on Smart Roads") Technical Annex. SAFE-SPOT Consortium, Brussels. (restricted)

[4]   CVIS Consortium (2005). *CVIS (Cooperative Vehicle Infrastructure Systems) Technical Annex*, CVIS Consortium, Brussels. (restricted)

[5]   COOPERS Consortium (2005). *COOPERS (CO-OPerative SystEms for Intelligent Road Safety) Technical Annex*, COOPERS Consortium, Brussels. (restricted)

[6]   FREILOT Consortium (2009). *FREILOT (Urban Freight Energy Efficiency Pilot) Technical Annex*, FREILOT Consortium, Brussels. (restricted)

[7]   DRIVE C2X Consortium (2011). *DRIVE C2X Technical Annex*, DRIVE C2X Consortium, Brussels. (restricted)

[8]   C-ITS Platform (2016). Platform for the Deployment of Cooperative Intelligent Transport Systems in the EU (E03188) C-ITS Platform Final Report, DG MOVE – DG Mobility and Transport, January 2016, Brussels.

[9]   C-ITS Platform (2017). Platform for the Deployment of Cooperative Intelligent Transport Systems in the EU (C-ITS Platform) Phase II Final Report, DG MOVE – DG Mobility and Transport, September 2017, Brussels.

[10] C-Roads Platform. (2018). Harmonised C-ITS Specifications for Europe, Release 1.2: C-ITS Infrastructure Functions and Specifications; Common C-ITS Service Definitions; Roadside ITS G5 System Profile. C-Roads Platform, Brussels.

[11] CCAM Partnership. (2024). *The Strategic Research and Innovation Agenda (SRIA)*. CCAM Partnership, Brussels.

[12] ITS America (2023). *ITS America National V2X Deployment Plan: An Infrastructure & Automaker Collaboration*. ITS America, April 2023, Washington, DC.

[13] Lu, M. (ed.) (2016). *Evaluation of Intelligent Road Transport Systems: Methods and Results*. Institution of Engineering and Technology (IET), London. DOI:10.1049/PBTR007E.

[14] Lu, M., and Wevers, K. (2007). Application of grey relational analysis for evaluating road traffic safety measures: advanced driver assistance systems against infrastructure redesign. *IET Intelligent Transport Systems*. 1(1): 3–14.

[15] Boardman, A.E., Greenberg, D.H., Vining, A.R., and Weimer, D.L. (1996). *Cost-Benefit Analysis: Concepts and Practice.* Prentice Hall, Upper Saddle River, NJ.

[16] Hauer, E. (2011). Computing what the public wants: some issues in road safety cost-benefit analysis. *Accident Analysis and Prevention*. 43(1): 151–164.

[17] Trilling D.R. (1978). A cost-effectiveness evaluation of highway safety counter-measures. *Traffic Quarterly*. 32:41–67.

[18] Lichfield, N. (1956). *The Economics of Planned Development.* Estates Gazette, ltd, London.

[19] Lichfield, N. (1964). Cost-benefit analysis in plan evaluation. *Town Planning Review*. 35:159–169.

[20] Hill, M. (1968). A goal-achievements matrix for evaluating alternative plans. *Journal of the American Institute of Planners*. 34:19–29.

[21] Saaty, T.L. (1980). *The Analytic Hierarchy Process.* McGraw-Hill, New York.

[22] Saaty, T.L. (1995). *Decision Making for Leaders: The Analytic Hierarchy Process for Decisions in a Complex World.* RWS Publications, Pittsburgh, PA.

[23] Yoon, K.P., and Hwang, C.L. (1995). *Multiple Attribute Decision Making: An Introduction.* Sage Publications, London.

[24] Hwang, C.L., and Yoon, K. (1981). *Multiple Attribute Decision Making: Methods and Applications.* Springer-Verlag, Berlin.

[25] Roy, B. (1968). Classement et choix en présence de critères multiples (la méthode ELECTRE). RIRO. 8:57-75. (French)

[26] Brans, J.P. (1982). L'ingénierie de la décision. L'élaboration d'instruments d'aide à la décision. Quebec: Université Laval. (French)

[27] Brans, J.P., Vincke, P, and Mareschal, B. (1986). How to select and how to rank projects: the PROMETHEE method. *European Journal of Operational Research*. 24:228–238.

[28] Brans, J.P. (1996). The space of freedom of the decision maker modelling the human brain. *European Journal of Operational Research*. 92:593–602.

[29] Zadeh, L. (1965). Fuzzy sets. *Information and Control*. 8(3):338–353.

[30] Deng, J. (1982). Control problems of Grey Systems. *Systems and Control Letters*. North-Holland Co. Publisher. 5:288–294.

[31] Guo, H. (1985). Identification coefficient of relational grade. *Fuzzy Mathematics*. 5:55–58. (Chinese).

[32] Studer, L. and Marchionni, G. (2016). Evaluation of the impact of ITS. In: Lu (ed.) *Evaluation of Intelligent Road Transport Systems: Methods and Results.* IET, London. 67–122.

[33]   Sintonen, H., Bolovinou, A., Guelsen, B., *et al.* (2023). *Hi-Drive Deliverable D4.3: Experimental Procedure.* Volkswagen Group Research, Wolfsburg, Germany.

[34]   Mrozek, J.R., and Taylor, L.O. (2002). What determines the value of life? a meta-analysis. *Journal of Policy Analysis and Management.* 21(2):253–270.

[35]   Viscusi, W.K., and Aldy, J.E. (2003). The value of a statistical life: a critical review of market estimates throughout the world. *Journal of Risk and Uncertainty.* 27(1):5–76.

[36]   De Blaeij, A., Florax, R.J.G.M., Rietveld, P., and Verhoef, E. (2003). The value of statistical life in road safety: a meta-analysis. *Accident Analysis & Prevention.* 35(6):973–986.

[37]   Böhm, M., Flechl, B., Aigner, W., and Visser, C. (2016). ITS evaluation policy – culture and needs. In: Lu (ed.) *Evaluation of Intelligent Road Transport Systems: Methods and Results.* IET, London.

[38]   N.N. (2021). Taxonomy and Definitions for Terms Related to Driving Automation Systems for On-Road Motor Vehicles. SAE Standard J3016, 27–46, USA.

[39]   Ritchie, H. (2020). Global Greenhouse Gas Emissions by Sector 2020. Our World in Data. Climate Watch, The World Resources Institute (WRI).

[40]   Belkhir, L., and Elmeligi A. (2018). Assessing ICT global emissions footprint: trends to 2040 and recommendations. *Journal of Cleaner Production.* 177: 448–463.

[41]   Lu, M., and De Bock, J. (eds.) (2015). *Sustainable Logistics and Supply Chains: Innovations and Integral Approaches.* Berlin: Springer. DOI:10. 1007/978-3-319-17419-8.

[42]   Lu, M. (2023). Sustainable mobility for flourishing communities. Paper ID 103. In *Proceedings of the 29th World Congress on Intelligent Transport Systems*, 16–20 October 2023, Suzhou, P.R. China.

Index

www.ingramcontent.com/pod-product-compliance
Lightning Source LLC
Jackson TN
JSHW012015201224
75816JS00003B/13